HOW TO USE, CALIBRATE, REPAIR AND UPGRADE
VACUUM TUBE TESTERS

IGOR S. POPOVICH

DISCLAIMER & COPYRIGHT NOTICE

The information contained in this book is to be taken in the context of general overview, not specific advice. You should not act on the information contained herein without seeking professional advice. Neither the author nor the publisher (or any other person involved in the publication, distribution or sale of this book) accepts any responsibility for the consequences that may arise from readers acting in accordance with the material given in the book. Professional advice about each particular case / instance should be sought.

Designs and circuit diagrams are intellectual property of their copyright holders and should not be used without their permission. They are discussed here from educational perspective under the "fair use" provision.

© **Copyright Igor S. Popovich, 2017**

All rights are reserved.

No part of this publication may be used or reproduced in any form or transmitted by any means, without the prior written permission from the publisher, except in the case of brief quotations in articles and reviews.

INDEMNITY NOTICE

Tube testers involve lethal voltages, high temperatures and other hazards. By reading this book you automatically agree to indemnify its author, publisher and retailer against any claims, of any nature, and for any reason.

The accuracy of drawings and schematics presented cannot be guaranteed and a particular tester could have been wired differently or modified over the years.

Published by Career Professionals in Australia

Second, revised edition, 2022

Bulk purchases

This book may be purchased in larger quantities for educational, business or promotional use. Please e-mail us at sales@careerprofessionals.com.au

National Library of Australia Cataloguing-in-Publication Data:

Popovich, Igor S.,

HOW TO USE, CALIBRATE, REPAIR AND UPGRADE VACUUM TUBE TESTERS

ISBN: 978-0-9806223-7-9

1. Electrical engineering 2. Electronics

I Igor S. Popovich II Title III Index

621.3

CONTENTS

1. HOW VACUUM TUBES WORK _____ Page 9

- ELECTRON EMISSION AND VACUUM DIODE
- VACUUM DIODE IN DC AND AC CIRCUITS
- ELECTRON TUBES - NAMING CONVENTIONS, BASES AND PIN NUMBERING
- TRIODE PARAMETERS AND CHARACTERISTICS
- CASE STUDY: STATIC PARAMETERS OF 12AX7 (ECC83) DUO-TRIODE

2. TESTING & MATCHING VACUUM TUBES _____ Page 19

- TUBE PARAMETERS AND THEIR VARIATION WITH TEST CONDITIONS
- TUBE TESTING PRINCIPLES AND PROBLEMS
- NEGATIVE GRID CURRENT - CAUSES AND DETECTION METHODS
- TESTING CASE STUDIES AND PRACTICAL EXAMPLES OF TUBE MATCHING
- HOW TO TEST A TUBE THAT IS NOT LISTED IN THE BOOK OR TUBE CHART
- EXPERIMENT: COMPARISON OF TRIPLETT 3444, B&K700 and MERCURY 1000 TEST VOLTAGES FOR A FEW COMMON PREAMP AND POWER TUBES

3. EMISSION TESTERS _____ Page 35

- HOW EMISSION TESTERS WORK
- EICO 625 & 635
- KNIGHT KG-600, HEATHKIT TC-1, TC-2, TC-3 & IT-17
- CONAR 221, 223 & 224
- PRECISION 640 (NRI 71) & PRECISION 660
- TRIPLETT 2413 & 3414
- ELETTRA PROVAVALVOLE

4. GRID CIRCUIT TESTERS _____ Page 47

- HOW GRID CIRCUIT TESTERS WORK
- SECO 78, 88 & 98
- SENCORE "MIGHTY MITE" TESTERS: TC114, TC130, TC-136, TC-142, TC-154
- SENCORE TC162 & TC28 ("THE HYBRIDER")
- B&K 600 & 606 DYNA-QUIK TUBE TESTERS
- B&K 607 & 667
- B&K 625 DYNA TESTER
- PRECISION APPARATUS COMPANY (PACO) 650 & T-62
- MERCURY 1101, 1101C & 1101CT
- AMERICAN SCIENTIFIC DEVELOPMENT COMPANY TV-20

5. DYNAMIC CONDUCTANCE TESTERS _____ Page 61

- THE OPERATING PRINCIPLE BEHIND DYNAMIC CONDUCTANCE TESTERS
- SICO MODEL 85
- SICO TV-12
- SYLVANIA 139 & 140
- SYLVANIA 219 & 220
- JACKSON TUBE TESTERS
- EICO 666 & 667

6. PROPORTIONAL MUTUAL CONDUCTANCE TESTERS _____ Page 79

- HOW PROPORTIONAL MUTUAL CONDUCTANCE TESTERS WORK
- WESTON 798
- TRIPLETT 3423
- TAYLOR 45D
- AVO VALVE CHARACTERISTIC METER (MK III)
- METRIX 310CTR
- SIMPSON 330

7. HICKOK-TYPE TESTERS _____ Page 95

- THE HICKOK BRIDGE CIRCUIT AND HOW IT MEASURES TRANSCONDUCTANCE
- HICKOK TESTERS
- B&K 500
- B&K 550
- B&K 650
- B&K 675
- B&K 700 & 707
- B&K 747
- MERCURY 1000, 1200 & 2000
- PRECISE 111
- PRECISE 116
- DYNAMATIC DM456
- SIMPLE FIXES & UPGRADES FOR HICKOK-TYPE TESTERS

8. TRUE MUTUAL CONDUCTANCE TESTERS _____ Page 133

- WESTON 981 & HEATHKIT TT-1
- TRIPLETT 3444
- SENCORE MU-140 & MU-150
- RCA WT-110A
- SECO 107, 107-B & 107-C
- MODERN TRANSCONDUCTANCE TESTERS

9. REPAIRING & UPGRADING VINTAGE TUBE TESTERS _____ Page 155

- CHOOSING & BUYING A TUBE TESTER
- SAFETY RULES and PRECAUTIONS
- TYPES OF FAULTS, COMMON CAUSES AND FAULT LOCATION METHODS
- COMPONENT TESTING
- ANALOG (MOVING COIL) METERS
- A CRASH COURSE ON PROPER SOLDERING
- SOCKET ADAPTERS

10. TESTING & MATCHING TUBES WITHOUT A TUBE TESTER _____ Page 171

- SIMPLE OHMMETER AND MULTIMETER CHECKS
- EMISSION AND TRANSCONDUCTANCE TESTS
- DIY CURVE TRACER FOR MATCHING TUBES

INTRODUCTION

Why a resurgence of interest in tube testers?

The revival of tube technology that started in the early 1990s has also led to renewed interest in vacuum tube testers and their ultimate purpose, testing and matching audio tubes.

The prices of vintage tube testers have risen to levels unimaginable just a few years earlier. Just as with vintage tube amplifiers, tube testers their owners could not give away then are now fetching many hundreds and some even thousands of dollars.

The ramp-up of vacuum tube production in countries such as Russia, China, and Slovakia, and large, mostly military stocks of those countries, has made such tubes relatively affordable, while the dwindling remnants of western production that stopped in the 1970s and early 1980s are so expensive that a matched quad of such lovelies costs as much as a small used car ... well, almost ...

When genuine NOS Mullard or Telefunken preamp tubes such as 12AU7 and 12AX7 sell for $200 or more, power tubes with brand name logos sell at ten times the cost of the currently produced ones tube dealers and resellers stand to gain small fortunes.

Thus, when decent money beckons, a proliferation of forgeries can be expected.As in any business, there are decent and honest tube (re)sellers, and there are crooks who reprint old Chinese or Soviet tubes with western logos and sell them at ten times the price.

Many eBay sellers also spruiking matched pairs and quads of expensive currently-produced Chinese and Russian tubes such as 300B, who sell factory rejects and unmatched tubes as "closely matched."

Who needs a tube tester?

High prices of suspect tubes from untrustworthy sellers are probably the main reason lots of audiophiles and guitar players feel that a tube tester would help them ascertain if they got what they paid for and if their old tubes needed replacing in the first place, before the purchase.

If you only use a few tube types (for one hi-fi amp and/or preamp, or a guitar amp), and if you have a trustworthy technician who services your gear, you do not need a tube tester. If you fix your amps yourself and use a reputable tube supplier, you should not need a tester either. You can tell if a tube is faulty or weak just by measuring the DC and AC voltages on its electrodes (in a circuit).

On the other hand, if you are a technician who fixes amps and other tube gear or a tube reseller, then a reliable and accurate tube tester will be beneficial to your business and improve your bottom line. In both cases, declared results of meaningful tests should help you get a higher price for your goods or services.

Buying the wrong tester, using it to do things it was never designed to do, damaging or even destroying tubes it was supposed to test are just some of the ways a tube tester can work against your interests.

Tube testers suffer from false positives and false negatives. A false positive is when a perfectly good tube is tested as "faulty" or "worn" or "marginal" and discarded. The word "positive" has the same meaning as your blood or pregnancy test. If diagnosed "positive" for a disease or a "marker" of a certain medical condition, it means you are sick or will get sick soon.

The false negative is when a faulty tube is passed by a tube tester as a healthy one, meaning the tester could not find anything wrong with it. Then, if such a tube is inserted into a piece of tube gear, depending on what's wrong with it, it may cause secondary faults in the actual device (amplifier, for instance)!

In both cases, the inadequacy of a tube tester will cost its owner money. In the first case, a perfectly adequate tube is discarded; in the second, an expensive amplifier repair may be needed.

Tube testers are like cars. Some look good but aren't that easy or pleasurable to drive (use); others look old-fashioned or too complicated to use but are quite capable. Just as with cars, some brands and models of tube testers are overpriced; others are under-appreciated and thus cheaper than they should be.

Ultimately, they all have particular strengths and weaknesses; one aim of this book is to identify such positive and negative aspects of various types or families of tube testers in a general sense, and pluses and minuses of a few specific models that often come up for sale.

Most vintage tube testers analyzed in this book originate in the USA, but many European testers are equally good, if not better than most American testers. The most notable are AVO and Taylor (UK), Metrix (France), Neuberger and Funke (Germany), and UnaOhm (Italy). There are also decent tube testers made in Poland (ELPO P508) and the former Soviet Union (L3-3), but these are relatively rare and will not be discussed here.

SYMBOLS USED

Many acronyms and symbols will be used in this book. Here are some of them, in no particular order: NOS (New Old Stock), DC direct current, AC alternating current, CG control grid, SG or G2 screen grid, L-N-E (Line-Neutral-Earth) mains power connections, COM (common terminal - not necessarily grounded), GND ground (earth), TUT Tube-Under-Test, TT tube tester, HV high voltage, SR silicon rectifier, SS solid-state, K cathode, A anode, HK heater-cathode, LC inductance-capacitance, FSD full scale deflection, RMS root-mean-square, SW switch, CW clockwise, CCW counterclockwise, Em emission, Gm mutual conductance, Q operating point of a tube, LED light-emitting diode, FET field-effect transistor, ...

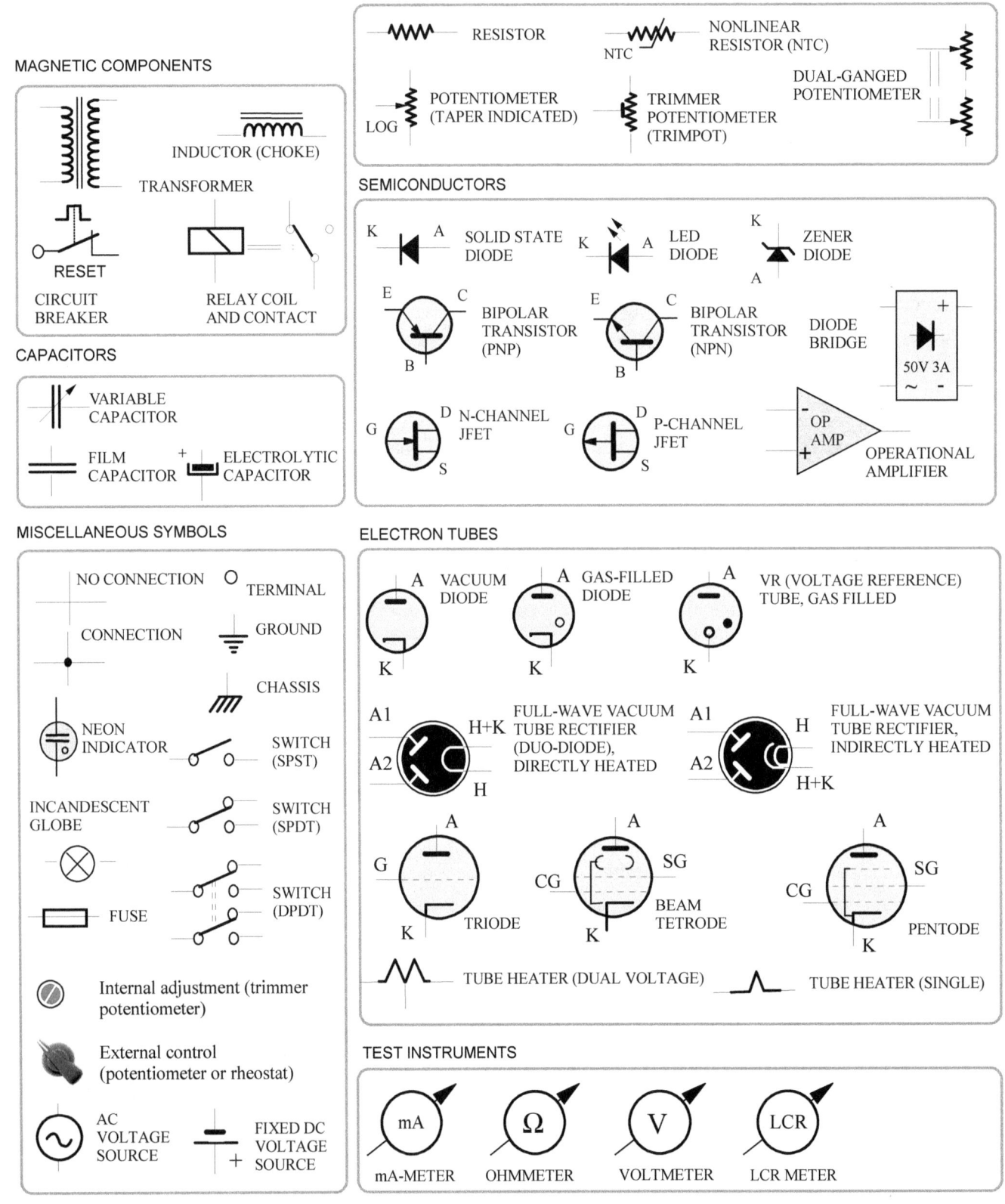

AUDIOPHILE TUBE AMPLIFIER BOOKS BY IGOR S. POPOVICH

Available from Amazon.com and all good online bookstores

Audiophile Vacuum Tube Amplifiers, Vol. 1:

- BASIC ELECTRONIC CIRCUIT THEORY
- ELECTRONIC COMPONENTS
- AUDIO FREQUENCY AMPLIFIERS
- PHYSICAL FUNDAMENTALS OF VACUUM TUBE OPERATION
- VOLTAGE AMPLIFICATION WITH TRIODES - THE COMMON CATHODE STAGE
- OTHER VOLTAGE AMPLIFICATION STAGES WITH TRIODES
- TETRODES AND PENTODES AS VOLTAGE AMPLIFIERS
- FREQUENCY RESPONSE OF VACUUM TUBE AMPLIFIERS
- IMPEDANCE-COUPLED STAGES AND INTERSTAGE TRANSFORMERS
- NEGATIVE FEEDBACK
- TONE CONTROLS, ACTIVE CROSSOVERS AND OTHER CIRCUITS
- PRACTICAL LINE-LEVEL PREAMPLIFIER DESIGNS
- PHONO PREAMPLIFIERS
- SINGLE-ENDED TRIODE OUTPUT STAGE
- PRACTICAL SINGLE-ENDED TRIODE AMPLIFIER DESIGNS
- PRACTICAL SINGLE-ENDED PSEUDO-TRIODE DESIGNS
- SINGLE-ENDED PENTODE AND ULTRALINEAR OUTPUT STAGES

Audiophile Vacuum Tube Amplifiers, Vol. 2:

- PRACTICAL SINGLE-ENDED PENTODE AND ULTRALINEAR DESIGNS
- PUSH-PULL OUTPUT STAGES
- PRACTICAL PUSH-PULL AMPLIFIER DESIGNS
- BALANCED, BRIDGE AND OTL (OUTPUT TRANSFORMERLESS) AMPLIFIERS
- THE DESIGN PROCESS
- FUNDAMENTALS OF MAGNETIC CIRCUITS AND TRANSFORMERS
- MAINS TRANSFORMERS AND FILTERING CHOKES
- POWER SUPPLIES FOR TUBE AMPLIFIERS
- AUDIO TRANSFORMERS
- TROUBLESHOOTING AND REPAIRING TUBE AMPLIFIERS
- UPGRADING & IMPROVING TUBE AMPLIFIERS
- SOUND CONSTRUCTION PRACTICES
- AUDIO TESTS & MEASUREMENTS
- TESTING & MATCHING VACUUM TUBES

Audiophile Vacuum Tube Amplifiers, Vol. 3:

- THE FRONT-END: SUPERIOR INPUT & DRIVER STAGES
- FROM SHOCKING TO SUBLIME: LESSONS FROM COMMERCIAL LINE STAGES
- DIY LINE-LEVEL PREAMPLIFIERS: $10,000 SOUND ON $500-$1,000 BUDGET
- THE STARS OF THE AUDION ERA: ANCIENT TUBES IN MODERN AMPS
- CHEAP & CHEERFUL: PREAMP & DRIVER TUBES FOR AUDIO EXPLORERS
- SLEEPING GIANTS: OUTPUT TUBES FOR THOSE WHO WANT TO BE DIFFERENT
- THE QUEEN OF HEARTS: SINGLE-ENDED AMPLIFIERS WITH 300B TRIODES
- TRIODES, PENTODES AND BEAM TUBES: MORE SINGLE-ENDED DESIGNS
- BIG BOTTLES: SET AMPLIFIERS WITH HIGH VOLTAGE TRANSMITTING TUBES
- THE WAY IT USED TO BE: VINTAGE PUSH-PULL AMPLIFIERS
- NEW? IMPROVED? MODERN PUSH-PULL AMPLIFIER DESIGNS
- CUTE, CLEVER OR CONTROVERSIAL? INTERESTING IDEAS FROM TUBE AUDIO'S PAST AND PRESENT
- THRIFTY TIPS & TRICKS: TIME & MONEY SAVING IDEAS
- OUTPUT AND INTERSTAGE TRANSFORMERS: FROM COMMERCIAL BENCHMARKS TO YOUR OWN DESIGNS
- MEASUREMENTS VERSUS LISTENING AND OTHER AUDIO DESIGN DILEMMAS

GUITAR TUBE AMPLIFIER BOOKS BY IGOR S. POPOVICH

Available from Amazon.com and all good online bookstores

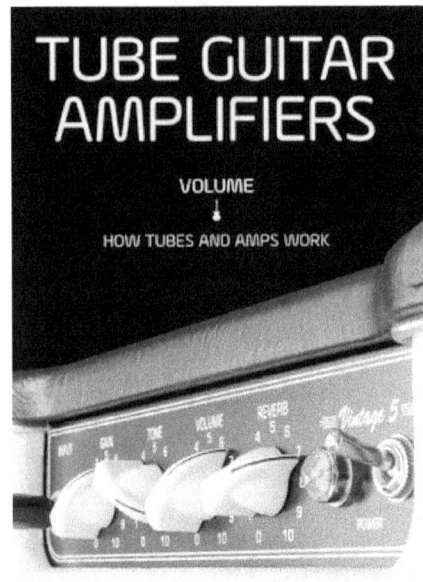

Tube Guitar Amplifiers Volume 1: How Tubes and Amps Work

- BASIC ELECTRONIC CIRCUIT THEORY
- AUDIO AMPLIFIERS
- ELECTRONIC COMPONENTS
- PHYSICAL FUNDAMENTALS OF VACUUM TUBE OPERATION
- TRIODES AS VOLTAGE AMPLIFIERS
- TETRODES, PENTODES AND BEAM-POWER TUBES
- INPUT CIRCUITS AND STAGES
- TONE CONTROLS
- ANALOG EFFECTS (TREMOLO, VIBRATO, REVERB) AND EFFECTS LOOPS
- POWER SUPPLIES FOR TUBE AMPLIFIERS
- SINGLE-ENDED TRIODE, PENTODE AND ULTRALINEAR OUTPUT STAGES
- PHASE SPLITTERS OR INVERTERS
- PUSH-PULL OUTPUT STAGES
- NEGATIVE FEEDBACK
- TRANSISTOR AND HYBRID GUITAR AMPLIFIERS

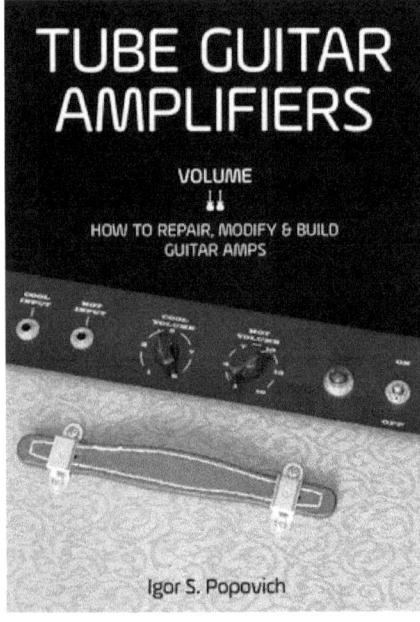

Tube Guitar Amplifiers Volume 2: How to Repair, Modify & Build Guitar Amps

- OUTPUT AND INTERSTAGE TRANSFORMERS FOR TUBE GUITAR AMPS
- LOUDSPEAKERS, OUTPUT ATTENUATORS & HEADPHONE CIRCUITS
- TROUBLESHOOTING AND REPAIRING TUBE GUITAR AMPLIFIERS
- WIRING, SOLDERING & MODIFICATION PRACTICES
- POWER SUPPLY MODIFICATIONS AND IMPROVEMENTS
- TONE TWEAKS
- MODERN PUSH-PULL AMPS
- DIY PROJECTS: CONVERTING SOLID STATE GUITAR AMPS TO TUBES
- DIY PROJECTS: ULTRA-SMALL AMPS
- REBUILDING COMMERCIAL AMPS IN A HANDWIRED (POINT-TO-POINT) FASHION
- DIY PROJECTS: QUIRKY & UNUSUAL DESIGNS
- CONVERTING VINTAGE TUBE GEAR INTO GUITAR AMPS

GETTING IN TOUCH WITH US

If you've liked the book and benefited from it, the best way to repay a favor is to recommend it to your friends and to write an online review.

Also, if you spot an error or an omission or should you have any constructive criticism of the book, I'd like to hear from you, so we can fix it together.

If you'd like to contribute rare or unusual tube testers for review, ideas or projects for the next edition or if you have ideas on how to make the book better, please let me know.

My e-mail is **igorpop@careerprofessionals.com.au**

HOW VACUUM TUBES WORK

1

- ELECTRON EMISSION AND VACUUM DIODE
- VACUUM DIODE IN DC AND AC CIRCUITS
- ELECTRONIC TUBES - NAMING CONVENTIONS, BASES AND PIN NUMBERING
- TRIODE PARAMETERS AND CHARACTERISTICS
- STATIC PARAMETERS OF A 12AX7 (ECC83) DUO-TRIODE

ELECTRON EMISSION AND VACUUM DIODE

Thermionic emission

In any material, atoms or molecules are at rest only at (the absolute zero) temperature of -273°C. At any other temperature, they are in a continual state of motion. Higher temperatures make this movement more vigorous and increase the velocity of electrons in the material. This "liberation" of electrons from a metallic electrode by virtue of its increased temperature is known as thermionic emission, and the emitter electrode is called a cathode.

When a metal electrode gets bombarded by ions or electrons of sufficient energy, some of that energy is absorbed by surface electrons which may be enough to exceed the work function of the metal and liberate the electrons. This phenomenon is called a secondary emission, meaning the released electrons are secondary electrons while the "primary" electrons provide the energy.

Oxide-coated indirectly-heated cathodes and filamentary type tubes

Early vacuum tubes were directly-heated or filamentary type. Typical examples still in use today (almost exclusively in hi-fi amps) are 2A3, 300B, and 211 power triodes. The heater filament acts as a cathode, i.e., it emits electrons. These tubes will indicate a short between a heater and cathode on tube testers, but that is normal since it is the same electrode!

Most modern vacuum tubes use oxide-coated cathodes. In such indirectly-heated tubes, an insulated heater wire is placed inside a cathode, a metal sleeve that gets energy as heat from the heater (conducted through a ceramic insulator), and the cathode then emits electrons. The cathode is made of nickel, nickel with a few percent of cobalt or silicon, platinum or Konal (an alloy made of nickel, cobalt, iron, and titanium), and then coated with a mixture of barium and strontium oxides.

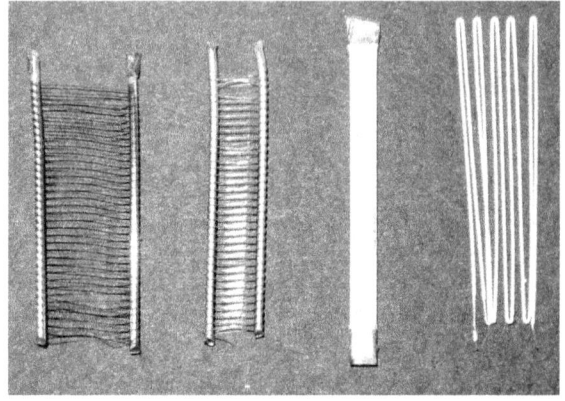

ABOVE: (L-R): The screen grid and the gold control grid with their vertical mechanical support rods. Next is the white oxide-coated cathode sleeve and the folded heater filament pulled out of the cathode sleeve of 6L6 beam tetrode.

The heater is a tungsten wire loop coated with a heat-resistant insulating material such as aluminium or beryllium oxide, placed inside the cathode sleeve. This coating conducts the heat to the cathode and insulates it electrically from the heater wire. The oxides provide electrons at temperatures as low as 750°C (dull-red heat).

Oxide-coated cathodes have many advantages over filamentary ones, the most important of which (from the user's perspective) is their much longer life of several thousand hours.

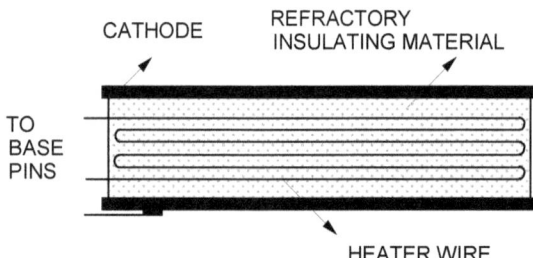

ABOVE: Construction of an indirectly-heated cathode-heater assembly

The space charge or "electron cloud"

A thermionic diode is the simplest vacuum tube with only two electrodes, a cathode and an anode. Assuming a filamentary diode, where the heater acts as a cathode (for instance, 5Y3 or 5U4), the filament heats up and emits electrons once the heater voltage is applied.

If there is no positive voltage on the anode, the emitted electrons disperse into the surrounding space and form a cloud or "space charge" around the cathode. The cloud is negatively charged and repels newly-arrived electrons back to the cathode. Soon an equilibrium is established, meaning that as some excited electrons leave the cathode, as many others are forced back to it. The density of the cloud is the greatest near the filament and reduces with the increased distance from it.

When connected to a source of positive voltage VB ("B" for battery), the anode will start attracting negative electrons from the cloud. Those closest to the anode will be swept to it first. As electrons reach the anode, others come from the cathode and take their place in the electron cloud. The cloud is not affected by the anode, which cannot disperse it.

The electron cloud reduces the inter-electrode potential from the linear distribution (dotted line on the next page) and causes the tube's nonlinear character. This flow of electrons is an electric current through the vacuum and is said to be "space-charge limited."NOTE: The adopted convention for current flow is the opposite of the electron flow.

Anode characteristic of a vacuum diode

If we connect a variable DC voltage source across a diode (+ to the anode and - to the cathode) and hook up an mA-meter in series to measure the DC current, with an increased voltage between the anode and cathode V_{AK} (or "across the tube"), the current will increase exponentially. The anode draws electrons from the space charge, and the cathode replenishes these electrons. The anode current is limited only by the space charge, and the cathode can produce more electrons than the anode can attract.

However, above a specific voltage V_C the rise will become slower, the current has reached the saturation level. The anode has depleted all the electrons from the space charge, and the cathode cannot produce enough electrons to reestablish the space cloud. The anode current is said to be temperature limited. Normally tubes operate in the space charge region.

VACUUM DIODE IN DC AND AC CIRCUITS

To illustrate the fundamental properties of a vacuum diode, let's connect it to a source of relatively high DC voltage, 200V in this case. Without any load connected, the internal resistance of the diode will limit the current in this circuit, and all the power would be dissipated into heat inside the diode.

However, the DC current through the diode may exceed the maximum value declared by the manufacturer, and the device may be damaged or destroyed, so we need to connect a load in series with the diode. It does not matter where we place it, between the anode and the DC source (as illustrated) or between the cathode and the source.

To enable us to perform the DC analysis of this simple circuit, numerical values have been chosen, 200V for the DC source and load resistance of 25,000Ω.

The basic concept of Ohm's law applies here, but Ohm's law assumes linear resistances and here we have a nonlinear diode, whose resistance we don't know. Moreover, the diode's resistance changes with the magnitude of the current flowing through it. Likewise, the voltage drop across the diode is not linearly proportional to the current.

There are two ways to analyze this kind of circuit. One uses the I-V curves for the specific diode given by its manufacturer to find the solution graphically. The other option is simplifying things prudently and "linearizing" the diode. We will do that soon while performing the AC analysis.

The circuit is a simple voltage divider. Depending on the static resistance of the tube and the load, the DC source's voltage V_B will be split in some proportion between the two: $V_B = V_L + V_A$ ("A" for anode-cathode voltage)!

The I-V curve for the diode is given (by its manufacturer), and on it, we draw a "load line" for the resistor R_L (next page). We only need two points to define a line. One is the battery voltage V_B (200V). We mark that point on the voltage or X-axis (point A). The second point can be chosen arbitrarily. Simply choose a voltage and calculate the corresponding current.

For instance, if we move 200V to the left of point A (right on the current or vertical-axis), the current in that point B would be $I_B = 200/25,000 = 8$ mA. We mark that value and have our point B. Now we draw the load line for $R_L = 25k\Omega$.

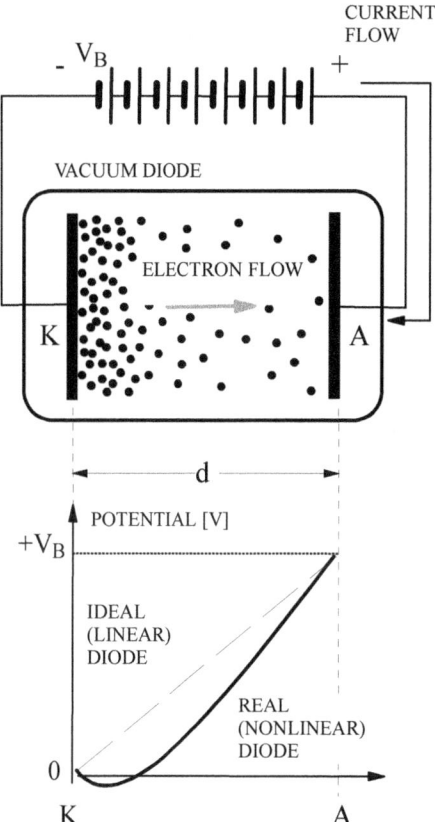

ABOVE: The physical behavior of a vacuum diode
BELOW: The anode characteristic of a vacuum diode for two different cathode temperatures (T2 > T1).

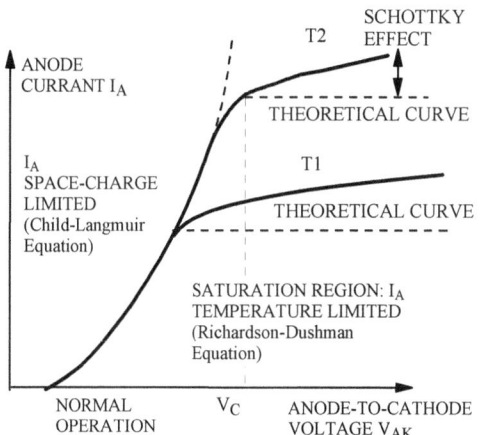

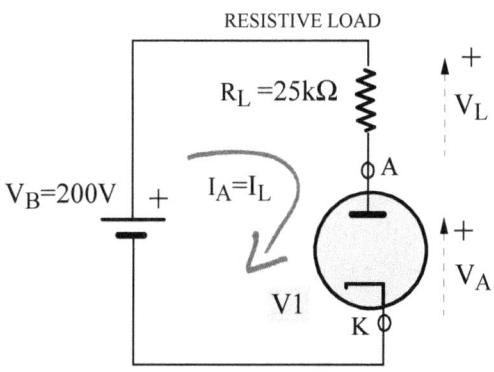

ABOVE: Vacuum diode V1 in a simple series DC circuit with a resistive load

Once we've drawn the R_L line, we get the intersection of the two I-V curves. That is the operating or "quiescent" point Q. Draw a horizontal line towards the vertical or current axis and read the value of the current in point Q as I_Q = 3.3 mA. The current through the diode tube ("anode current") and the load is the same current $I_Q = I_L = I_A$, so we can call it simply I.

In a similar fashion, draw a vertical line down to the voltage axis and read V_A = 120V. That is the voltage drop on the diode tube. The rest of the 200V battery voltage is dropped across the load resistor (V_L = 200-120 = 80V). We have "solved" our circuit, we know all its currents and voltages.

The "static" or DC resistance of the tube R_I in point Q is the voltage drop across it divided by the current:

$R_I = V_A/I_L$ = 120V/0.0033A = 36,364Ω or approx. 36kΩ.

For each value of the load, the operating point Q will be different. Two additional cases are illustrated, for the load resistor of twice the previous value or 50kΩ and half the previous value or 12.5kΩ (steeper curve).

For increased load resistance the current decreases and the operating point moves to the left, meaning less and less voltage is dropped across the tube and more and more voltage is applied across the load. As an exercise, calculate the internal DC resistance of this tube in Q1 and Q2.

"Plate resistance" is a term that may be confusing to some. It is not the electrical resistance of the material the anode or "plate" is made of, but the parameter used to model the behavior of a tube as a variable resistance, when viewed from the outside, from the circuit perspective. So, instead of "plate" or anode resistance of a tube we will use "internal resistance".

Vacuum diode in AC circuits

A vacuum diode V1 is connected across a source of AC voltage v, and a load resistor RL is connected between its cathode and the ground. With the same topology as before, we have added a source of a sine-voltage. What will be the waveform and amplitude of the AC voltage on the load resistor?

Again, we can solve this circuit graphically in a tedious, point-by-point fashion. For each point on the sine wave of the input voltage, we would find a projected point reflected of the diode's I-V curve.

The positive halves of the sine voltage are passed on by the vacuum diode. The valve is fully open, and the AC current flows. The diode does not conduct negative during negative peaks because the AC signal makes its anode negative with reference to its cathode. The valve is closed, and no current flows through it.

This unidirectional property of the diode makes it a rectifier because it "rectifies" or "straightens" the input AC signal whose average value is zero (positive peaks cancel negative peaks) into a rectified pulsating signal where there are no negative currents, so there is a positive average value of the rectified signal.

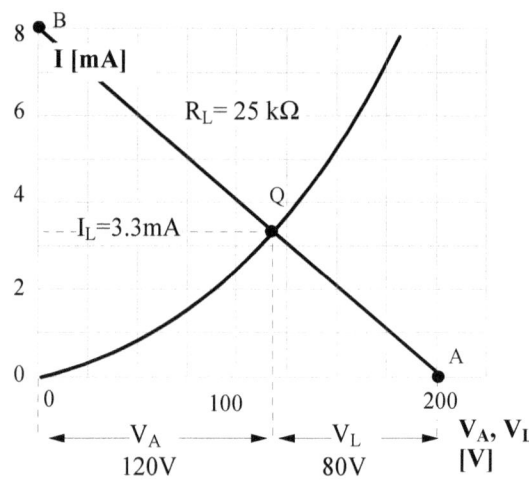
ABOVE: Graphical analysis of the series DC circuit with a vacuum diode

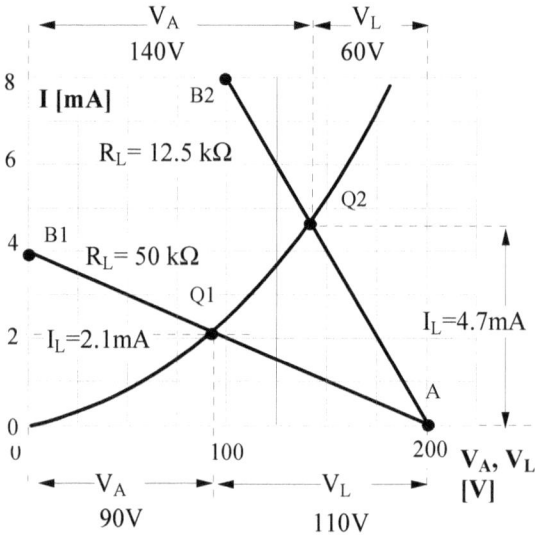
ABOVE: Graphical analysis of the series diode DC circuit for two different load values

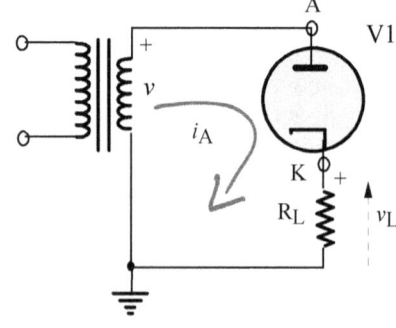
ABOVE: Vacuum diode V1 and its load connected to an AC voltage source. The load is the cathode circuit.

As tube tester manufacturers do in their user manuals and promotional literature, we can loosely call this rectified signal a DC current or voltage. Still, it is far from ideal, steady DC voltage. Nevertheless, it is a crucial first step in AC-DC conversion.

Notice a few important things. The waveform thus constructed is the current through the circuit, not the voltage. The voltage waveform would be of the same form (shape) for a resistive load, proportional to this current.

HOW VACUUM TUBES WORK

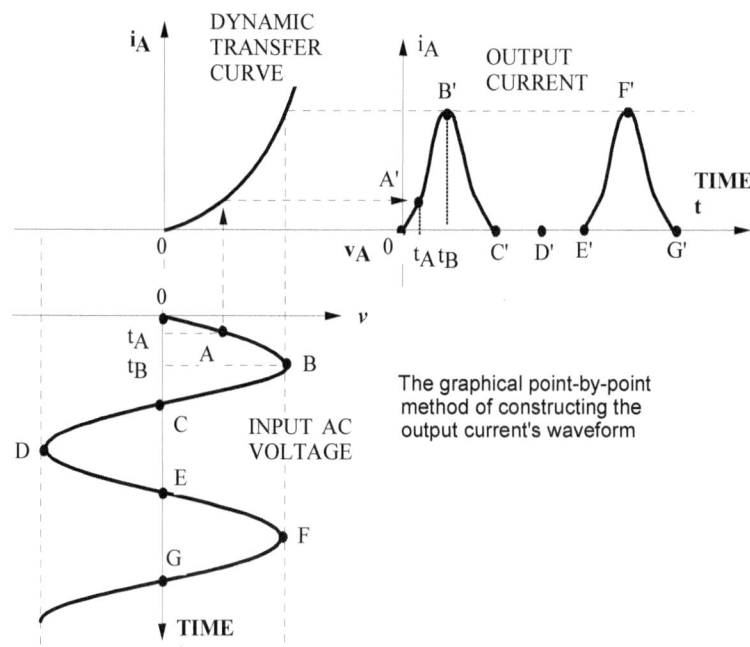

The graphical point-by-point method of constructing the output current's waveform

Also, the output signal is distorted due to the curvature of the diode's transfer curve. A linear transfer curve (a straight line) would result in no distortion of the rectified signal.

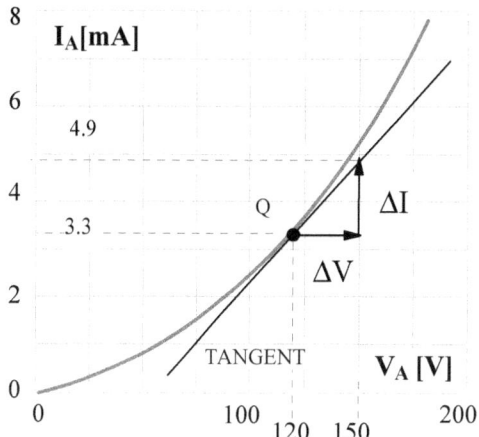

ABOVE: Determining the internal dynamic resistance of a diode from its I-V curve

Dynamic resistance of a vacuum diode

The vacuum diode's resistance to AC currents, or impedance, differs from its resistance to DC currents. The dynamic resistance at any point along the I-V curve is the slope of the tangent at that point. Once we draw the tangent, we arbitrarily choose DV and DI and read their two axis values. Greek capital letter "delta," symbol Δ, stands for "difference." We have $\Delta V = 150-120 = 30V$ and $\Delta I = 4.9-3.3 = 1.6$ mA, so the dynamic internal resistance of the tube is $r_I = \Delta V/\Delta I = 30/0.0016 = 18,750\ \Omega$ or $18.75 k\Omega$.

The symbol for dynamic or AC resistance is a lower case letter r. In this case, the AC internal resistance of the diode is about half its DC resistance R ($18k\Omega$ versus $36k\Omega$).

ELECTRON TUBES - NAMING CONVENTIONS, BASES AND PIN NUMBERING

Vacuum tubes, as they are called in the United States, have also been called "electronic tubes" or "thermionic valves." The term "valve" is used mainly in the UK and Australian English, and in other languages, like "valvola" in Italian or "válvula" in Spanish. In some countries, the term for the electronic tube is "lamp," such as "lampa" in Serbian and "lampe" in French. Germans call it "Röhre."

The term electron "tube" probably comes from the tubular shape of the glass bulb, and "valve" most likely originated in the mechanical-electrical analogy behind the operating principle of the triode. Just as turning the handle on a mechanical valve controls the flow of water through it, changing the voltage (or bias) on the tube's control grid modulates the flow of electrons through it.

Americans use the colloquial term "plate" instead of the proper term "anode." We will use "vacuum tube" and "anode" in this book.

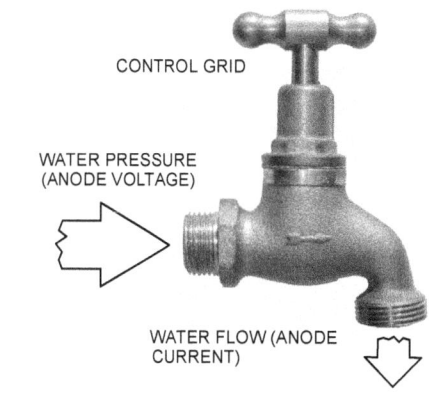

ABOVE: The water tap analogy of a triode explains why vacuum tubes are also called "valves".

Tube naming "conventions"

European tube makers tried to follow some kind of naming system. The same cannot be said for their American competitors. The European system uses four letters followed by 2 or 3 digits. Let's look at a few examples.

ECC83 means a double triode (CC) with 6.3V heating (E) and using a Noval socket (8).

ECC40 is a double triode (CC) with 6.3V heating (E) and uses a RimLock socket (4).

EL34 is a power pentode (L) with a 6.3V heater (E) and an Octal socket (3).

PL519 means a TV output pentode (L) with 300mA heater current (P), "5" indicates a Magnoval socket.

ECF86 is a 6.3V (E) voltage amplifier triode (C) and preamplifier pentode (F) using a Noval socket (8). PY88 is a single diode (Y for "half-wave rectifier"), with a 300mA heater (P) using a Noval socket (8).

Special quality tubes were marked SQ or named by reversing the order of two of the letters and numbers. For instance, E82CC is the SQ version of ECC82, while E86C is an SQ version of EC86.

The KT prefix (KT66, KT88, etc.) means "Kinkless Tetrode." The number has no meaning, apart from the fact that the higher the number, the higher the anode dissipation of the tube.

The American naming convention is less consistent and not at all self-explanatory. The first number indicates heating voltage, so 12AX7 is a tube with a 12.6V heater, while 6L6 has a 6.3V heater.

The last digit indicates the number of electrodes + heater, so the 2A3 triode has three elements; one electrode is a heater as well (directly-heated cathode). 6L6 has five electrodes + a separate heater, for a total of six elements. 12AX7 has six electrodes (two triodes) and a common heater, thus 7 in total.

Then there are "numerically-named" tubes such as 6922, 5687, and many others, which are usually the "industrial" versions of their consumer-aimed equivalents. These numbers have no meaning, and that is why I titled this section naming "conventions" instead of naming "standards." As different tube manufacturers developed new tubes, they simply named them at their convenience without any regard for standardization.

Not to be outdone, the USA military established their own naming, where numbers are usually preceded by the acronym JAN, which stands for "Joint Army-Navy."

Tube bases and pin numbering

No matter what base a tube uses, the pins are always numbered in a clockwise fashion if viewed from underneath (pins facing you). The same view is of their sockets if viewed from the lug side (where components are soldered).

1st number	
3	Octal
4	RimLock
5	Magnoval (large 9-pin)
8	Noval (small 9-pin)
9	Miniature 7-pin
1st letter	
A	4.0V heater
B	180mA heater
C	200mA heater
D	Battery-powered tubes, 1.25-1.4V heater
E	6.3V heater voltage
F	13V heater
G	5.0V heater
P	300mA heater current
U	100mA heater current
O, Z	Cold cathode tubes
2nd, 3rd & 4th letter	
A	Single diode
B	Duo-diode
C	Triode (voltage amplifier)
D	Power triode
E	Tetrode
F	Preamplifier pentode
H	Hexode
L	Output (power) pentode
M	Indicator tube
N	Thyratron
X, W	Gas-filled tube
Y	High voltage half-wave rectifier
Z	High voltage full-wave rectifier

PIN NUMBERING DIRECTION

NOVAL   OCTAL

TRIODE PARAMETERS AND CHARACTERISTICS

Triode "curves" in 3D

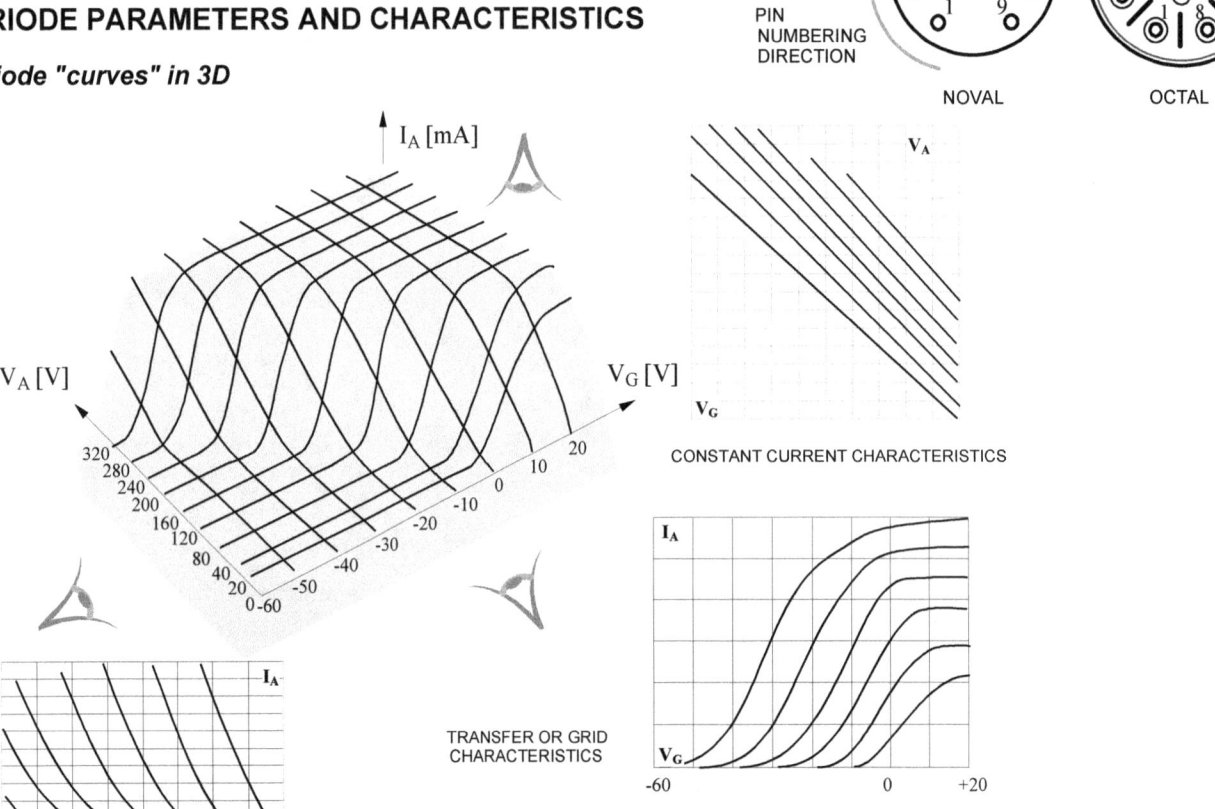

ABOVE: To visualize and understand triode's behavior, a contour needs to drawn in a 3-D space. Alternatively, three projections can be used, so the characteristics can be drawn in two dimensions.

HOW VACUUM TUBES WORK

Since there are three variables of interest, grid voltage, anode voltage, and anode current, the triode's behavior is a contour in 3-dimensional space. Dealing with 3D curves on 2D paper is awkward; drawing two variables in two-dimensions is much easier, with the third variable as a parameter.

There are three possible triode graphs. The I_A vs. V_A graph with V_G as a parameter is also called anode or plate characteristics. The I_A vs. V_G graph with V_A as a parameter depicts one or more transfer curves, and V_A vs. V_G graph with I_A as a parameter is called "constant current characteristics" (since I_A is kept constant, which is the meaning of "parameter")!

Current control by triode's grid, amplification factor and "reachthrough"

We have seen that the current through a vacuum diode is a function of one variable, the anode voltage: $I_A = KV_A^{3/2}$. In triodes, apart from depending on the anode voltage, the anode current is a function of another variable as well, the voltage on the control grid: $I_K = K(V_G + DV_A)^{3/2}$

Since $I_K = I_G + I_A$ (cathode current is a sum of the grid and anode currents), and in most circuits (but not all!) the grid current is negligibly small compared to anode current and can be considered zero, this equation is usually written as $I_A = K(V_G + DV_A)^{3/2}$, but that is only valid if $I_G = 0$!

The constant K is called perveance, as in a diode, and the constant D comes from the German noun *Durchgrief*, meaning "reach through" since the anode voltage and its electric field have to reach through the grid to "get" electrons from the cathode. It is a measure of the degree of anode's control over the current, its "effectiveness".

The concept is not used in English literature at all; Anglo Saxons, for some reason, exclusively use the inverse concept called voltage amplification factor, for which the Greek letter μ (pronounced mju) is used.

Since $\mu = 1/D$, they write this equation as $I_A = K(V_G + V_A/\mu)^{3/2}$, or, more often, they use the alternative way of expressing anode current, $I_A = A(\mu V_G + V_A)^{3/2}$, where constant A loses its meaning, it is not perveance any more.

For instance, the D-factor for LS50 power pentode is 20% or 0.02, meaning its amplification factor is 1/0.02 = 5. For 12AX7 triode, with μ=100, D is 1/100 = 0.01 or 1%. These differences can be interpreted in the sense that the anode current in 12AX7 is much more sensitive to grid voltage changes (20 times more sensitive, since its μ is 20 times higher!) than for the LS50 pentode.

Using the reach through figures for interpretation, in LS50 with D=20%, 20% of anode current changes are due to the anode voltage influence and 80% as a result of grid voltage changes, while in the 12AX7 triode, only 1% of anode current changes come from changes in the anode voltage and 99% result from the grid action, meaning that the grid is 20 times more effective in the 12AX7's case.

How to read tube data sheets

Most USA-made tube testers were designed so their users need not refer to the tube's datasheet to test it. All necessary tester settings were given in a roll-chart or tube setup book (often a separate document from the user's manual). However, to understand tube testing principles and methods, we need to understand the information given in their datasheets.

This Raytheon data sheet (next page), or rather the 1st page of the datasheet, was chosen to illustrate acceptable deviations in tube parameters. Usually, datasheets from other tube makers did not include such figures. Some were very basic, almost useless, others very detailed, going for 30 or so pages, with most falling somewhere in between.

It pays to consult more than one datasheet for the same tube type, so if you are confused, or something is not clear, simply download a few other datasheets from online sources and compare them.

The "Mechanical data" (1) and "Mechanical ratings" (2) sections are of no particular interest to us, the verbal description of the tube (3) isn't of much interest either, but the "Basing" and "pinout" info is very important (4).

Of most interest is the "electrical Data" section, in particular, three columns, "Normal operation," "Test limit or design minimum," and "Test limit or design maximum." It is not clear what the difference is between "Normal operation" and "Normal Test Conditions"; all specified figures are identical.

Some parameters, such as plate voltage, don't have a minimum, only a maximum, in this case, $330V_{DC}$.

Some information is misleading. For instance, under "plate dissipation," while the maximum allowed is 2.5 Watts, the normal operation is not 2.5 Watts; it can be any dissipation that the circuit designer chooses, up to 2.5W.

Starting with heater current, notice that the allowable tolerances are from 142 to 158 mA (for 12.6V heating), so -9.5% and +10.5% from the nominal 150mA.

The amplification factor μ (mju) is allowed to vary from 50 to 70, or from 83.3% to 116.7% of the nominal 60, so +/-16.7% or a total of 33.4%.

Technical Information

12AT7WA

RELIABLE MINIATURE DOUBLE TRIODE

❸ The 12AT7WA is a heater-cathode type, high-mu double triode of miniature construction, suitable for use in amplifier, mixer, oscillator, multivibrator and computer circuits. The high amplification factor (60) makes it ideal for audio amplifier service where a gain of 40 is easily realized. The low plate resistance plus the ample peak plate current make this type ideal for pulse amplifier and low power servo amplifier service. This type is controlled for cathode interface and is designed for dependable operation under conditions of shock and vibration encountered in mobile, aircraft and missile applications. A heater center tap is provided to permit operation at 12.6 volts or parallel operation at 6.3 volts.

MECHANICAL DATA

ENVELOPE	T 6½ Glass
OUTLINE	JEDEC (6–2)
BASE	9 Pin Miniature
BASING	9A
MOUNTING POSITION	Any

❶ PHYSICAL DIMENSIONS

MECHANICAL RATINGS:
❷
- Maximum Impact Acceleration (Shock-Test-Note 3) .... 630 G
- Maximum Vibrational Acceleration (96 hour Fatigue Test-Note 4) .... 2.5 G
- Maximum Bulb Temperature .... 165 °C
- Altitude .... 60,000 ft.

ELECTRICAL DATA

❺

Ratings and Normal Operation	MIL-E-1 Symbol	Test Limit or Design Minimum	Normal Test Conditions	Normal Operation	Test Limit or Design Maximum	MIL-E-1 Units
Ratings						
Heater Voltage Series	Ef:	11.4	12.6	12.6	13.9	V
Heater Voltage parallel		5.7	6.3	6.3	6.9	V
Plate Voltage	Eb:	---	250	250	330	Vdc
Grid Voltage	Ec1:	−55	0	0	0	Vdc
Plate Dissipation (per Plate)	Pp/p:	---	---	2.5	2.5	W
Heater-Cathode Voltage	Ehk:	---	---	---	±100	V
Grid Resistance Fixed Bias	Rg/g:	---	---	---	0.25	Meg.
Grid Resistance Cathode Bias	Rg/g:	---	---	---	0.5	Meg.
Tests						
Plate Current (1) (per plate)	Ib/p:	7	---	10	14	mAdc
Cathode Resistance (per cathode)	Rk/k:	---	200	200	---	ohms
Transconductance per plate	Sm/p:	4500	---	5500	6500	μmhos
Amplification Factor	Mu/p:	50	---	60	70	---
Heater Current (Series)	If:	142	---	150	158	mA
Plate Current (2) Cut-off (Ec1 = −20 Vdc)	Ib/b:	---	---	---	100	μAdc
Plate Current (1) difference between sections	ΔIb:	---	---	---	3.2	mAdc

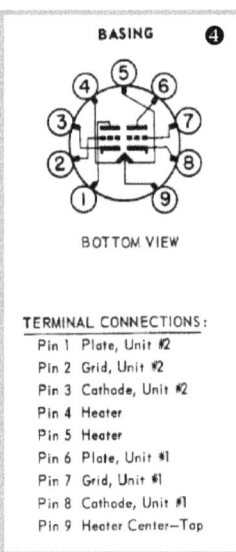

❹ BASING

BOTTOM VIEW

TERMINAL CONNECTIONS:
- Pin 1 Plate, Unit #2
- Pin 2 Grid, Unit #2
- Pin 3 Cathode, Unit #2
- Pin 4 Heater
- Pin 5 Heater
- Pin 6 Plate, Unit #1
- Pin 7 Grid, Unit #1
- Pin 8 Cathode, Unit #1
- Pin 9 Heater Center-Tap

Likewise, mutual conductance Gm is specified as 5,500 micromhos (5.5mA/V) at the anode voltage of 250V and anode current of 10mA. Assuming that "test limits" mean "reject limits", presumably Raytheon would reject tubes with gm below 4,500 and above 6,500 micromhos in that test point!

Percentage-wise, these are 81.8% and 118.2% or -18.2% and +18.2%, or a whopping 36.4% total variation. They certainly allowed themselves a lot of wiggle room, didn't they?

CASE STUDY: STATIC PARAMETERS OF 12AX7 (ECC83) DUO-TRIODE

Transfer characteristics

Since 12AX7 is not just the most common preamp tube in guitar amps but is also used in hi-fi amps and preamps, let's use it as a practical example to learn about the three most important triode parameters and three types of triode characteristics.

Transfer characteristics show how anode current I_A changes with a DC bias on the control grid (V_G), with anode voltage V_A as a parameter (kept constant). There is a different curve for each value of anode-to-cathode voltage V_A, so we are talking about an infinite number or a family of curves.

Manufacturers usually publish at least two or three of them. We have six here (next page), from 50V to 300V.

TUBE PROFILE: ECC83 (12AX7)

- High μ duo-triode, Noval socket
- Heater: 6.3V/300mA or 12.6V/150mA
- V_{AMAX}= 300V_{DC}, V_{HKMAX}=180V_{DC}
- P_{AMAX}=1W, I_{KMAX}=8 mA
- TYPICAL OPERATION:
- V_A=250V, V_G=-2.0V, I_0=1.2 mA
- Gm = 1.6 mA/V, μ=100, r_I = 62.5 kΩ

ABOVE: In my books on tube amplifiers I use this kind of "tube profile" to outline its most important parameters. Apart from typical operation in a circuit and static parameters <u>in that operating point</u>, maximum anode & screen voltages, anode dissipation and cathode current are also of interest to circuit designers.

HOW VACUUM TUBES WORK

As V_A is increased, a higher negative grid bias is needed to keep the anode current I_A at the same level (draw a horizontal line through all six curves). The anode voltage has a large impact on the triode's anode current. That is not the case for tetrodes and pentodes, as we will see later. Notice the word "average" (1) printed by the manufacturer. That means that each tube's curves will be slightly different, so take these as an approximation only!

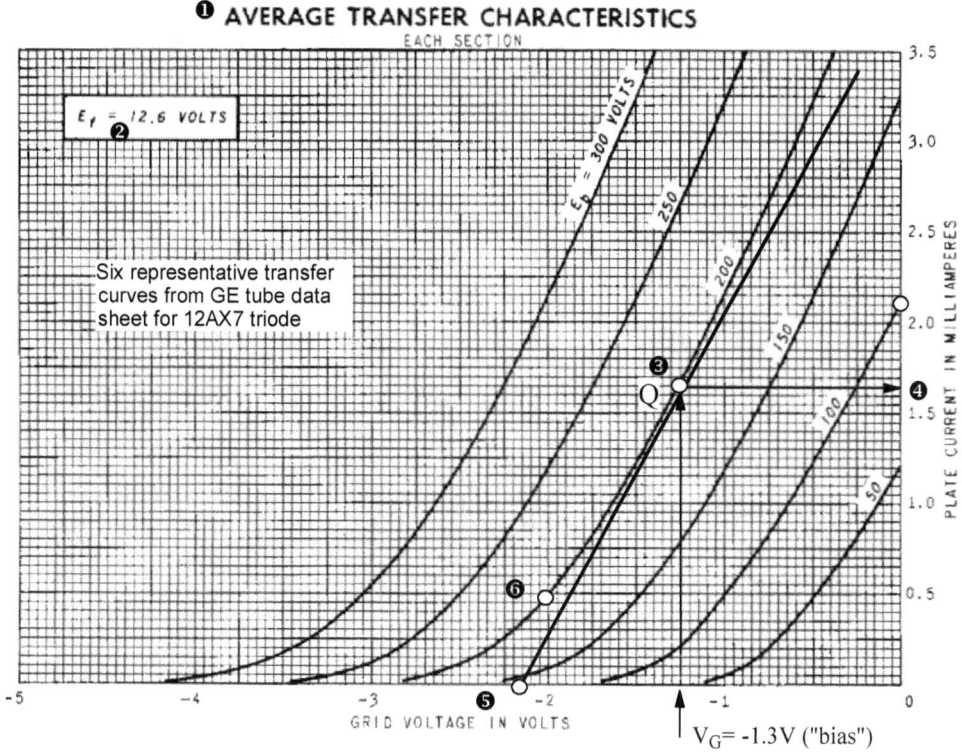

The heater voltage is also specified (2); in this case, E_F=12.6V. "F" is for filament (heater).

That indicates that the curves for the same tube (in this case, the "bogey" or "average" tube) will also vary with heater voltage changes!

Let's take one TX curve now as an example, V_A=200V, and position our operating "idle" or quiescent point at the grid bias of -1.3V (3). That "bias" will determine the anode or "plate" current, which we read by drawing a horizontal line to the right until we reach the vertical or anode current axis, I_A= 1.65 mA (4).

Transconductance or mutual conductance

A derivative "d" of a function is the slope of a tangent to that function in a particular point. In the first approximation, the derivative "d" can be replaced by a small difference, for which the Greek capital letter Delta is used (Δ).

Transconductance or *mutual* conductance between the control grid and anode (Gm in English) is defined as a change in I_A (anode current) caused by a small change in V_G (DC voltage on control grid or "bias" voltage): Gm=$\Delta I_A/\Delta V_G$ That is the slope or steepness of the transfer curve in any chosen point. In general, conductance is inverse of resistance: G=1/R, so the unit for conductance is 1/Ω or S (Siemens), where S=A/V ("ampere per volt").

You can use any two points to read Gm in that point. For instance use the "projected cutoff" point (5) and the Q point (3). So we get a change in bias from -2.2V to -1.3V, so ΔV_G = 0.9V. At the same time the anode current increased from zero in point (5) to 1.65mA in (4), so ΔI_A=1.65mA! Now we can calculate Gm=$\Delta I_A/\Delta V_G$ =1.65 mA/0.9V = 1.83 mA/V. This means that for each Volt of control grid voltage the anode current jumps 1.83mA! That is actually quite a low transconductance for a vacuum tube. Some tubes have Gm in the 20-30 mA/V range.

Americans use "micromho" as a unit for Gm, "mho" being a "unit" for conductivity, a "reverse" of "ohm," a unit of resistance, but that should be "inverse" (1/Ω) and not "reverse"! Prefix "micro" means 10-6 or one-millionth part of "mho." So, 1mA/V = 1,000 "micromhos".

The upper part of the TX curve (higher anode currents affected by lower bias voltages) is quite linear, meaning that its slope is constant, and thus Gm is constant. However, the bottom part of the TX curve is very nonlinear; the slope reduces. Draw a tangent and read Gm in point (6) for V_G= -2.0V, and you will get a much lower figure.

Internal resistance and voltage amplification factor of a triode

As we have already seen with vacuum diodes, the anode curves describe the relationship between anode current and the voltage drop across the tube (between its anode and cathode). The slope of the curves is inversely proportional to the internal resistance of the tube, and we can determine such resistance graphically from the curves.

The slope is low at low anode currents, meaning the internal resistance is high, and as anode current goes up, the slope rises, and the internal resistance falls. Let's choose a couple of points at the same anode voltage (200V).

Notice that there is no curve for V_G= -1.3V, only for -1V and -1.5V. However, the curve is there; it's just that manufacturer did not publish it on this graph that we redraw from the datasheet. We can approximately position it by copying the -1V curve and pasting it slightly closer to the -1.5V than the -1V curve.

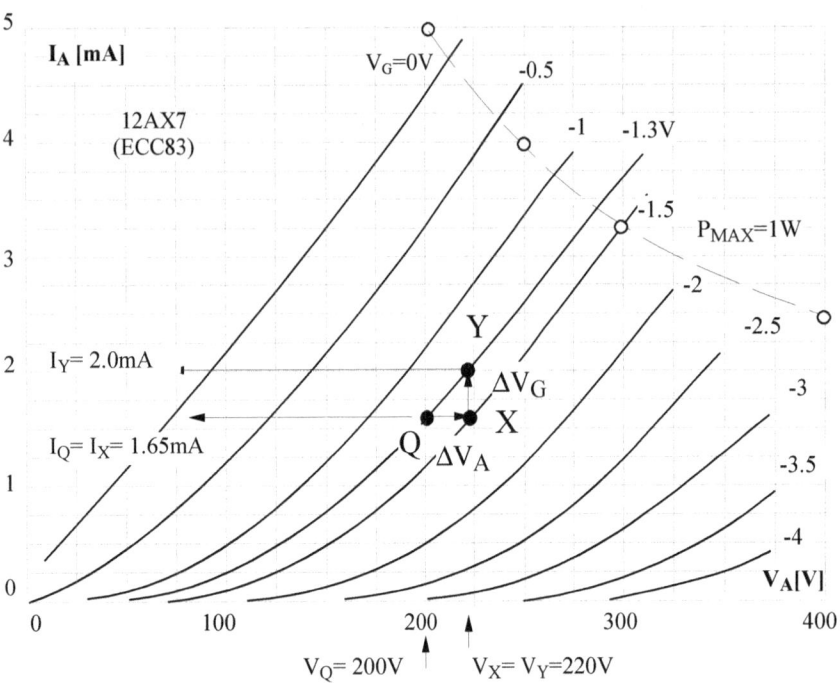

To determine the tube's internal resistance in point Q (I_A=1.65mA), move to the right until we hit the V_G= -1.5V curve. Read the anode voltage in that point (X), 225V, and move upwards to the -1.3V curve (point Y). The anode current is I_Y= 2.0mA.

ΔV_A= 220-200 = 20V, and that jump in anode voltage caused anode current to jump ΔI_A = 2.0-1.65 = 0.35 mA. The tube's internal resistance in point Q is r_I= $\Delta V_A/\Delta I_A$ = 20/0.00035 = 57,143 V/A or 57.14 kΩ.

Tube's voltage amplification factor is defined as the change of anode voltage divided by the change of grid voltage that caused it: $\mu=\Delta V_A/\Delta V_G$! The two identical units (volts) cancel one another, so m is dimensionless. Since ΔV_G=1.5-1.3=0.2V, $\mu=\Delta V_A/\Delta V_G$ = 20V/0.2V = 100.

ABOVE: Estimating 12AX7 triode's internal resistance r_I in points A and B, and its amplification factor in point C, using its anode characteristics.

BELOW: How the three static parameters of 12AX7 triode vary with anode current levels

The Barkhousen's equation

Now that we know how to read the tube parameters from the graphs, the good news is that we usually don't have to bother doing it at all. In most cases, at least for tubes intended for audio applications, manufacturers had published the μ-Gm-r_I graphs.

Of course, there will be a different graph for each anode voltage. This one (right) is for V_A=100V.

We cannot even position our point Q here since the horizontal or current axis only goes up to 1.6mA, and we had 1.65mA, due to the much higher anode voltage of 200V we used in the previous example.

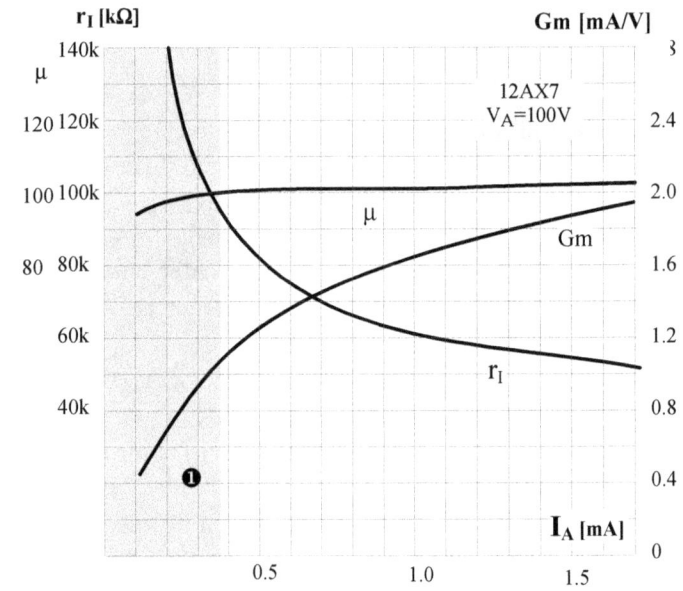

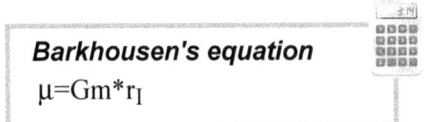

Barkhousen's equation
$\mu=Gm*r_I$

Notice how μ (amplification factor) is constant for almost the whole range of anode currents, except the very low current range. In that region, the internal resistance of the tube shoots up, while Gm drops rapidly off as the anode current drops. Normally, an audio designer would avoid using a tube in a low anode current regime (1).

The visual relationships of the three curves indicate that Gm and r_I change inversely to one another, so we could speculate that their product would remain more or less constant, and sure enough, another gentleman had done it all a long time ago. In his honor, the simple equation that links the three parameters, $\mu=Gm*r_I$, was named Barkhousen's equation.

We estimated the tube's Gm at point Q as 1.83 mA/V and its internal resistance at the same point as rI=57,143 ohms. r_I=57,143 ohms. So, the voltage amplification factor μ should be $\mu=Gm*r_I$ = 1.83 * 57.143 = 104.6, and we got μ=100 from our third graphic estimation.

That is incredibly close for such imprecise graphs and "eyeball" readouts from a computer screen (in my case) and a small printed graph (in your case). Again, don't lose your sleep even if you get +/-10% discrepancies, this is an estimation tool only, and such accuracy is usually adequate for design or troubleshooting purposes.

TESTING & MATCHING VACUUM TUBES

2

- TUBE PARAMETERS AND THEIR VARIATION WITH TEST CONDITIONS
- TUBE TESTING PRINCIPLES AND PROBLEMS
- NEGATIVE GRID CURRENT - CAUSES AND DETECTION METHODS
- TESTING CASE STUDIES AND PRACTICAL EXAMPLES OF TUBE MATCHING
- HOW TO TEST A TUBE THAT IS NOT LISTED IN THE BOOK OR TUBE CHART
- EXPERIMENT: COMPARISON OF TRIPLETT 3444, B&K700 and MERCURY 1000 TEST VOLTAGES FOR A FEW COMMON AUDIO POWER TUBES

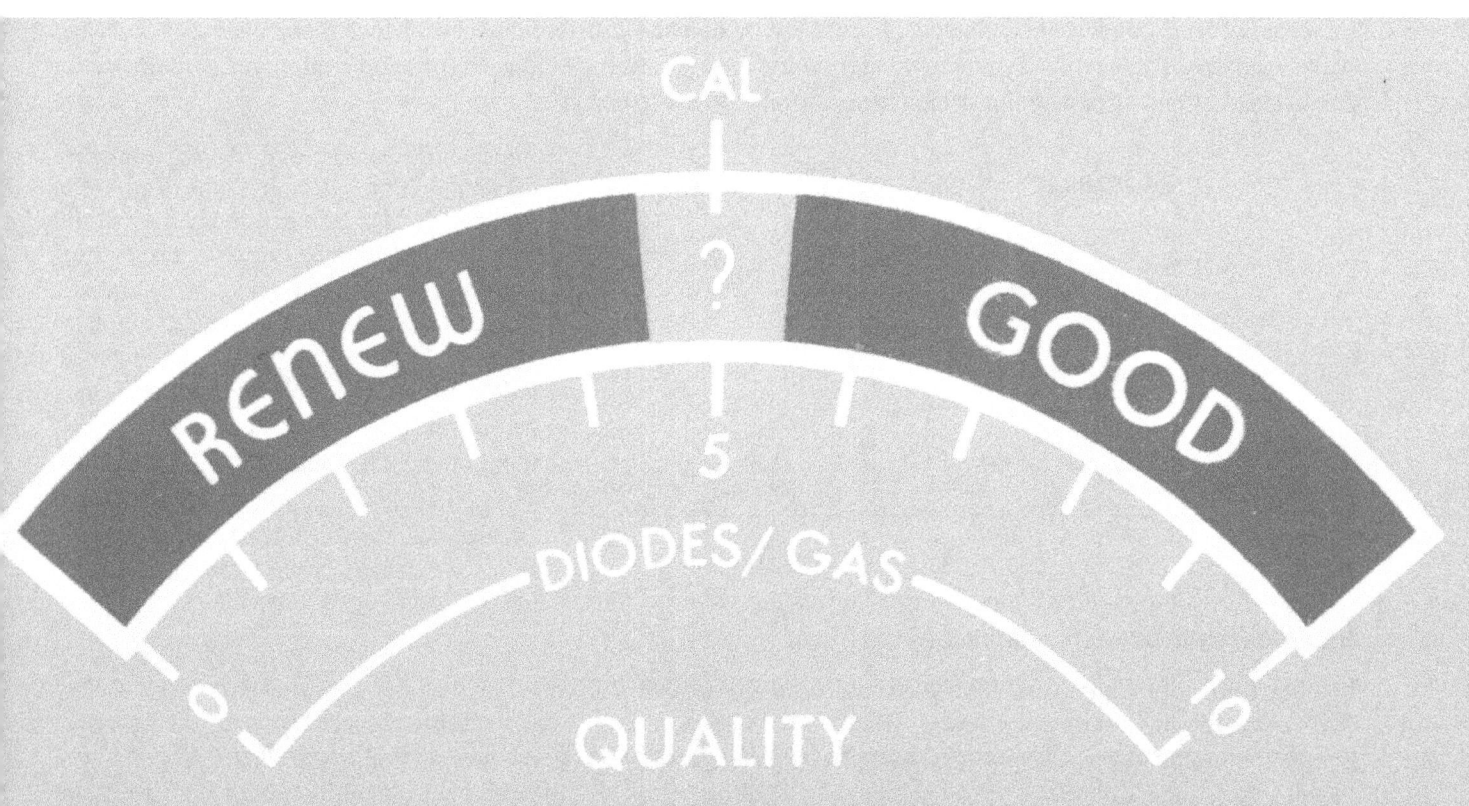

TUBE PARAMETERS AND THEIR VARIATION WITH TEST CONDITIONS

Variations in tube parameters

One major source of error or discrepancy between estimated or calculated figures (during the design of a circuit) and real, measured figures after an amplifier is built is the variation in tube parameters.

The published data sheets should be used with reservation, only as a starting point. Actual tube characteristics could be significantly different, especially for certain brands and types of tubes.

Tubes are made of many mechanical parts (grid, cathode, heater, anode, supports, insulators, etc.) that are manually assembled, so the spacing between components cannot be fully controlled. That is the leading cause of the varying electrical parameters even amongst the same batch of tubes.

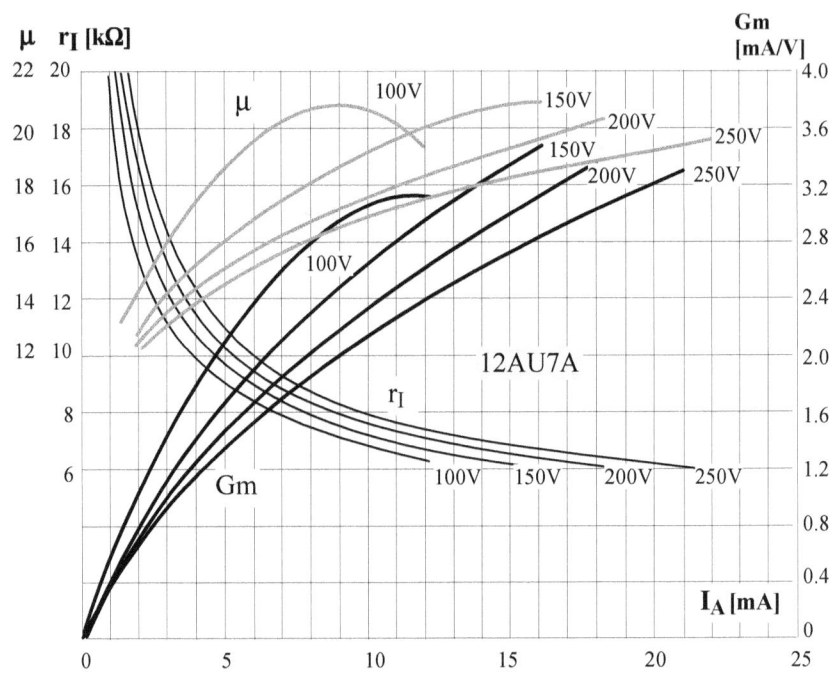

ABOVE: How triode's parameters vary with anode voltage and current (12AU7A)

Factors such as slightly different materials used in different batches, metal for the heater, cathode, and other electrodes, and the chemical composition of the getter and the glass bulb itself, also impact the ultimate measurements of a tube.

The three dynamic parameters (Gm, μ, and r_I) can vary by as much as +/-40%! That means if you buy twenty unmatched tubes of the same type, whose nominal Gm is 2 mA/V, the 40% variation would be 2.0*0.4 = 0.8 mA/V. Therefore, the "normal" or acceptable range would be from 1.2 to 2.8 mA/V!

Anode currents can vary +/-20%, for the same type of tube and made by the same manufacturer! Heater currents vary a bit less, "only" +/-10%. To put things into perspective, without going too deep into the mathematics (or rather the statistics) of it all, we need to understand the normal or standard distribution curve, which mathematically describes such variations. The curve is symmetrical around the mean or average value ($\sigma = 0$) and continuous, with inflection points at $\sigma=+1$ and -1 (where the shape changes from convex to concave).

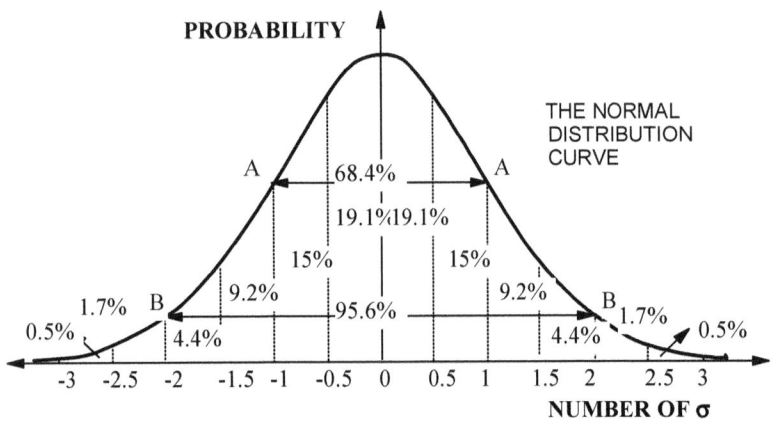

The distribution curve is a probability density function; it specifies the probability that a measured value would fall within certain limits expressed as multiples of standard deviation σ (sigma). These limits are marked on the graph as +/-σ/2, +/-σ, +/-2σ, etc. For instance, 68.4% of samples will fall within the one sigma range (+/-σ), and 95.6% will fall within the +/-2σ range.

If a picture is worth a thousand words, an experiment is worth at least ten times more, so let's calculate σ for a batch of NOS tubes.

How tube manufacturers grade tubes

Manufacturers of phono cartridges use a similar screening method to tube manufacturers. For instance, in their 2M series, Ortofon has four quality levels: "Red" is the cheapest model, followed by "Blue," "Bronze," and finally, top-of-the-range, "Black." The retail price ratio between the Black and Red models is $795/$129 or 6 to 1!

I cannot be sure about Ortofon, I've used them as an illustrative example only, but generally, the phono cartridges ("engines") of the same series are all identical.

The diamond tips may be different, and the housings are of different color/shape/design for marketing purposes, but due to manufacturing tolerances and other factors, some measure much better than the others and are selected for the top quality levels. The subsequent lower-spec cartridges are progressively downgraded into cheaper models. Tube selection follows the same methodology.

Once a series or batch of vacuum tubes is made, after the period of activating, maturing, and stabilizing (or "training"), the best tubes are selected based on the lowest noise, best-matched halves (for duo-triodes), lowest grid current, and other factors. These are labeled "premium" or SQ (Special Quality) tubes. In the old days, some of those had their pins plated in gold or their name reversed (E88CC would be a high-quality ECC88).

Gold plating does not carry the same weight as it once did; it is often simply a marketing gimmick aimed at unsuspecting users. Since very few end-users possess a laboratory- type tube tester that can test for noise, mA of grid current, heater-cathode leakage, and other parameters that ordinary tube checkers cannot detect, such audacity by some tube makers will continue unchallenged.

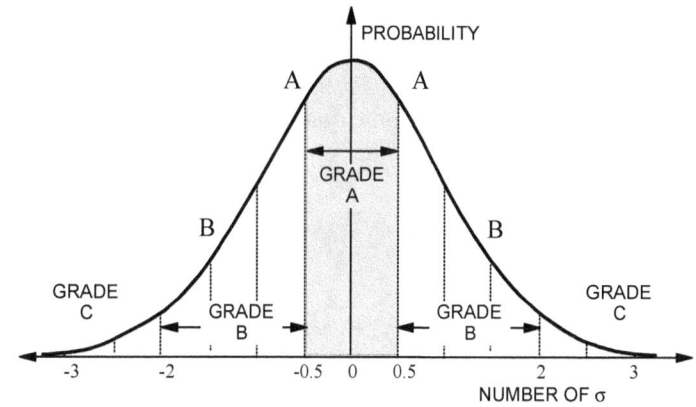

In vintage test equipment, you may see tubes with evacuation tips marked with heat-resistant red, blue, or green paint. These have been specially selected for that particular purpose or service based on specific criteria (low noise, balanced halves, high gain, wide bandwidth, etc.).

The next quality level tubes, let's call it "Grade B," are boxed individually and sold to OEM (original equipment manufacturers) and service centers.

Tubes from the tails of the distribution curve, those testing very high or very low, are sold in bulk, often even without the manufacturer's markings (so as not to compromise their reputation). These are then sold to resellers who flog them off to ignorant buyers.

ABOVE: One of many ways of grading vacuum tubes.
BELOW: Typical variations of mutual conductance of a batch of thirty 6L6-GA tubes.

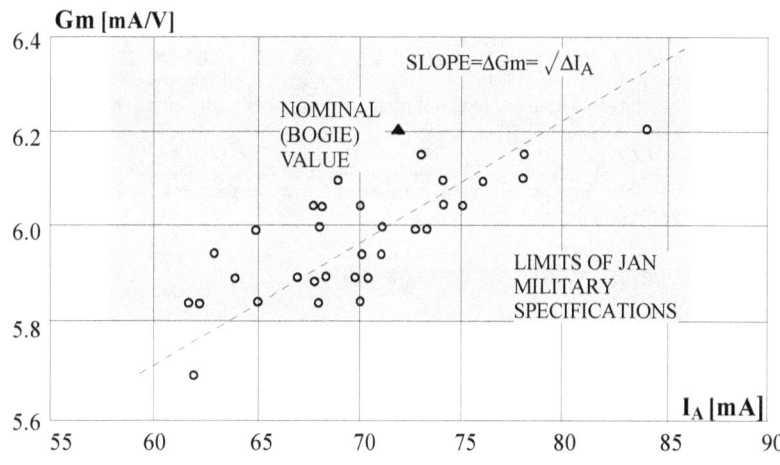

It may seem strange to label very strong-testing tubes from the positive tail of the distribution curve the lowest grade, but that is what they are.

If the nominal anode current of a power triode at $V_A=250V$ on the anode and -12V on the grid is 40mA, you don't want to plug a replacement tube with an anode current of 68 mA, do you? That would unbalance your push-pull output transformer and possibly even take it into saturation.

How Gm varies with anode and screen voltage changes

This hard-to-find information is handy for estimating the actual value of the tube's transconductance, primarily if your tester tests tetrodes and pentodes as triodes or if it tests them at an unusually low voltage way below manufacturer's nominal values, as most commercial testers do.

Say you've just bought a quad of EL12 tubes from an eBay seller in Germany, and you are testing them on your Mercury 1000 tester. Mercury testers don't have the socket for EL12, EL156, and other German power tubes, but you have made an adapter (see page 168).

All four measure around 10,500 micromhos, while the RSD (manufacturer) datasheet specifies 15 mA/V or 15,000 micromhos. Are the tubes defective, used, or weak (low in emission)? Remember, low emission would manifest itself as low transconductance.

The answer is no! Your tester measured the EL12 tube at a much lower plate voltage of $100V_{DC}$ instead of the nominal $250V_{DC}$ (specified in the datasheet). So, it shows the Gm to be $Gm_R = Gm_N * \sqrt{(100/250)} = Gm_N * 0.63 = 9.45$ mA/V, instead of the declared 15 mA/V, meaning there is nothing wrong with the tubes.

THE DIFFERENCE IN MUTUAL CONDUCTANCE WHEN TUBES ARE TESTED AT REDUCED INSTEAD OF NOMINAL VOLTAGES	THE DIFFERENCE IN MUTUAL CONDUCTANCE WHEN TUBES ARE TESTED AS TRIODES
$Gm(V_{RED}) = Gm(V_{NOM})\sqrt{(V_{RED}/V_{NOM})}$	$Gm(TRIODE) = Gm(PENT.) + Gm(SCREEN)$ $Gm(TRIODE)/Gm(PENTODE) = I_K/I_A$

Now, you may think that these tubes are unusually strong, since they measured 10.5 mA/V, while the nominal value at 100V on the anode is around 9.5 mA/V (as we have just established). Is that the case?

Again, no. Look at the second rule-of-thumb formula, which shows that the transconductance when pentodes are tested as triodes is slightly higher because it is an algebraic sum of anode transconductance and screen grid transconductance. The ratio of triode Gm and pentode Gm (always higher than one) is the same as the ratio of cathode current and anode current, since the cathode current is always higher than anode current, the difference being the screen grid current ($I_K = I_A + I_S$).

This applies to testers such as Mercury 1000/ 1200/ 2000 and Hickok 799, which test all amplifying tubes as triodes. That is why your tester tested the quad at around 10,500 micromhos instead of 9,500 micromhos. So, these tubes test around the nominal value and are most likely to be unused or NOS (New Old Stock) as declared by the seller.

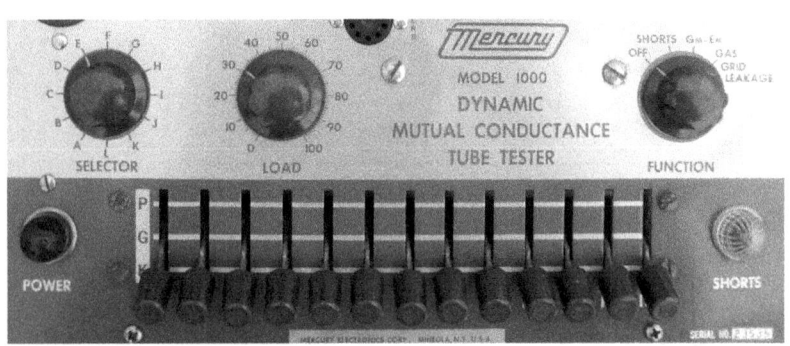

ABOVE: Mercury mutual conductance testers (models 1000, 1200 and 2000) connect all tube's electrodes to either P (plate), G (grid) or K (cathode), so all amplifying tubes are tested as triodes. The user does not have to worry about configuring heater pins which are pre-wired on sockets.

BELOW RIGHT: The not-often-published Ip (plate voltage) - VG (grid bias) graph with transfer cirves for a few typical anode voltages and half-a-dozen or so curves of constant transconductance

EXAMPLE: KT66 is a beam tetrode, close in parameters to 6L6. The published specifications are Gm=7.0 mA/V and r_I=22.5 kΩ, meaning that its amplification factor is $\mu = Gm*r_I = 157.5$

As a triode, $Gm_T = 7.3$ mA/V and $r_{IT} = 1.3$ kΩ, so the triode amplification factor is $\mu_T = Gm_T * r_{IT} = 9.5$!

The myth that tube testers that test pentodes and tetrodes as triodes are not as good as testers who test them as pentodes or tetrodes is just that - a myth. This is a perfectly sound testing method, providing you understand that the real pentode Gm is a few percentage points lower than the triode figure displayed on the meter.

How Gm varies with grid voltage ("bias")

The graphs on the right show two types of characteristics. The first is a family of five transfer curves for a typical triode (almost vertical lines). These are usually published by tube makers. Unfortunately, the graphs of Gm (six shown) are almost never published.

In the first approximation (or, if you wish, for an ideal triode), the Gm curves can be considered straight horizontal lines, meaning the Gm is constant over the range of grid bias voltages.

Also, Gm rises with anode DC current, although not in a linear fashion, but in proportion to its cube root. This can be concluded from the shape of the Gm curve on the Rp-Gm-μ graph.

So, in the "quick and dirty" approximation (a technical term that very few non-engineers understand), Gm is independent of the combination of electrode voltages that give rise to that anode current (anode voltage, screen voltage & grid voltage).

This fact enabled tube tester manufacturers to claim that emission testers, most of which display anode current under static conditions, are perfectly capable of distinguishing between strong or good tubes and weak or bad ones. If the anode current cannot reach the nominal level (due to the low emission capability of a tube), its mutual conductance will also be low.

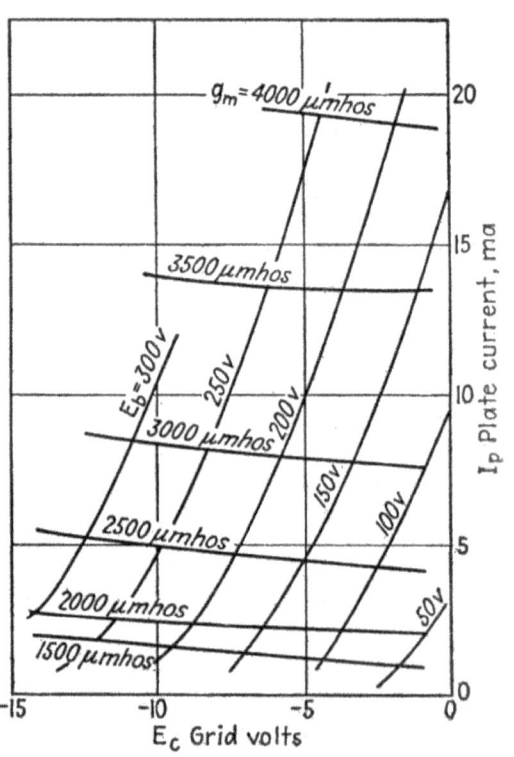

TUBE TESTING PRINCIPLES AND PROBLEMS

Should we match tubes on anode current or transconductance?

The answer to this question is simple (although the analysis behind it isn't): If you can, match them for *both*! If you cannot, match them based on the most important parameter for the application. For instance, match preamp (voltage amplification) tubes for Gm (mutual conductance) and output or power tubes for I_A (anode or plate current).

In single-ended (SE) amps, the use of unmatched tubes is not critical, and unless the difference is huge, you are unlikely to notice the imbalance in volume between the two channels, for instance. However, matched pairs are required for single-ended amps that use two or more output tubes in parallel connection (PSE)! Likewise, in push-pull (PP) and especially in parallel push-pull stages (PPP), you should use matched pairs, quads, octets, etc.

But, as always, there are other factors in play. Very few vintage tube electronics books pay any attention to the issue of tube matching. Let's see what "The Radiotron Designer's Handbook", IV edition, says about that:

"The matching technique varies with the class of operation for which the valves are intended. Matching valves for class A_1 service is not as critical as for valves intended for class AB_1, AB_2 or class B. For class A_1 service it is usually sufficient to match for zero signal plate current only.

Triodes for class AB_1, AB_2 or class B service should be checked at a number of points on the plate current grid bias curve. The points usually taken are (a) zero signal condition (b) a bias corresponding to the maximum permissible plate dissipation. The plate currents so measured should agree at all points within 2%. Triodes intended for class AB_2 or class B service should also be matched for amplification factor. When matching tetrodes and pentodes it is usually sufficient to match for zero signal plate current and power output."

The authors distinguish between triodes, which need better matching, and tetrodes/pentodes, for which matching isn't as critical. Secondly, plate currents should be matched within 2%, meaning that if one tube draws 66 mA, the other can draw between 64.7 and 67.3 mA. They don't say how they arrived at this 2% figure. Tubes for class AB and B amplifiers (large signal swings) must be matched not just in one point, but all along their transfer curves, in at least 3 or 4 points, and also matched for amplification factor μ (or Gm, which is related to μ)!

Accuracy? What accuracy?!?

Tube testers are like watches. A person with four watches is never sure what the right time is. Likewise, a poor soul with two or more tube testers can never be sure which one is correct! The same tube will test weaker on one unit and stronger on another. The lesson: buy one tester and stick with it or you may go mad.

Whether you go mad or stay sane, don't expect too much from your tube tester. Most were very crude and inaccurate devices even when new (due to the "el cheapo" design philosophy behind them). Now, when they are at least 50-60 years old, they are even less accurate.

You will not find many references to accuracy, repeatability and stability in tube testers' manuals. At least Weston had the decency to put things into perspective by declaring this in model 798 manual: "The tube testing section is rated with an accuracy of 15% for all mutual conductance ranges and 2% for the voltage regulator section."

What does this tell you? Well, we don't know if Weston meant +/-15% or +/-7.5% (15% overall) accuracy. Even assuming a higher accuracy of +/-7.5%, a tube's measured Gm of 5,000 micromhos could actually range anywhere from 4,625 to 5,375! It does not instill much confidence in the whole tube tester business, does it?

Scale resolution and accuracy

The accuracy of testing depends on so many factors (the testing principle used, calibration, component drift, mains voltage fluctuations, hum and interference, regulation of testers' mains transformers) that it cannot be ascertained at all. What we can determine is the resolution of the tube tester's scale. No matter how accurate the measurement may be, it cannot exceed the accuracy & resolution of the scale!

Let's agree that one division is the smallest unit of resolution. We can guess if the needle is somewhere between divisions, but that is another story.

The scale of TV-10/U military tester made by Hickok (RIGHT) has four Gm ranges (3,000 - 6,000 - 15,000 - 30,000). Let's determine the resolution of these scales (the Gm value of one minor division).

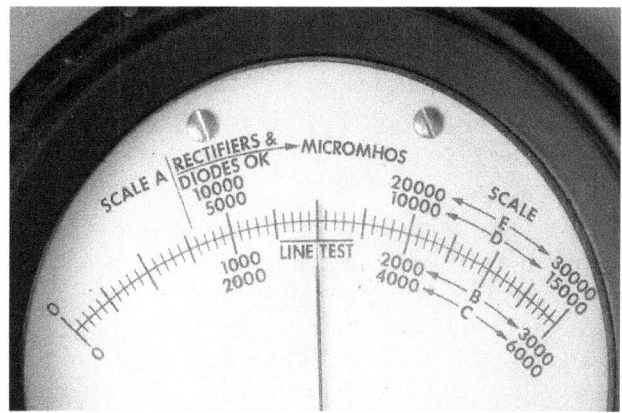

ABOVE: Many users find the scale of TV-10/U tester confusing

On the 3,000 micromhos scale, there are four major divisions between 2,000 and 3,000 micromhos. Each major division has five minor divisions, that is 4 x 5 = 20 in total. 20 minor divisions are equivalent to 1,000 micromhos, so 1 mindiv = 50 micromhos.

On the 6,000 scale, 20 mindiv is equivalent to 2,000 micromhos, so the resolution is 100 micromhos! Using the same reasoning, we get that the resolution of the 15,000 scale is five times lower than that of the 3,000 scale, or 5,000/20 = 250, and the resolution of the 30,000 scale is 500 micromhos (10,000/20 = 500).

How meaningful are tube test "certificates"?

I put certificates in quotation marks since these flimsy pieces of paper with minimal information could hardly be called certificates. At best, they are more of a marketing tool designed to instill confidence in the buyer. Even genuine ones have a very limited value since only one or two parameters are specified. At worst, they may be forgeries.

Once we bought a "matched" pair of current production Shuguang 2A3 tubes from a Chinese seller on eBay. Upon receipt, we tested the tubes on Triplett 3444 tube analyzer. The test "certificates" said I_P=45mA at $250V_{DC}$ anode voltage and -45V bias for both.

We replicated the test voltages and got 36mA of plate current for one and 47mA for the other! These were not matched in any sense; we returned them and lost $30 on return postage.

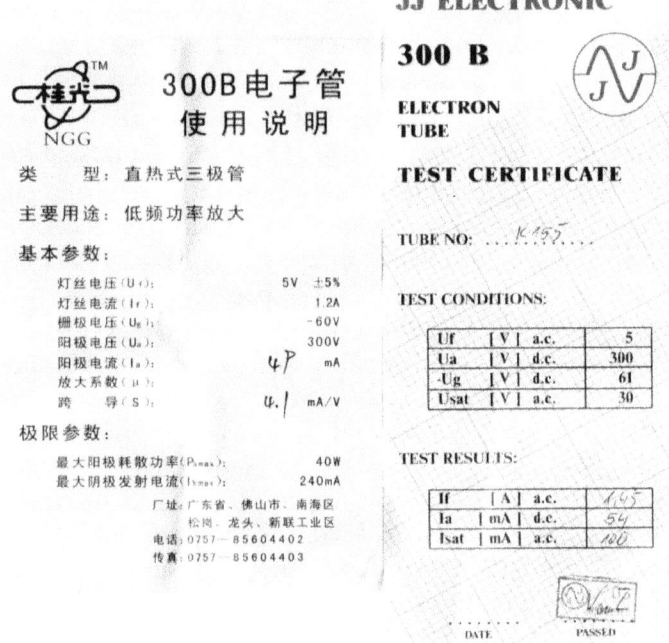

ABOVE: NGG specifies both Gm (4.1 mA/V) and anode current (49mA) for their 300B, one of better certificates. Most other manufacturers only specify one of those. Shuguang for instance, only lists the anode current. For some strange reason JJ does not specify Gm either, but includes heater current and I_{SAT} (anode current saturation level) instead, two parameters of minor or even no importance to tube buyers.

Calibration issues

Since there are so many types of tube testers and design variations, there is no standard way to calibrate tube testers. Apart from testers that belong to the same family, so their calibration procedures are similar (for instance, testers based on Hickok Gm bridge), each tester is different and requires a specific approach. They were all riddled with compromises and cut corners even when new, so what they are measuring is open to an endless debate.

The only proper way to calibrate a tube tester would be against a factory standard, and such a thing does not exist. Just as the same type tubes (say 6L6) from various makers are different, so are their standards of quality and parameter variations.

Arguably, emission testers don't need any calibration. It's all in relative comparisons between one tube and another anyway. Gm testers certainly do if they display true Gm on a calibrated scale (and very few do)! If they measure Gm but display on a % scale without actual Gm values, calibration may not matter much since you cannot be sure what the meter indicates.

The best we can do is calibrate various circuits in a tester, each in turn (the signal amplitude or frequency, bias voltage, shunt resistance, meter full-scale deflection, and so on) and then hope (or pray if you are a religious kind) that the overall indication will be meaningful.

Calibration instructions by testers' manufacturers are another sore point. In most cases, they weren't even included in testers' user manuals, and if they were, they were a sordid read, short on substance, and with little or no explanation of the background behind the steps you are performing.

People who wrote them didn't explain why certain steps and adjustments should be performed or what the implications of a certain tolerance are. For instance, what if a signal voltage isn't $1.5V_{AC}$ but $1.3V_{AC}$ (which is all you can get by tweaking trimpots without changing components), or what if bias voltages in your tester are 15% higher than the specified figures. Perhaps they didn't know themselves, didn't want to confuse users with too much mathematics and physics, or they wanted users to pay them to calibrate their testers!

The other group of testers (or manufacturers) base their calibration procedures on using a tube with a known Gm (obviously tested on another tester) as a calibration standard.

The main problem is that each tester tests tubes in a different operating point, so if the "standard" 6L6 was tested on a Hickok Tester with a pulsating 170V "DC" on its anode and screen is used to calibrate a tester that uses true (filtered) DC voltages of 250V, should you calibrate for the same Gm reading? I don't think so! Plus, the Hickok's signal was $2.5V_{RMS}$, and your tester uses $0.25V_{RMS}$. The list of such issues goes on and on, so I'm sure you get a picture. Such calibration is meaningless.

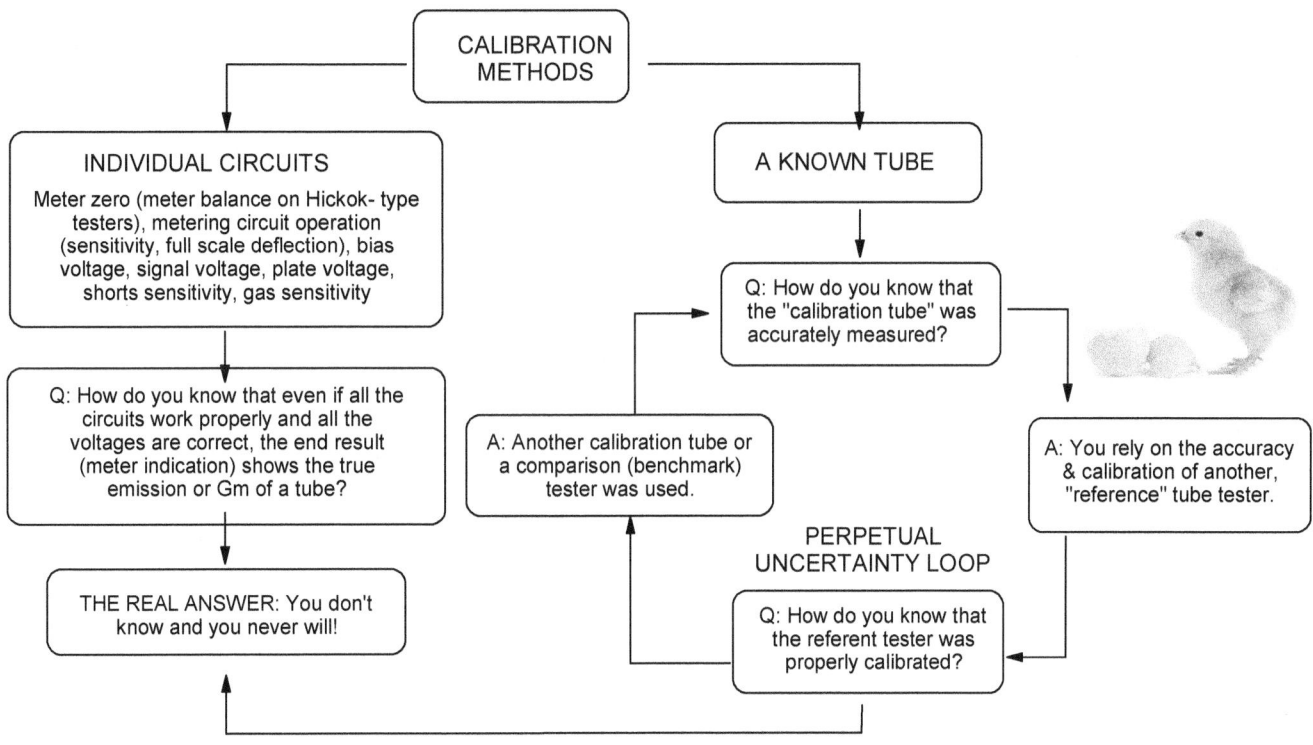

ABOVE: The disturbing truth or "The Chicken & Egg problem" in tube tester calibration

NEGATIVE GRID CURRENT - CAUSES AND DETECTION METHODS

Normal and abnormal grid currents

An mA meter in tube tests may show zero grid current when the grid voltage is zero, but that does not mean that no current flows at all. A more sensitive instrument such as a mA-meter would indicate a positive current of one or two microamperes.

The grid is acting as a mini anode and is collecting some "lazy" electrons, those that didn't have enough energy for a long journey to the actual anode. This is quite a normal situation, and we should not be concerned about it.

However, the grid current may flow in the opposite direction as well. This "negative" grid current develops a positive voltage drop on the external grid resistor (between grid and ground). It makes the grid less negative with respect to the cathode (reduces the negative bias), thus increasing the anode current.

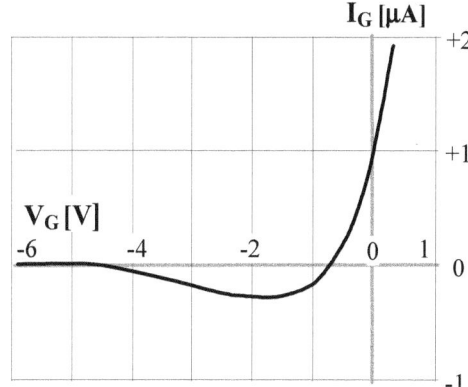

ABOVE: "Magnified" grid current versus grid voltage curve. This is a normal condition.

FAR LEFT: Positive grid current (from the external circuit into the grid). The grid behaves as a mini-anode and collects some electrons from the cathode. This is a normal situation.

LEFT: Negative grid current (from the grid into the external circuit). The grid behaves as a mini-cathode and emits electrons itself, which flow to the anode and through the power supply back into the grid. This is not a normal occurrence.

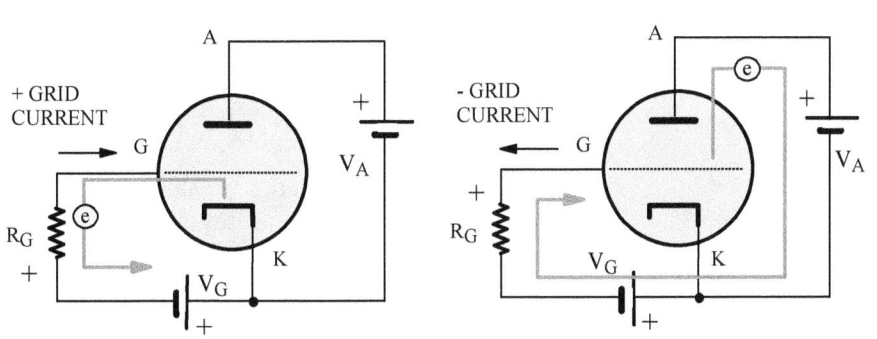

Grid emission

The oxide coating applied to the surface of the cathode during tube manufacture may vaporize and deposit itself on the grid, which then becomes a mini-cathode and starts emitting electrons itself. This condition, called "grid poisoning" or "grid emission", gives rise to the negative grid current.

Gassy tubes

The most common reason for the negative grid current is the residual gas ionization inside the tube. No matter how good the vacuum pumps are and how effective the "getter" material is inside a tube (its job is to "get" and bond as many ions and gas molecules as possible), there is always some residual gas left.

Also, as a tube ages, air leakage into the tube through glass seals increases, and the occluded gasses in the tube's elements are also gradually released with age.

With quality western-made tubes such as those by Mullard, Phillips, Telefunken, Valvo, etc., gas effects aren't usually present even after 60 or 70 years, so the most common reason for tubes to go "gassy" is improper operation.

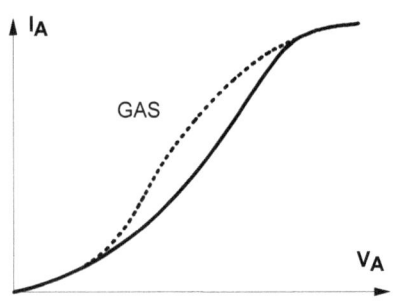

ABOVE: The effect of gas on I_A-V_A triode curves

That happens when tubes are underbiased, so their anode and screen currents are too high, and tubes get too hot. The presence of gas in a tube decreases the effectiveness of its grid bias and causes high, erratic space currents. In very gassy tubes, the control grid can completely lose its effectiveness.

High energy electrons traveling between the cathode and the anode collide with such molecules and knock one or more electrons out of them, creating positively charged ions. These positive ions are attracted to the negative grid, where they collect back the missing electrons and become neutralized molecules again.

This loss of electrons makes the grid more positive, and such positive bias increases the anode current. In some cases, a cumulative process happens, where such a high anode current increases the number of collisions between cathode electrons and gas molecules and raising the number of positively charged ions, which in turn makes the grid more and more positive, and results in continuously rising anode current and thermal runaway that eventually destroys the tube if power is not removed.

Detecting abnormal grid currents

Tube testers usually detect this negative grid current condition (almost always called a "gas test") by connecting a high value (typically 8.2-10 MΩ) resistor R_G into the grid circuit. The small DC voltage drop the positive grid current develops across R_G is amplified by a simple DC amplifier (a triode or JFET stage), which also acts as an impedance buffer. The amplifier drives an analog mA meter, calibrated in resistance units (MΩ).

Connecting a low resistance moving coil meter directly across the large grid resistor would render the circuit inoperative, so an impedance buffer is necessary. The amplifier is biased into cutoff when no grid current flows.

Most tube testers have only a qualitative scale for "Grid leakage - Gas." Better testers feature quantitative scales for "leakage," as in Triplett 3444, usually not calibrated in mA but in MΩ, so they effectively measure DC resistance between tubes' electrodes in the "hot" state.

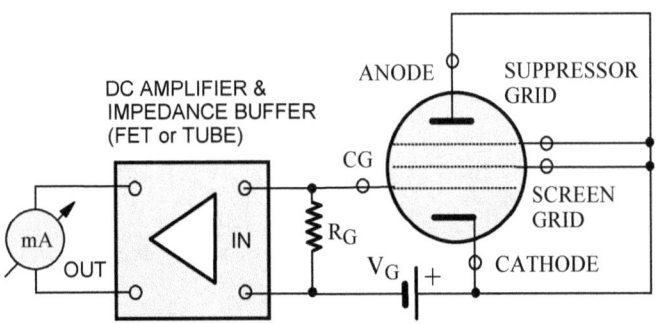

ABOVE: The operating principle behind the "gas test" or "grid leakage" of many tube testers. A pentode is shown to indicate the connection of the screen- and suppressor-grids, but the same circuit is also used for triodes and tetrodes.

A simpler approach used on testers such as Sylvania 220 and 620 is the grid shift gas test method. A large value resistor is switched into the grid circuit, and the grid signal is removed. In a healthy tube, that must result in a significant decrease of anode current, as indicated by the meter.

However, in a gassy tube, a negative current flowing through such a resistor will bias the grid more positively (just as in the previously described method), so the anode current will not drop during the gas test.

In fact, in many cases, the anode current even increases, which is a sure sign that the tube-under-test is either gassy or its grid is poisoned.

TESTING CASE STUDIES AND PRACTICAL EXAMPLES OF TUBE MATCHING

Now that we understand tube parameters such as Gm and a few important issues regarding testers' calibration and accuracy, let's look at a few examples of test results obtained by testing the same tubes on a few common mutual conductance testers. We still haven't covered the operational principles behind various types of tube testers. That will be done throughout the rest of the book, so here we will not focus on how those testers work, only on the relative comparisons between their results.

Experiment: Testing a batch of 12SN7GT triodes on Triplett 3444 tube tester

We had ten identical 12SN7 duo-triodes (identical to 6SN7, except the heater voltage) from the same series, all made one day long way back when I was still a baby, by our friends at GE. We tested them for Gm and anode current on Triplett 3444 Tube Analyzer. Luckily, this tester uses exactly $250V_{DC}$ as anode test voltage so that we can compare our test results directly with those published by GE in their datasheets.

Since 12SN7 is a duo-triode, we have 20 triodes in our sample, enabling us to compare the deviation between the pairs (two triodes in the same glass bulb) and between tubes. In many applications, you only need one duo-triode, but the two triodes inside it must be matched for the balanced circuit operation. Thus, matching the two triodes inside one tube is often more important than matching different tubes (for instance, between the left and right channel).

Once the average values and standard deviations have been calculated with the help of spreadsheet software, we got the following results: Gm_{AV}= 1.425mA/V, $I_{A\,AV}$=2.6 mA, $Gm\sigma$=0.13, $I_A\sigma$=0.513

Notice a smaller deviation of Gm, only 0.13, which means that Gm does not vary much. On the other hand, the anode currents varied widely, mainly because of five triodes (No 6, 12, 13, 14, 16). Number 6 is triode 3b, 12 is 6b, and so on. If those triodes were excluded, the standard deviation would drop significantly.

Triplett 3444 does not have the facility to precisely set grid bias while monitoring it on the main meter (like Precise 111 and some Hickok testers). Instead, it relies on the graduated scale of its "B-bias" control pot (photo below). However, if the knob is misaligned, the indication could be wrong.

In this photo, the pot is at its minimum, and the bias voltage is zero, yet the scale is not aligned correctly (1), so when the scale shows zero bias, the bias voltage will have already risen to between one and two volts.

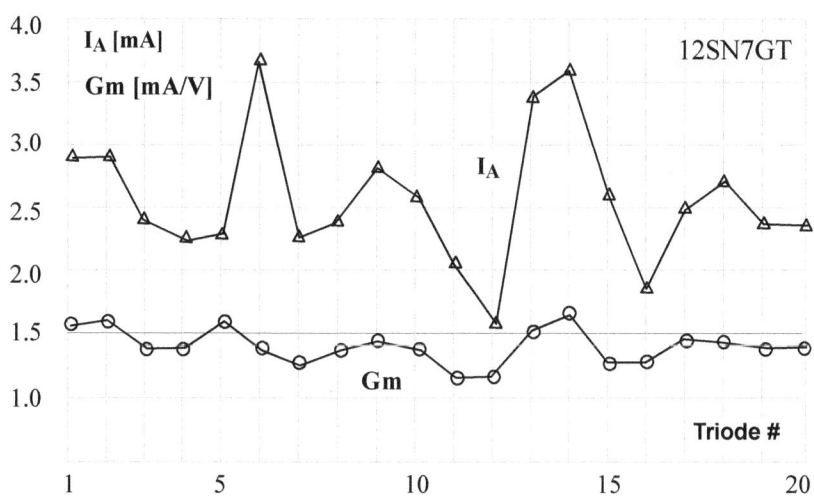

LEFT: Measured results in a visually-friendly form. Notice the uniformity of Gm results, and wide fluctuations of anode current!

BELOW: Test settings and actual element voltages of Triplett 3444 tester

Bias setting	12
Bias with no tube	-12.23 V_{DC}
Bias with a tube plugged in	-11.90 V_{DC}
Anode voltage with a tube plugged in	249 V_{DC}
Grid signal (4 kHz)	0.137 V_{AC}

Another issue is voltage sag when tube-under-test is plugged in. With the mechanical scale aligned and set at "12", the bias voltage was supposed to be -12.0 Volts. Without any tube plugged in, we measured -12.23V, which dropped to -11.90V once a 12SN7 duo-triode was plugged in for testing. This is a poor reflection on a supposedly decent tester such as Triplett 3444.

While a 0.33V bias discrepancy does not seem much, remember that with most preamp tubes (and even some low bias power tubes such as EL84), that can mean a significant difference in both Gm and anode current test results.

So when the tube tester's chart says what an average "good" figure for Gm should be, does that mean we should set the bias as per the listing and ignore the actual bias voltage, or the other way around?

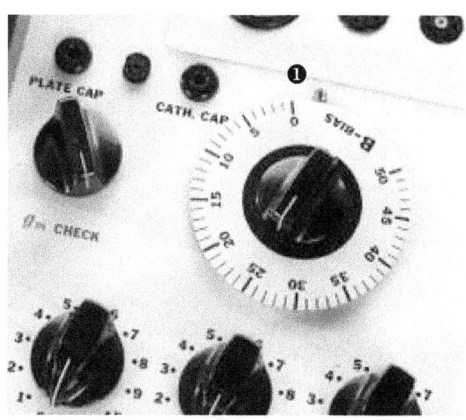

ABOVE: Before electrically calibrating any tester, make sure the knobs, scales, analog meters and other indicators are mechanically aligned first. This Triplett 3444 bias control is at its "electrical" zero, but way below its "mechanical" zero!

Experiment: Plotting the Gm - I_A curve

In this experiment, we chose tube #10 (because it is close to the average values on both gm and I_A) and used it to investigate how Gm varies with anode current to compare it with the curve published in the GE datasheet.

The shape of our measured Gm curve (dots) is the same as the published average or bogey curve, but our results are 15-20% higher.

Do not place too much faith in the published graphs and figures, and don't lose too much sleep if your predicted results (say, an amplification of a tube stage) differ significantly from the measured figures. Again, the published curves and figures are only for "ballpark" estimation and prediction purposes.

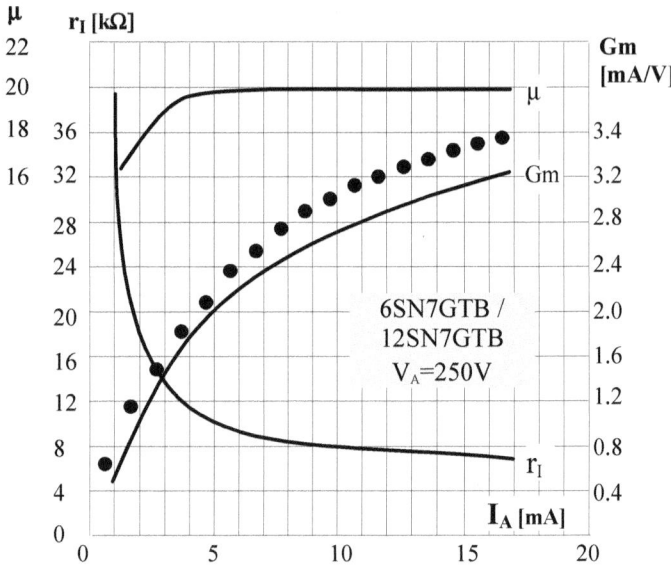

RIGHT: How the average tube's mutual conductance deviates from the Gm curve published by its manufacturer (GE).

Experiment: How to plot a tube's transfer characteristics on your tube tester

The worst way to "match" tubes is based on anode current in one steady point only, yet this is what tube sellers and even manufacturers do. The next best option is to match them for both anode current and mutual conductance, albeit in only one operating point (one anode voltage, one bias voltage).

Short of manually plotting the whole family of anode characteristics, which requires a range of anode voltages, or using a complex and expensive curve tracer, the quickest way to match tubes is to plot a transfer curve for each tube (by varying the bias control on your tester), superimpose them on top of each other and see what you get.

An experienced operator could do that in a surprisingly short time. Instead of nine test points as in this example (1V steps), three of four would be enough since the curve is always "smooth" and thus "predictable."

A transfer curve shows the relationship between DC plate current as a dependent variable (Y-axis) and negative grid bias voltage as an independent variable on X-axis at a fixed plate voltage. As we select 3-5 typical plate voltages, we get a family of transfer curves manufacturers specify in their datasheets.

In this experiment, we measured transfer curves on Triplett 3444 tube analyzer of two brands of 6L6 tubes, the current production JJ and NOS Chinese production from the 1960s, labeled "Tube Art."

A 100V plate/screen voltage was used to stay below the 50mA current limitation of the tester. With $250V_{DC}$ anode/screen voltages, the anode currents would quickly shoot toward 100mA or even higher.

The results were drawn on the graph taken from the GE catalog for the 6L6GC tube to see how much variation there is between the two brands and the vintage datasheet specification. Notice the difference between the published curve by General Electric and the two tubes tested. Would these tubes perform differently in an amplifier? Probably.

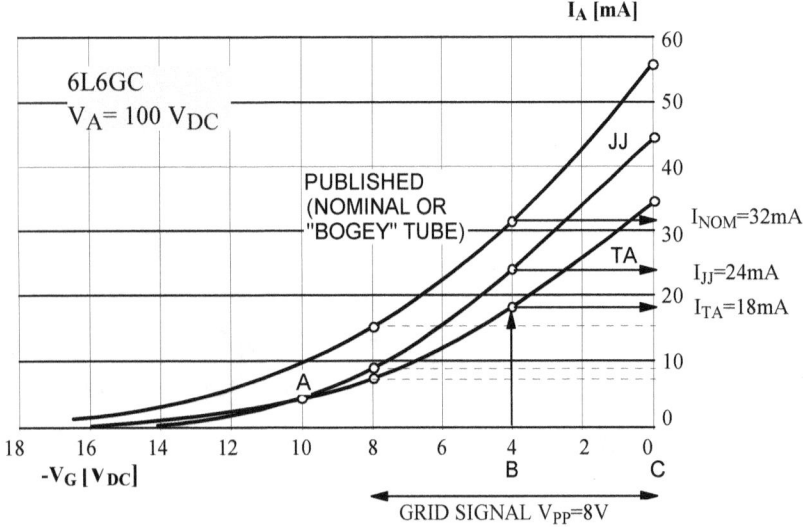

It is evident that both the Tube Art (TA) and JJ tubes pull significantly lower currents than the GE data sheet's nominal, average, or "bogey" tube. Also, below point A (where two transfer curves cross), the TA tube pulls more current than the JJ tube.

JJ's TX curve is steeper than TA's, meaning JJ tube has a higher mutual conductance. To the right of point A, JJ's TX curve is steeper (the slope of the tangent at any point) than that of the TA tube, and the anode currents increase faster, so JJ's transconductance is higher than TA's across the whole range of operation.

TESTING & MATCHING VACUUM TUBES

What would happen if we plugged these tubes into an amplifier? Let's assume that our amp is biased in point B (V_G=-4V_{DC}). While GE datasheet specifies that an average tube should pass 32mA of anode current in that point (V_G=-4V_{DC} and V_{AK}=+100V_{DC}), the JJ tube only draws 24mA and TA pulls even less, 18mA!

And finally, let's look at the maximum anode currents reached in point C, the point of zero bias (V_G=0V_{DC}). We are assuming a symmetrical grid signal around point B or -4V bias, a sine wave with its peaks at V_G=-4V_{DC} and V_G=0V_{DC} or 8V peak-to-peak. The TA tube's anode current will vary from a low of 8mA to a peak of 34mA, a swing of 26mA. JJ's tube will swing between 9mA and 44mA, a total of 35mA. Finally, the bogey GE tube should swing between a low of 16mA at -8V bias to a high of 57mA at zero bias, a total swing of 41mA.

You probably realize now where this discussion is heading. The TA tube will produce the lowest audio power of the three tubes. Assuming identical anode voltage swings, due to its larger current swing JJ's tube will produce a 35/26=1.35 or 35 % higher power output. However, it will still fall significantly short of the vintage GE tube that would produce 41/35 = 1.17 or 17% higher output.

Wait, you may protest, we cannot make meaningful conclusions based on such a tiny sample, a single tube from each maker, and you'd be right. We cannot proclaim that all Tube Art tubes are "weaker" than all JJ tubes or that all JJ tubes are "weaker" than the vintage GE-made tubes. This was simply to illustrate that variations between brands and tubes made in the same factory can be significant. Swapping such tubes around in your amp will change operating conditions, output power, and your amp's tonal voicing.

Investigation: The Groove Tubes performance rating system

Groove Tubes is a California-based tube reseller/rebrander. They source tubes from Russia, Slovak Republic (JJ), and China, test them, "grade" them, and reprint them with their logo. According to their website and info gained by us from various photos, after the 2-hour burn-in and stabilizing period, preamp tubes are tested for hum and microphonics in guitar amps (their main market). The gain test follows this in customized testbeds, where the signal frequency is swept over the audio range, and the gain-vs-frequency curve is plotted using PC-based "Audio Precision" hardware and software.

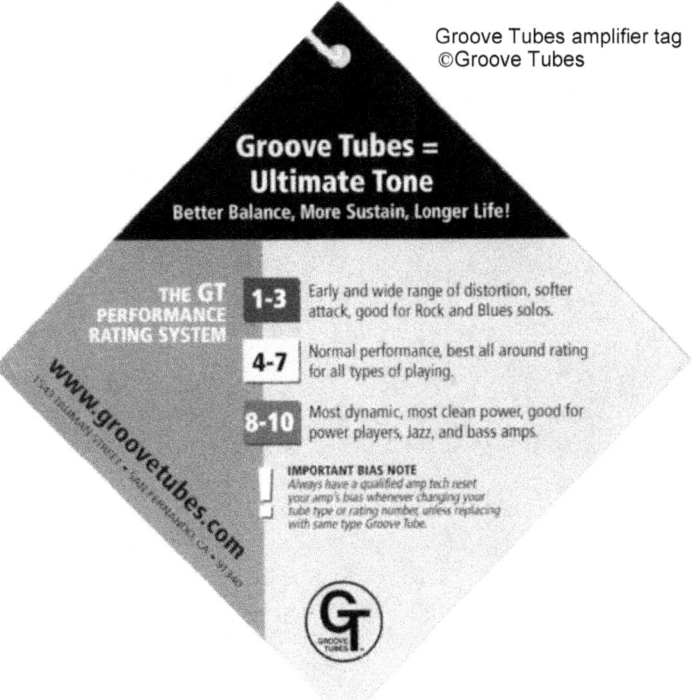

Groove Tubes amplifier tag
©Groove Tubes

Tube parameters are tested using the Amplitrex AT-1000 tester. Based on the results, the best (most balanced) tubes are selected for the "Special Applications Group" and further tested on Hagerman Vacu-Trace curve tracer. The tubes with dynamically matched halves are named "Matched Phase Inverter" (MPI) tubes and sold at a premium.

The photo on the GT website shows only one curve used for matching, not a whole family.

It is unclear if that is one of the anode curves or a transfer curve at a specific fixed anode voltage since transfer curves also have the same shape (for triodes only).

Apart from the similar basic tests, power tubes are "dynamically energized" and matched in terms of distortion levels or "gain-to-distortion ratio" and graded on a 1-10 scale. The tag from the Crate Palomino V8 amplifier hints at the differences between the three categories of tubes but does not explain what that means.

The "early distorters" have the least amount of headroom, meaning they start clipping the signal's peak early. Their "gain-to-distortion ratio" is the lowest of the three groups. These are marked as grades 1-3. "Normal" or average tubes are graded as levels 4-7, and late distorters (our names) have the widest headroom and thus distort at much higher signal levels.

Experiment: Testing a pair of "Test 4" and a pair of "Test 7" GT-6L6C tubes

In our stash of different lightly used tubes, we found two pairs of Groove Tubes GT-6L6C, one pair marked as "Test 7", the other as "Test 4", so we plugged them into Triplett 3444 analyzer and recorded the anode currents with every 1V change in bias.

Results for 2V jumps in bias voltage are shown so the table can fit onto the page width here. The test was done with 250V on the anode and 100V on the screen. Notice that all test points do not perfectly fit the smooth curves, but again that is due to the manual and error-prone (subjective reading on an imprecise scale) nature of the test.

The two Grade 4 tubes were well matched, with almost identical test results, certainly within the margin of error (the manual setting of bias pot). The two used Grade 7 tubes were not identical, perhaps they were matched when new, but now they have aged differently.

Groove Tubes claim that their tubes will age identically, but that is questionable. Tube aging is a complex process and depends not just on tubes but also on the amplifier they are used in. Even if two tubes were perfectly matched initially, they are unlikely to stay matched after a few years of operation!

BIAS [V_{DC}]	0	-2	-4	-6	-8	-10	-12	-14	-16	-18	-20	-22
Grade 4 tube #1	41	31	25	20	16	11.5	8	4.6	2.3	1	-	-
Grade 4 tube #2	41	31	25	19.5	15.5	11.2	8	4.6	2.3	0.8	-	-
Grade 7 tube #1	49	39	33	28	22	16	12	7	3.7	1.7	0.6	-
Grade 7 tube #2	46	36	29	24	20	15	10.5	6	3.2	1.35	0.3	-

Of course, the test results shown here and our conclusions are still valid. Assuming a fixed bias of -6V, in idle state (operating point Q4), G4 tubes will pull 18mA, and G7 tubes will be biased much hotter, pulling 28mA.

Assuming a maximum input grid signal of $12V_{PP}$ (peak-to-peak), centered around -6V bias, the anode current of the G4 tube will swing between the minimum of 8mA to a maximum of 41mA, a total peak-to-peak value of 33mA. The anode current of the G7 tube will vary from 12mA to 49mA, a total of 37mA. So, a G7 tube will amplify the signal more and distort at a higher power level (later) than the G4 tube.

Notice the identical positive halves for both tubes, around 21mA (identical positive headroom), but the negative peak of the G4 tube is smaller, 12mA compared to 16mA for the G7 tube (smaller negative headroom for the G4 tube). Since there is a more significant difference between its positive and negative halves, the G4 tube will distort more. Remember, if + and - halves of the output signal were identical, the harmonic distortion would be zero.

So, although GT groups these two pairs into the same 4-7 category, there is a noticeable difference in their characteristics and behavior.

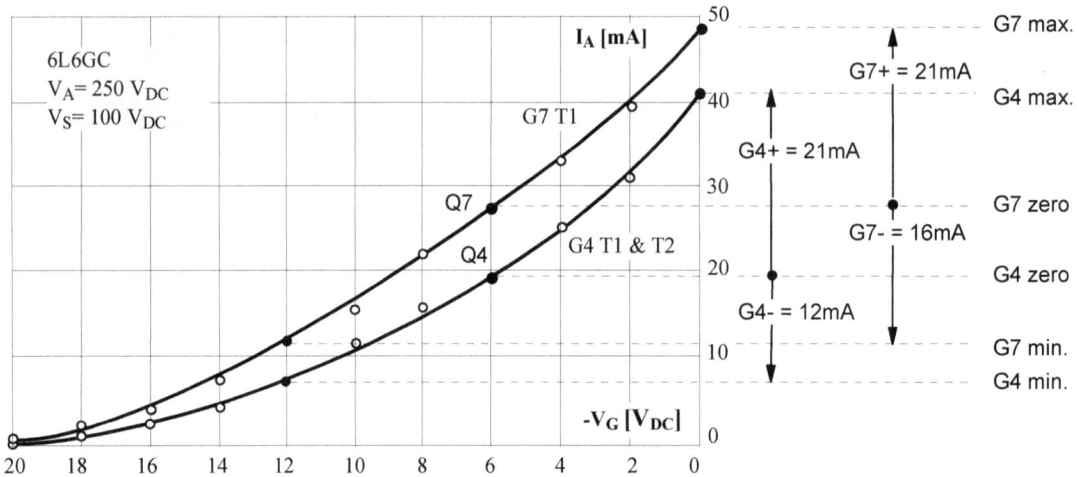

HOW TO TEST A TUBE THAT IS NOT LISTED IN THE BOOK OR TUBE CHART

We have used 6L6 as the base tube in both of the following experiments since even older testers from the 1940s and 50s have an octal socket and 6L6 test data listed. An adapter will need to be made or bought if the tubes you want to test don't use one of the sockets available on your tester. We used Eico 666 dynamic conductance tester (the results apply to model 667 as well) and Hickok 533A, a mutual conductance tester. Test terminals were installed on both testers so anode DC current could be measured by an external True-RMS multimeter.

Example #1: PL508 power pentode

PL508 (and EL508) are robust, long-lasting, cheap (no major amplifier manufacturer uses them) power tubes, great sounding in audiophile and guitar amps. Mullard, Philips, and Telefunken-branded tubes are still available. Since PL508 uses a Magnoval (Large Noval) or B9D socket, and only the latest tube testers made in the mid-to late-sixties have such a socket, we made a hardwired Magnoval-Octal adapter for this test. Eico 667 has a Magnoval socket, so no adapter is needed.

TESTING & MATCHING VACUUM TUBES

The PL508 tube datasheet by Phillips specifies a maximum cathode current of 100mA and maximum anode dissipation of 12 Watts. The heater voltage is 17V.

The "Typical characteristics" section specifies that with anode/screen voltage of +190V and -17V bias, 60mA of anode current should flow (plus 5mA screen current, so 65mA cathode current) and that mutual conductance in that point should be 9 mA/V.

The best settings to use on EICO 666/667 were found to be Plate: 90, Grid: 90, V switch: position 2, S switch: position 1. The lever bank and push-button bank test settings are as for 6L6. Hickok's bias was set at "30," and the Gm range was 6,000 (6mA/V). How did we arrive at those settings? Well, since Eico is a dynamic emission tester and Hickok tests mutual conductance, we adjusted the Hickok "Bias" setting to get anode currents of around 60mA, as per the datasheet, the same ballpark as Eico currents.

On Eico 666, we had four settings to play with, the "Plate" meter shunt, the grid signal amplitude "Grid," the voltage switch "V," and the coarse shunt switch "S." In position "2" the V-switch selects $15V_{AC}$ for the control grid, $45V_{AC}$ for the screen grid and $90V_{AC}$ for the anode.

Tubes 1-9 were made in the UK by Mullard, tubes 10-14 were marked "Westinghouse" USA, of unknown manufacturer.

Correlations between various measurements are very illustrative. EICO's readings almost perfectly correlate with anode current I_A. The conclusion is that 666 is primarily *an emission tester*, at least when it comes to testing power tubes. Its readings are moderately positively correlated with Gm (0.32), meaning that a low Gm tube should test low on that tester well.

Remember, the strongest negative correlation would be -1.0, 0 (zero) means no correlation at all, 1= perfect, 100% correlation.

Surprisingly low is the correlation of 0.27 between the two testers' anode currents. The two data sets look very close, 61-64 mA range for Eico and 56.5-65.5 mA for Hickok 533A.

The test results track each other, all except tube #9. Eico showed it as a healthy tube, above average, while Hickok showed both low Gm and the lowest I_A of all 14 tubes tested. Go figure! This anomaly could have lowered the correlation factors.

Notice also the very low Gm indications on 533A, around 3mA/V, a mere 1/3 of the expected figures.

Notice how uniform the test results are, considering that these tubes were not selected or matched in any way! This scenario is usually only seen with SQ (Special Quality) tubes.

Tube #	EICO 666 I_A (mA)	EICO 666 reading	Hickok 533A I_A (mA)	Hickok 533A Gm
1	62	114	65.5	3,200
2	63	116	63.5	3,350
3	62.5	114	64.5	3,280
4	63	114	59.5	3,150
5	64	114.5	60.5	3,150
6	65	117	63.5	3,200
7	64	116	61.5	3,050
8	64	115	62.5	3,120
9	64	114	56.5	2,470
10	69.5	124	63	3,450
11	61.5	111	61.5	3,300
12	61	109	58	3,220
13	61	110	57	3,050
14	62	111	61	2,950

Correlation coefficients	
Currents (I_A) between two testers	0.27
Tester readings (column 3 vs. col. 5)	0.32
EICO current vs. reading	0.95
Hickok current vs. Gm	0.62
EICO reading vs. Gm	0.32

ABOVE: Testing a batch of 14 PL508 power pentodes on Eico 666 and Hickok 533A
BELOW: Testing a quad of EL12 power pentodes on Eico 666 and Hickok 533A

Example #2: EL12 power pentode

EL12 is a predecessor of the highly acclaimed (and very expensive) F2a and F2a11 power tubes. I think EL12 is an even better sounding tube, and certainly much, much cheaper.

The design data for a single-ended output stage specifies that with anode/screen voltage of +250V and -7V control grid bias, the anode current of 72mA should flow (with 8mA screen current, so 80mA cathode current) and that mutual conductance in that point should be 15 mA/V.

The best settings to use on EICO were found to be Plate: 75, Grid: 75, V switch: position 1, S switch: position 1. Again, the lever bank and push-button bank test settings are as for 6L6.

Hickok's bias was set at "10" and Gm range was 15,000 (15mA/V).

Compared to PL508 results, the correlation between two testers' readings is higher (0.47 versus 0.32), as is Eico's reading versus Gm, also 0.47 versus 0.32 (by pure coincidence, I assure you). Hickok's current versus Gm tracking is surprising, showing a very strong negative correlation of -0.79!

Tube	EICO 666 I_A (mA)	EICO 666 reading	Hickok 533A I_A (mA)	Hickok 533A Gm
1	75	100	36.5	7,000
2	77	101	36.5	7,150
3	78	103	37	6,800
4	72	92	38.5	6,750

Correlation coefficients	
Currents (I_A) between two testers	-0.77
Tester readings (column 3 vs. col. 5)	0.47
EICO current vs. reading	0.97
Hickok current vs. Gm	-0.79
EICO reading vs. Gm	0.47

EXPERIMENT: COMPARISON OF TRIPLETT 3444, B&K700 AND MERCURY 1000 TEST VOLTAGES FOR A FEW COMMON AUDIO TUBES

To give you an idea and a feel for test results, we have done a few experiments for comparison purposes to see how the same pairs and quads of common audio tubes test on three different vintage mutual conductance testers: Triplett 3444, B&K 707, and Mercury 1000.

The table below outlines the major technical differences of these testers; we will discuss their functionality and other aspects later in the book. Note that the voltages given in the table are as specified by the manufacturers; the actual or measured voltage may be different in the actual testers!

	TRIPLETT 3444	MERCURY 1000/1200/2000	B&K550	B&K707
PLATE VOLTAGE(S)	12, 30, 100, 250 V_{DC}	110 V_{AC}	203 V_{AC}	193 V_{AC}
SCREEN VOLTAGE(S)	12, 30, 45, 100, 250 V_{DC}	Same as plate (triode connection)	148 V_{DC}	148 V_{DC}
BIAS VOLTAGES	0-5 and 0-50 V_{DC} variable	-1.7, -5.9 V_{DC} (fixed)	-0.2 -2.5 -7.5 - 19.5 (fixed)	-0.2 - 19.5 (fixed)
SIGNAL V_{AC} (RMS)	0.033, 0.1, 0.333 and 1.0	1.0 V_{RMS} (2.82 Vpp)	1.5 V_{RMS} (4.23 Vpp)	1.5 V_{RMS} (4.23 Vpp)
SIGNAL FREQUENCY	5,000 Hz	50 or 60 Hz (mains)	50 or 60 Hz (mains)	50 or 60 Hz (mains)

The B&K 550 tester was included in this table to illustrate that B&K reduced the number of bias voltages from four on the 550 and 650 models to only two on the later 700/707 models and reduced the plate voltage! The main functional difference is that models 550 and 650 display true Gm in micromhos, while later units (models 700, 707, and 747) only display on scales of 0-100 (presumably in percentages). Also, because it uses a much lower plate voltage of 110 V_{AC}, Mercury didn't have to include a bias higher than -5.9 V, as B&K had to. Both use the Hickok bridge circuit, although Mercury uses a lower grid signal voltage.

Since they didn't have such capability, anode current measuring circuits were installed in the B&K and Mercury testers. The figures after Gm are anode currents in mA (measured with a True-RMS digital multimeter).

6K6 testing

The specified test voltages on Triplett 3444 were $V_A = V_S = 100 V_{DC}$, $V_G = -8 V_{DC}$. Using 100V to test a power tube instead of the 250V available is strange. It goes to show that even reputable tester makers such as Triplett made some strange decisions and an occasional gross mistake.

The plate currents were 5-6 mA, and at such feeble power levels, all tubes seemed generally matched, tubes #1 and #2, and the second pair, tubes #3 & #4 in particular. Due to such a low test voltage, the mutual conductance was very low, 1.4 - 1.6 mA/V.

We increased the plate voltage from 100V_{DC} to 250V_{DC} and got more meaningful results. Due to poor regulation of Triplett tester's transformer, instead of 250V under no load, the anode and screen voltage drooped to only 222V with only 40 or so mA flowing!

Note **: Tube #2 got damaged due to a high heater voltage during testing on a faulty B&K 700 tester (after Triplett test), so the results are not comparable to Triplett 3444 results, but both B&K 700 (after repair) and Mercury 1000 picked up its weakness!

TEST VOLTAGES	PLATE VOLTS (V_{DC})	GRID BIAS (V_{DC})	SIGNAL (V_{DC})
Triplett 3444 revised	222	-8.5	0.2
B&K 700	120	-8.0	1.46
Mercury 1000	92	-1.55	0.95

6K6 Sylvania (Gm - Ip)	#1	#2**	#3	#4
Triplett 3444 specified	1,610 - 6.0	1,610 - 6.0	1,420 - 5.0	1,450 - 5.0
Triplett 3444 revised	3,700 - 42.5	3,450 - 41.5	3,500 - 42.0	3,550 - 40.5
B&K 700	94 - 16.1	48 - 10.4**	95 - 15.7	92 - 15.0
Mercury 1000	2,420 - 24.2	1,200 - 15.1 **	2,420 - 21.8	2,220 - 22.5

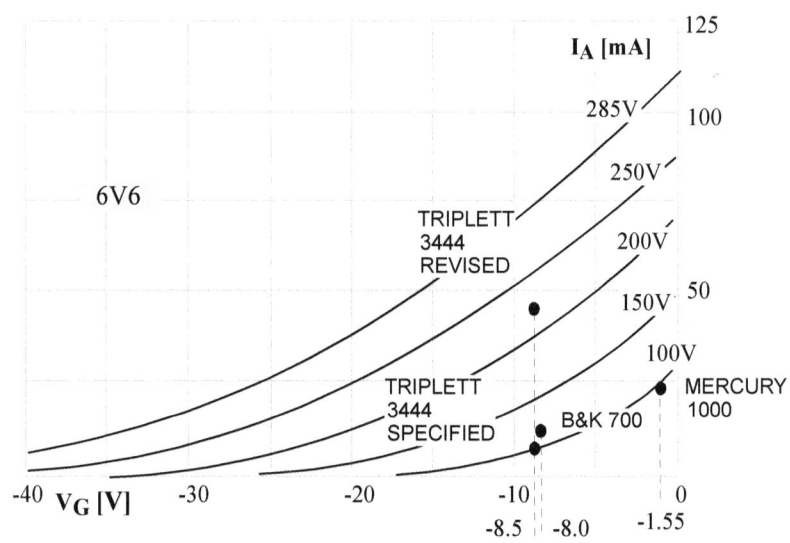

6L6 testing

The specified bias on Triplett 3444 was -18V, but plate currents at that setting were over 50 mA, so we increased the bias to 25 on the dial, which we measured as -27V.

Mercury 1000 specifies the "Load" as 95, but the readings were off the scale, so we reduced that setting to 85. For some strange reason, its setting for the 6L6 tube is for Em (emission) testing only, while the settings for 5881 are Gm settings!

So, if Mercury owners want to test 6L6 for transconductance, they need to use 5881 settings.

The published TX curve at Triplett revised test point (-27V bias and 250V anode volts) indicates an anode current of around 20-23mA, but we measured 39-46 mA!?

Another reminder not to take things too seriously unless you want to go mad!

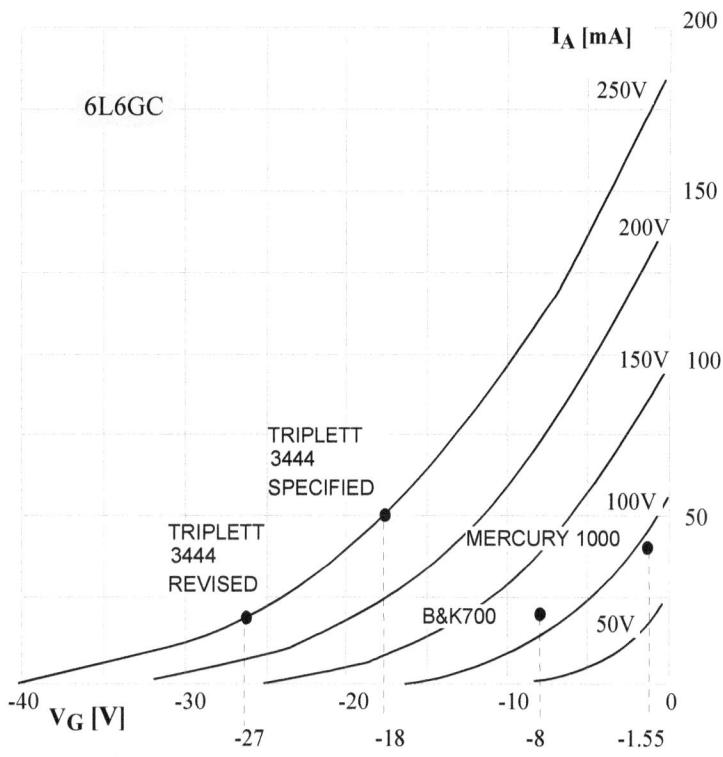

TEST VOLTAGES	PLATE VOLTS (V_{DC})	GRID BIAS (V_{DC})	SIGNAL (V_{AC})
Triplett 3444 revised	249	-27	0.2
B&K 700	120	-8.0	1.46
Mercury 1000	92	-1.55	0.95

6L6 (Gm - Ip)	#1 JJ	#2 JJ	#3 Tube Art	#4 Tube Art	#5 Sylvania	#5 Sylvania
Triplett 3444	7,000 - 42.5	7,000 - 42.5	4,600 - 39.0	4,600 - 39.0	5,500 - 46.0	5,600 - 44.0
B&K 700	107 - 34.0	105 - 34.8	74 - 26.7	78 - 27.9	91 - 33.3	90 - 34.0
Mercury 1000	4,800 - 28.3	4,750 - 28.5	3,600 - 22.0	3,700 - 23.0	4,100 - 26.2	4,000 - 26.5

6BQ5 (EL84) testing

For some reason, the Triplett 3444 anode voltage did not sag from the nominal 250V with 6L6, but it did drop to 234V with EL84.

To compare the Triplett and B&K results, if 10,400 is 88% (tube #1), then 7,850 for tube #2 should be 66.5%, yet B&K tested it as 78%. Tube #3 at 82% on B&K is close to 80% that it should show, and tube #4 at 89% is close to 90.5% conversion from the Triplett result.

Mercury and B&K bias levels are too high, so anode currents are too low, esp. in Mercury's case (8-10mA)!

Note **: Instead of testing low as on Triplett and B&K, these two tubes tested very high on Mercury 1000. The reason remains a mystery.

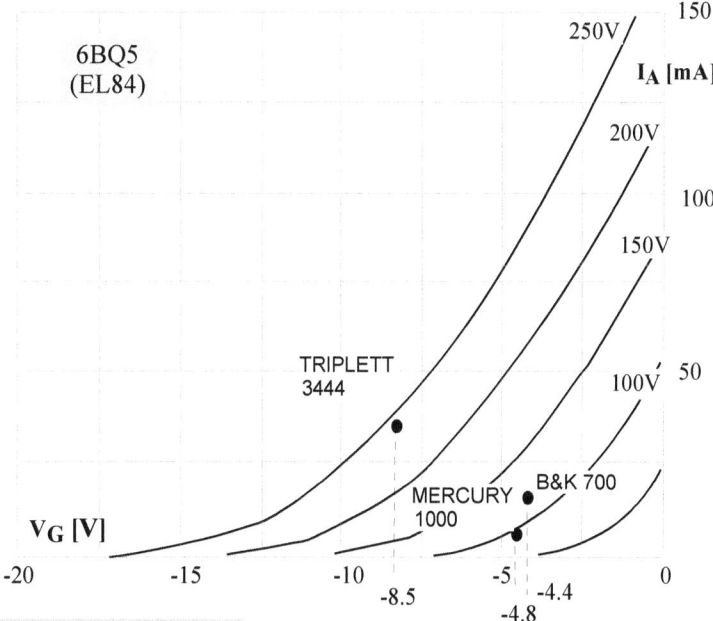

TEST VOLTAGES	PLATE VOLTS (V_{DC})	GRID BIAS (V_{DC})	SIGNAL (V_{AC})
Triplett 3444	234	-8.50	0.2
B&K 700	120	-4.4	1.46
Mercury 1000	92	-4.8	0.95

6BQ5 (Gm - Ip)	#1	#2	#3	#4
Triplett 3444	10,400 - 32.0	7,850 - 21.5	9,400 - 26.0	10,700 - 28.0
B&K 700	88 - 20.2	78 - 16.5	82 - 17.5	89 - 18.5
Mercury 1000	8,200 - 10.2	16,800 - 8.2 **	18,000 - 10.0 **	8,100 - 8.1

12AX7 (ECC83) testing

Four random 12AX7 tubes by four different manufacturers were tested on the same three Gm testers. Due to their high grid signal (1.1V and 0.95V), B&K 700 and Mercury 1000/1200/2000 testers obviously cannot test 12AX7 properly, let alone be used for matching such tubes. What surprised us was that even Triplett 3444 used a relatively high 0.4 V_{RMS} signal to test this tube when smaller grid signals were available. $0.4V_{RMS}$ is $1.13V_{PP}$!

12AX7 test settings on Triplett 3444:
Bias: 18 (0-5V) or -1.8V_{DC}
Plate voltage: 1 (250V_{DC})
Reject: 1.0 mA/V (1,000 micromhos), nominal: 1.3 mA/V (1,300 micromhos)

TEST VOLTAGES	ANODE VOLTS (V_{DC})	GRID BIAS (V_{DC})	SIGNAL (V_{AC})
Triplett 3444	249	-1.85	0.4
B&K 700	204	-1.5	1.1
Mercury 1000	103	-1.4	0.95

Gm - Ip	#1a RCA	#1b RCA	#2a Philips	#2b Philips	#3a JJ	#3b JJ	#4a Ei - YU	#4b Ei- YU
Triplett 3444	1,650 - 0.8	2,400 - 1.9	2,120 - 1.5	2,150 - 1.5	1,750 - 0.9	1,750 - 0.9	1,570 - 1.2	1,660 - 1.3
B&K 700	35 - 1.5	21 - 0.9	23.5 - 1.1	24.0 - 1.2	21 - 0.8	21 - 0.9	22 - 0.9	22 - 0.9
Mercury 1000	1,120 - 0.7	1,600 - 1.2	1,120 - 0.8	1,250 - 0.9	850 - 0.6	1,020 - 0.7	1,040 - 0.7	1,040 - 0.7

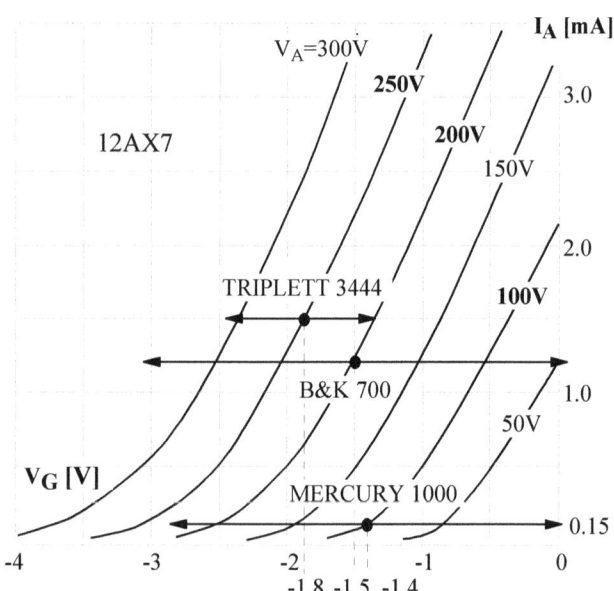

The visual comparison of test bias and anode voltages and anode currents is illustrated on the transfer characteristic of the 12AX7 triode.

Mercury 1000 uses a low anode test voltage of 100V. Even worse is the choice of the bias voltage, -1.4V, which at such low anode voltage is too high (too negative), so during the negative half-wave of the grid signal, the tube is beyond cut-off and does not conduct! The bias should be changed to around -0.5V.

Don't be confused by the fact that the graph says the anode currents in Mercury's case should be around 0.15mA. That would be the case if 0.1V or lower test signals were used. Since peak-to-peak grid signals for Mercury and B&K testers are 2.7 and 3.1V, respectively, their measured currents are the average values over the whole swing, from cut-off to zero bias and beyond.

"The 12AX7 problem" or why low bias tubes aren't tested properly on most vintage testers

The transconductance of tubes such as 12AX7 and ECC88 (6DJ8) isn't accurately measured on most tube testers because the amplitude of their grid test signal is too large compared to the DC bias voltage. The transfer curve for ECC88 illustrates what happens.

A 0.1 V_{PP} signal would measure Gm in one point (Q). When a grid signal of $1.13V_{PP}$ ($0.4V_{RMS}$) is applied to a tube biased at -1V (point Q), as in Triplett 3444, the measured value is an average over the X-Y range. Not so bad, since Gm is relatively constant in that linear range, that result will be close to the actual gm in point Q.

However, most testers apply a much larger signal voltage. For instance, B&K 550 grid signal is $1.5V_{RMS}$. Its peak value is 1.5 * 1.41 = 2.115 V_P. The peak-to-peak value is twice that or the whopping 4.23 V_{PP}!

If tested at 90V plate voltage, that would take 6DJ8 almost to the -3.6V cutoff, through the area of rapidly changing Gm (curvature). The story at the other end is even worse. The test signal takes the tube above the 0V bias into a region of positive grid current (beyond point W). Again, the reading will be some kind of dubious average value between points Z and W.

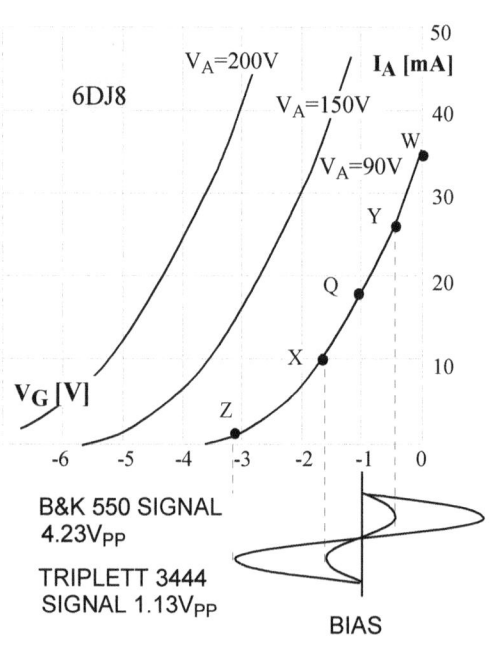

ABOVE: Most testers' signal voltages are way too high for low bias tubes such as 6DJ8 and 12AX7!

EMISSION TESTERS

3

- HOW EMISSION TESTERS WORK
- EICO 625 & 635
- KNIGHT KG-600, HEATHKIT TC-1, TC-2, TC-3 & IT-17
- CONAR 221, 223 & 224
- PRECISION 640 (NRI 71) & PRECISION 660
- TRIPLETT 2413 & 3414
- ELETTRA PROVAVALVOLE

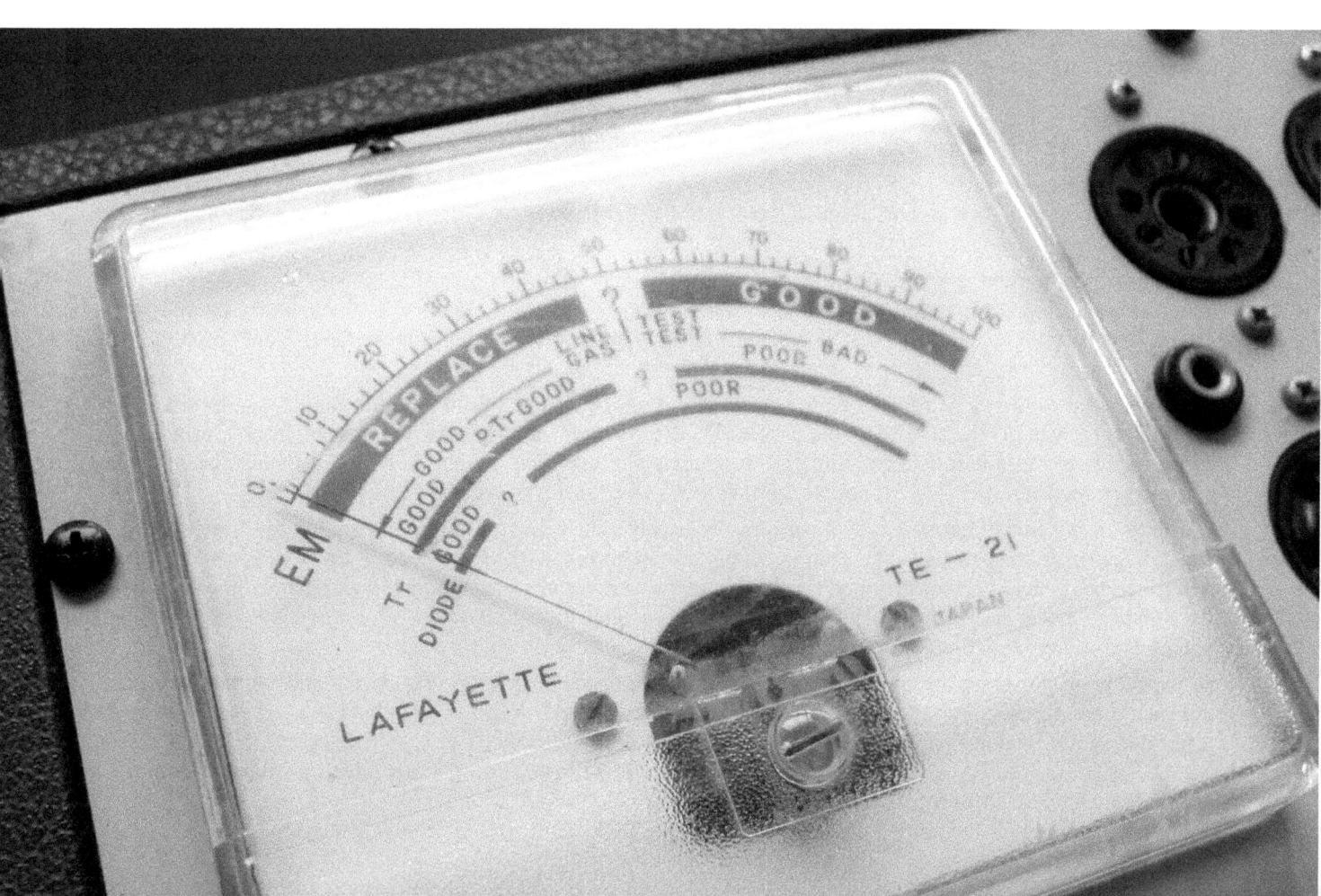

HOW EMISSION TESTERS WORK

During the emission test, there are only two circuits. One is the heater connection between the 0V or COM terminal and the switch-selectable heater voltage (H2). The other circuit is formed by connecting all the tube electrodes together (except cathode) to a source of low AC voltage (20-40V), the power transformer's secondary winding.

The meter (with a series resistor and variable adjustable shunt control) is connected between the cathode and the COM terminal, and once the "Test," "Merit," or "Value" switch is closed, the circuit is closed, and current flows through the meter.

The higher the emissive capacity of the tube's cathode, the higher the current flowing. Most of that current flows from the cathode to the closest electrode, the control grid (G1). Since grids are wound using very thin wire, too much current would burn out the wire just like a fuse. That is why voltages of 30V or less are used; higher test voltages would result in higher current levels.

For that reason, some testers strap G1 to the cathode, so electrons pass through G1, and most of the current flows to the screen grid instead (wound with a much thicker wire).

The tube under test rectifies sinusoidal mains voltage into a series of sinusoidal pulses whose RMS value is 1/2 of its peak value. However, moving coil meters respond to the average value, which in this case is $V_{AV}=0.318V_{MAX}$.

Dynamic conductance and proportional Gm testers also feature the same waveform as their anode current and even as their "DC" bias voltage! Thus, it is very important to understand this waveform.

RIGHT: The "DC" current through the meter is a series of half-wave rectified sinusoidal pulses! Tube-under-test is the rectifier.

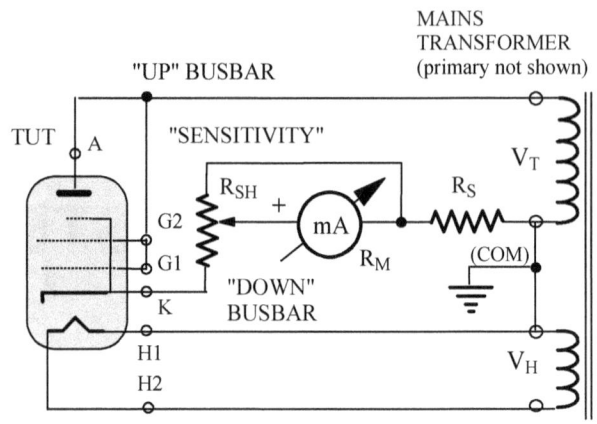

ABOVE: A typical emission test circuit

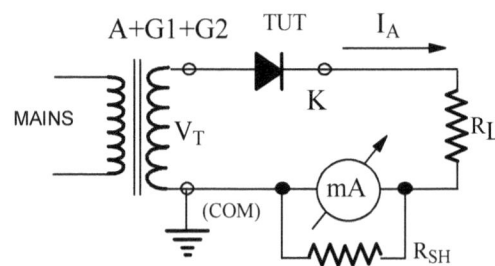

ABOVE: The simplified test circuit. R_L is the total load, meter's internal resistance R_M, plus the series resistor R_S and the lower part of the "Shunt" or "Sensitivity" rheostat.

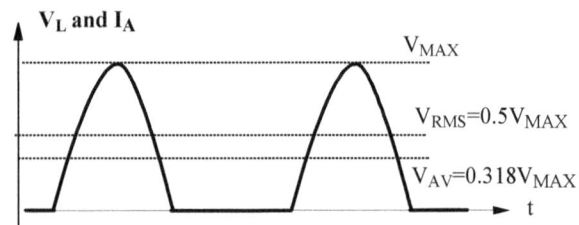

Case study: Same tubes, same model tester, different anode currents

You sell a matched pair of NOS Telefunken EL34 tubes on eBay. In a week, the buyer lodges a complaint with PayPal claiming that goods aren't "as described," and PayPal removes the money from your account and refunds the buyer, no questions asked!

You measured the plate currents on your Triplett 3423 tester using the precision resistor method (in line with the anode circuit) and got the readings of 48.6 mA and 48.2 mA, which was as close a match as one could hope for. He had the same circuit installed in his 3423 tester, re-tested the tubes, but his plate currents were only 34 mA. He claims that it is way too low, that the tubes are not new. What would you do?

The clue is not in the calibration difference between your two testers but in the multimeters used to measure plate currents and how this type of tester operates. The TUT (tube-under-test) has AC voltages on its anode and grid, so the tested tube works as a rectifier, providing its own anode and bias "DC" voltages. However, since the whole tester is a half-wave rectifier., the "DC" is not the "proper" DC, but a series of rectified sine pulses.

A true RMS meter will measure any waveform accurately. The buyer used a cheap analog multimeter (non-RMS), which measures average values of the signal, but is calibrated in the RMS values for a sine wave, which is an average multiplied by 1.11 (11% more). When they measure any other waveform (square, triangular, or half-wave rectified AC pulses as in this case), these meters introduce an error since their calibration is for pure sine waves only and does not apply to other waveforms.

For this type of pulsating waveform, the RMS value is 1/2 of the peak value V_{MAX}, and the average value is 31.8% of peak. The same applies to the current. A true-RMS meter will measure $0.5I_{MAX}$, but the average-responding meter calibrated in RMS values will measure $I_{AV}=0.318I_{MAX}$, but will show $0.318*1.11*I_{MAX}= 0.353I_{MAX}$, so the ratio will be $0.353I_{MAX}/0.5I_{MAX} = 0.71$! So, the non-RMS meter will only show 71% of the true value (an error of -29%), and that is what happened to your buyer.

EMISSION TESTERS

EICO 625 & 635

Eico 625 is one of the best emission testers, with a few features usually only found on more advanced models, "Line adjust" (1), "Circuit overload" (2), acting as a primary fuse, and an illuminated roll chart (3).

There is even an internal tube, a duo-diode (4), but it's only used as a rectifier to drive the analog meter during line adjustment.

All tubes are tested using $30V_{AC}$ test voltage (5). R4 and R5 are fixed resistors (7), in series with the wiper of the "Shunt" control potentiometer (6) and the meter.

Resistors R1, R2, and R3 are load resistors in series with the meter (8).

Older sockets for directly-heated triodes such as 2A3 and 300B are included. There is a blank cutout (9) for an installation of a socket of your choosing, a handy feature.

Talking about line adjustment, most testers that use a rheostat in parallel with a section of the primary winding were designed to provide nominal secondary voltages with 90V or, in this case, 100V on the primary. The higher the mains voltage above that level, the more voltage the rheostat has to "kill" and the hotter it gets! Rheostats in tube testers are rated at 25 Watts and fall in the 200-300 Ω range. China-made replacements are available on eBay.

Although it may seem that Eico 635 would be a newer or more advanced version of the older model 625, that is not the case.

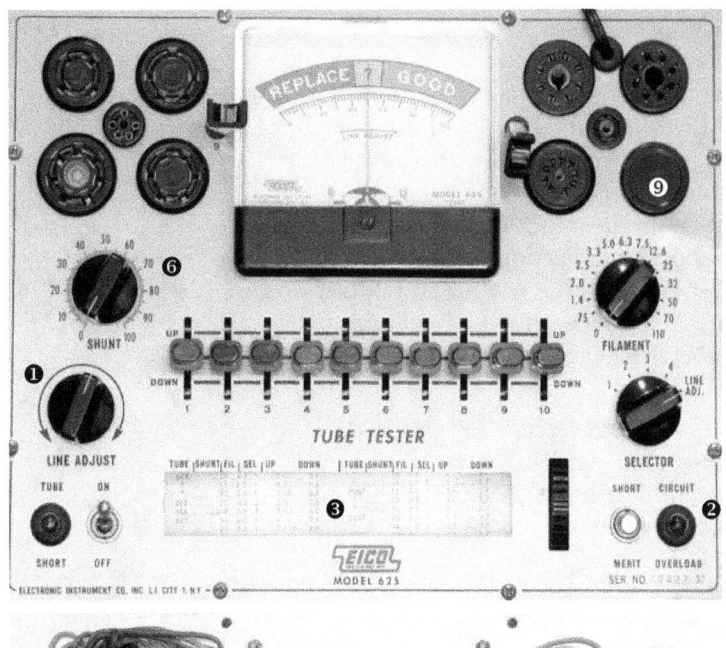

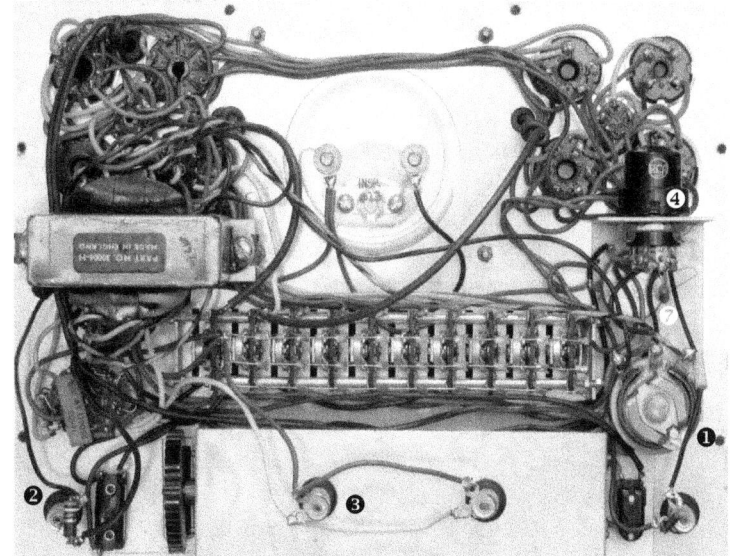

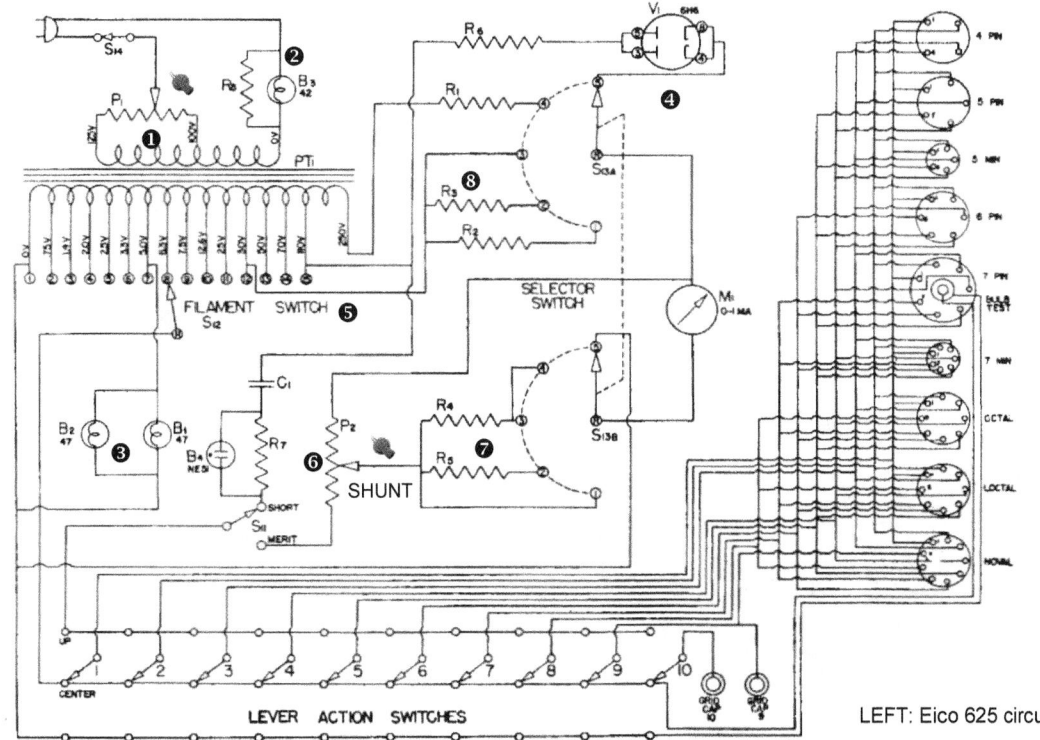

On the other hand, Eico 635 looks good but is a very basic, even primitive checker, electrically and functionally identical to Mercury 1101 grid circuit checker; see its circuit diagram on page 59.

There is no comparison between the two; model 625 is miles ahead of 635 in its design and testing capabilities.

LEFT: Eico 625 circuit diagram

KNIGHT KG-600, HEATHKIT TC-1, TC-2, TC-3 & IT-17

KnightKit was an attempt by Allied Radio to steal some business from hugely successful Heathkit, just as National Radio Institute tried to do with their "Conar" kit version of the same tester. KnightKit 600A and 600B are almost identical to Heathkit's tube checker series (TC-1, TC-2, TC-3, and IT-17).

Test voltages are the same, 30, 100, and $250V_{AC}$, and while Heathkit uses 14 heater voltages, Knight adds two more, 4.2V and 19.6V.

Apart from naming variations ("TOP-CENTER-BOTTOM" instead of "UP-CENTER-DOWN"), Heathkit TC-2 does not have the internal "Calibration control" 1k2 pot, using a fixed resistor instead.

The resistor added in series with the meter in circuit "2" is 360R instead of 470R, while the load resistors are 3k6-820R-0R-2k5 instead of 5k1-1k-0R-2k5 in KnightKit's case.

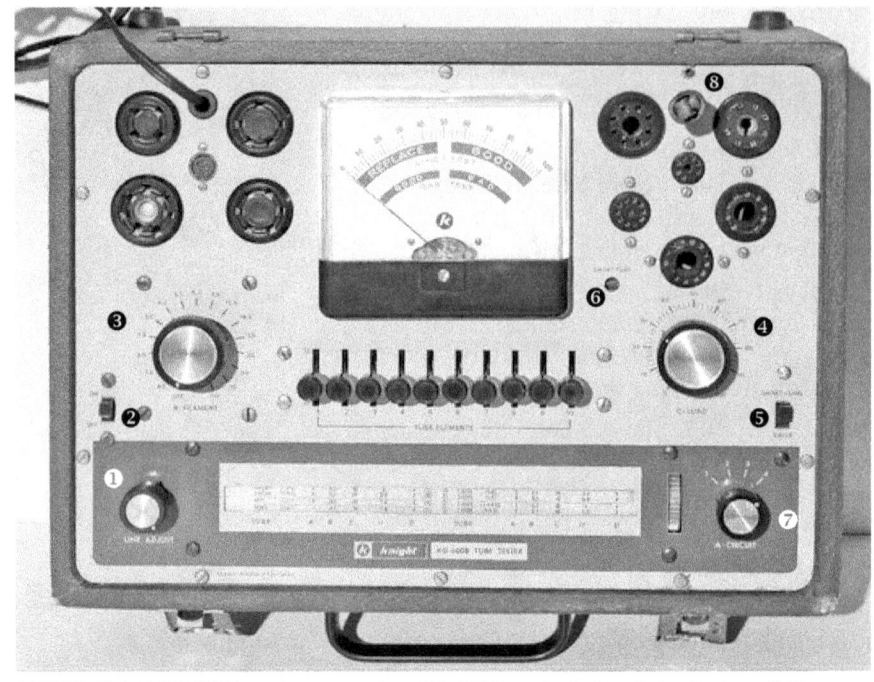

ABOVE: Knight KG-600A and more modern KG-600B are typical emission testers, 10 levers (one for each tube pin) in up or down position (all electrodes connected either to cathode or anode) and a choice of 4 test voltages (A-circuit) selector.

RIGHT: The assembly manual was well laid out and in no way below the high standards set by Heathkit with their literature.

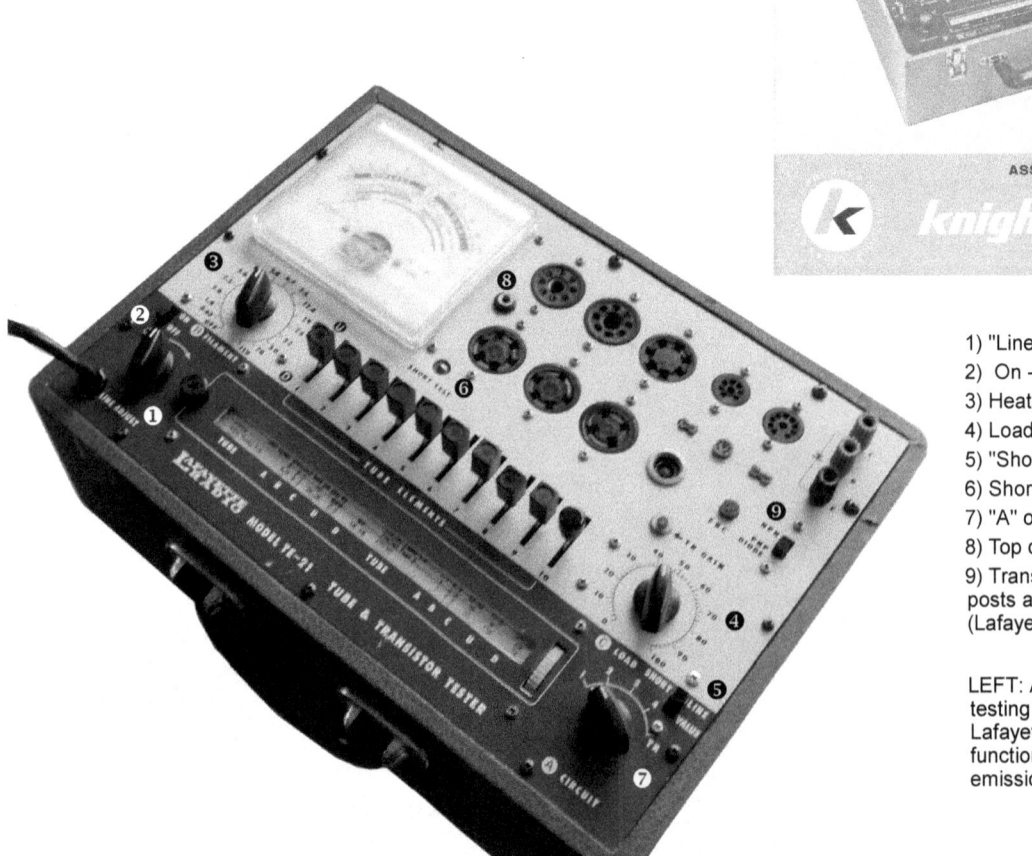

1) "Line adjust" rheostat
2) On - Off switch
3) Heater voltage selector
4) Load potentiometer control
5) "Shorts/Line-value" test switch
6) Shorts indicator (neon)
7) "A" or "Circuit" selector switch
8) Top cap connector
9) Transistor test sockets, binding posts and NPN-PNP selector switch (Lafayette TE-21 only)

LEFT: Apart from the added transistor testing section, the elegant looking Lafayette TE-21 (made in Japan) was functionally identical to Knight/Heathkit emission testers

EMISSION TESTERS

In Heathkit's tube checkers, the "Load" potentiometer, called "Plate," is connected directly between the "Bottom" busbar and the 0V point on the upper secondary winding, without the 51R resistor as in the KG-600 case.

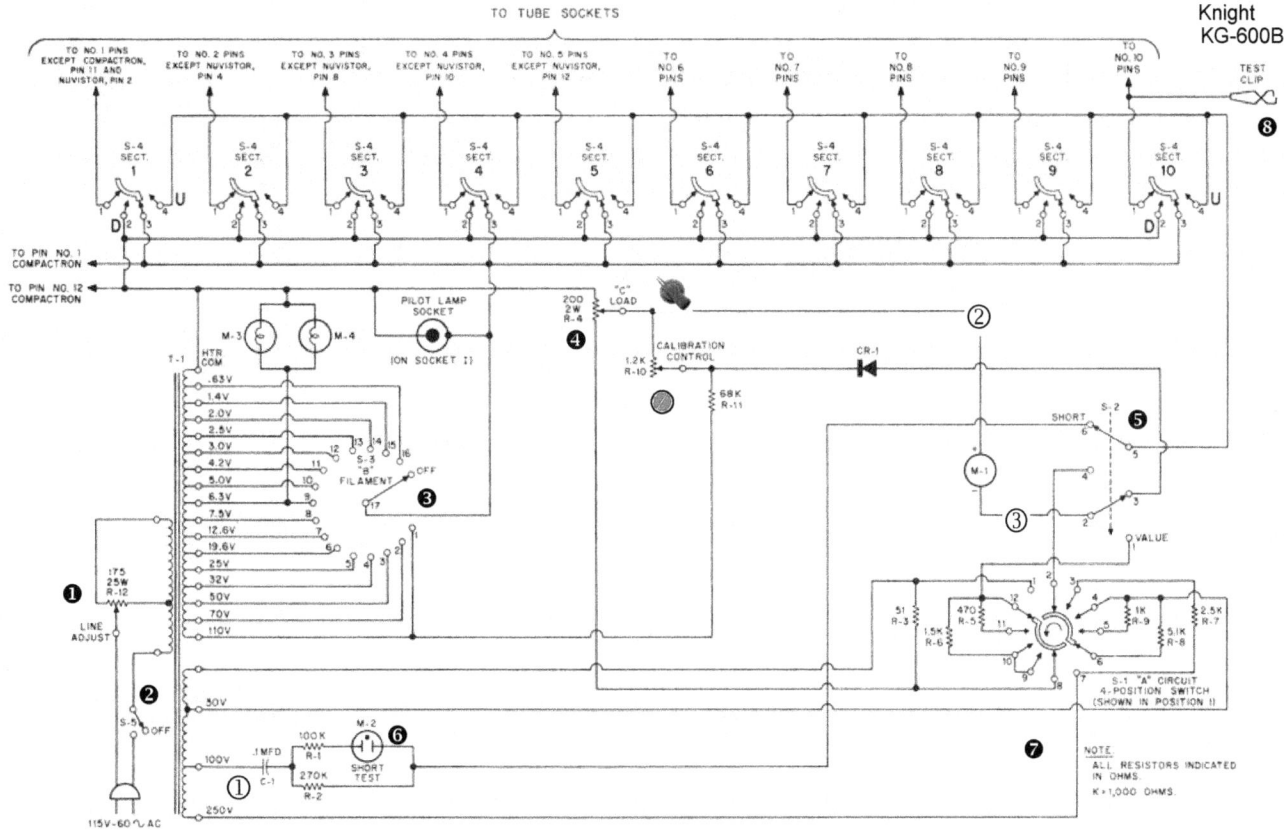

Knight KG-600 (ABOVE) and Heathkit TC-2 circuit diagram (BELOW)

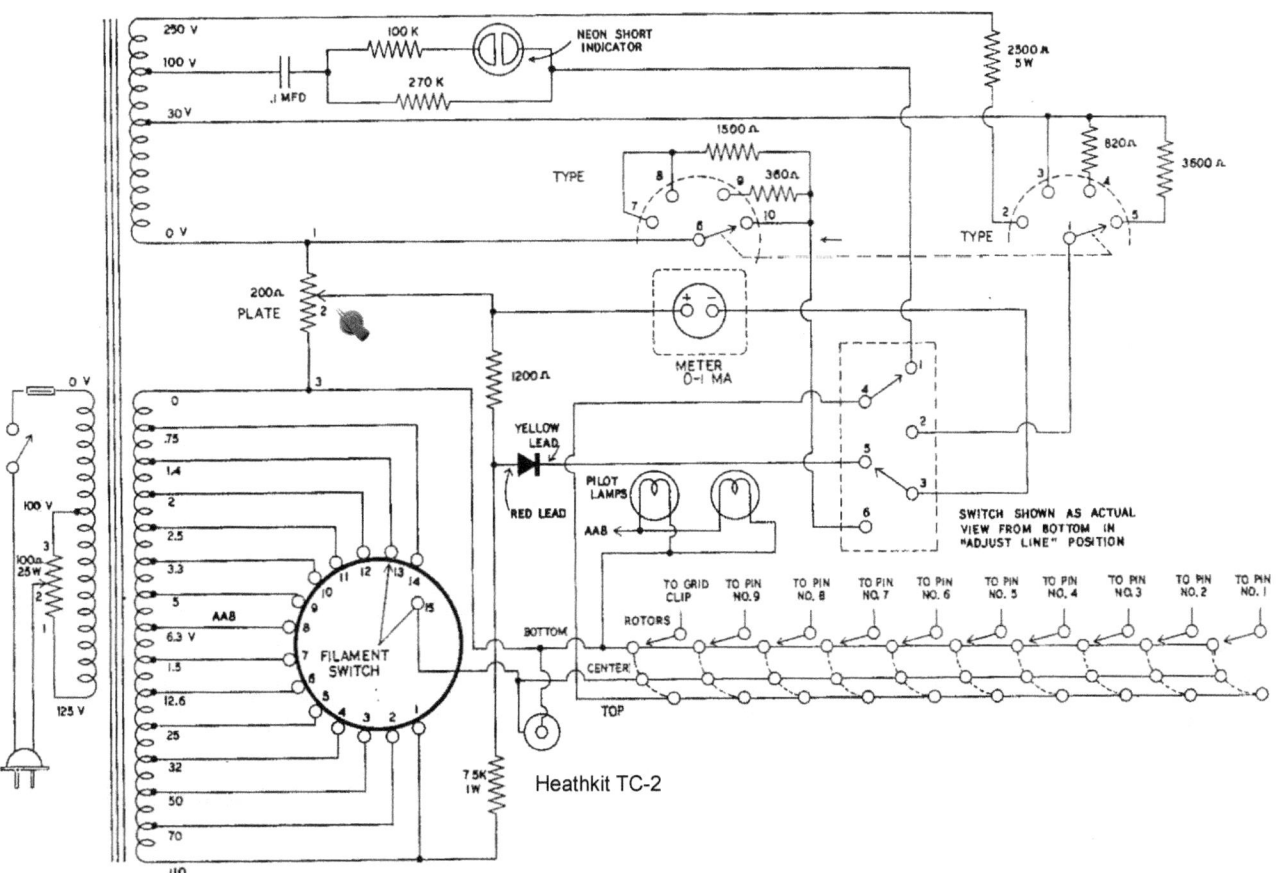

The operating principle behind the emission test

The first three positions (1, 2 & 3) of the "A" or "Circuit" test switch use $30V_{AC}$ test voltage. While the 51Ω resistor is always part of the circuit, the other series resistor's value varies. In Position "1," the resistance is 5k1; it is lowered to 1k in position "2" and to zero (short circuit) in position "3".

Another bunch of resistors is switched in and out within the 'LOAD' or shunt loop, 1k5 in position "3", 470R in position "2," and no resistance in position "1". In position "4," a much higher test voltage is used, full $250V_{AC}$.

The example below illustrates the way 6L6 is tested. The anode and screen are strapped together in the "UP" position, while the control grid, cathode, and one side of the heater are strapped together in the "DOWN" position. The metering circuit is connected between the other side of the heater and the 0V (COM) terminal.

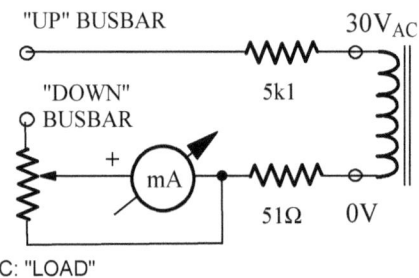

"A" CIRCUIT: Position "1"

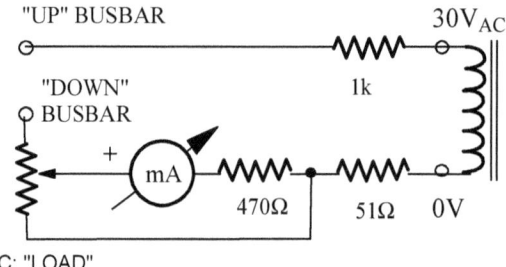

"A" CIRCUIT: Position "2"

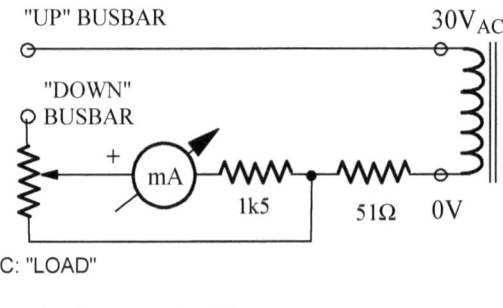

"A" CIRCUIT: Position "3"

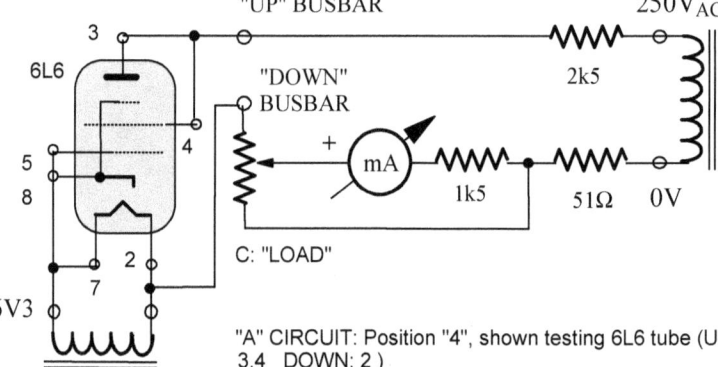

"A" CIRCUIT: Position "4", shown testing 6L6 tube (UP: 3,4 DOWN: 2)

Switches for pins 5,8,7 stay in the center position, meaning those electrodes remain connected to the selected heater voltage, 6.3V in this case.

Shorts & leakage neon circuit

This is the most common shorts/grid leakage circuit used, mostly on lower-spec (cheaper) emission testers. Either C_2 or R_2 is used, with the resistive option being more common. Typically, the C_1 capacitance is between 10 and 50 nF, and R_2's resistance varies from 680 kΩ to 2 MΩ.

The test voltage V_T is usually 80 - 117 V_{AC}. R_2 determines the circuit's sensitivity (when the neon globe will turn on). The higher its value, the more sensitive the detection circuit is. Heathkit/Conar and Knight testers use 270kΩ; others use 330kΩ.

The switching arrangement (not shown) for each test switch position isolates one electrode while all others are bundled together. Here the leakage between the heater and all other electrodes is checked.

C_1 is a blocking capacitor, which prevents the Tube-Under-Test from rectifying the AC test signal, so only AC current can flow if there is less than infinite resistance between the electrodes under test (RHK in this case). R_{HK} is not an external resistor but an internal resistance of the tube.

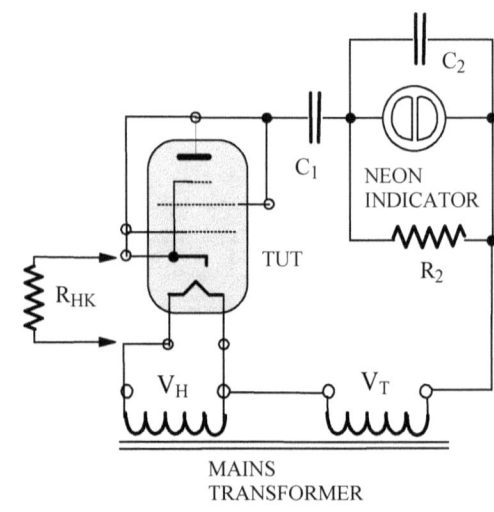

ABOVE: A typical shorts & leakage neon test circuit

Replacing the primitive "shorts & grid leakage" neon circuit with an analog megohmmeter

IMPROVEMENT PROJECT

This simple modification to any tester with a fixed (neon-type) shorts test circuit enables precise measurement of inter-electrode leakage instead of the primitive on-off neon indication.

The same AC supply point is used; in this case, it is a $100V_{AC}$ tap on the power transformer.

The capacitor, two resistors, and neon globe are removed, and an additional latching DPDT switch (either sliding or rotary) is added (SX).

The megohmmeter is shown added to Knight KG600 tester. The wiring is identical for Conar and Heathkit tube checkers, the only difference being the terminal numbers on switch S2, the "Short-Value" switch.

In the LA ("Line Adjust") position of the SX switch, the meter is connected as before, between terminal 2 of S2 (- of the meter), and the meter's positive connection goes to the "Load" control pot (its slider).

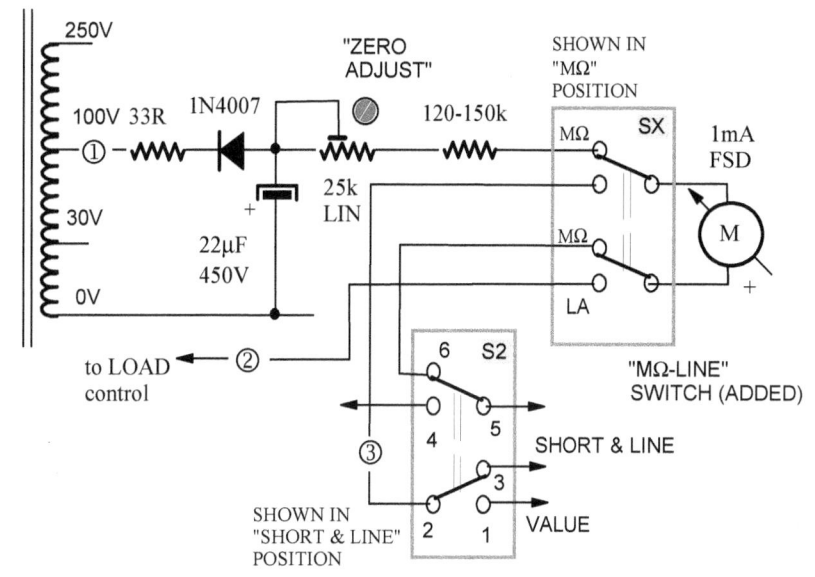

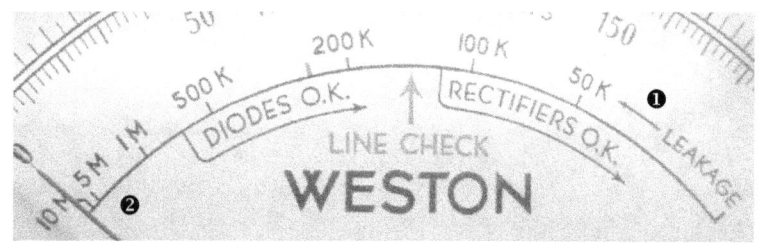

ABOVE: Ohmmeter scales are highly nonlinear; the sensitivity and resolution are very high at lower resistance values (1) and reduce towards higher resistances, where the scale gets highly compressed (2). Your scale will look like this example from Weston 981 tester.

The first calibration step is to make a short circuit between any two tube socket pins and adjust the series trimmer pot for the zero Ω reading of the meter. The resistor values are not critical.

The second calibration step is to mark a few resistance values on one of the tester's scales (or draw a dedicated MΩ scale if you prefer. Insert various resistors (100k, 200k, 500k, 1M, 3M, etc.) and mark the meter's readings on the scale.

CONAR 221, 223 and 224

Conar was the brand name of the National Radio Institute in the USA. After initially using a rebadged Precision 640 (NRI 71) tester in their technician training courses, in the mid-1960s, they released their versions of Knight and Heathkit tube checkers. Models 221, 223, and 224 use the same test settings and are all but functionally identical (except two internal components of different values, which do not affect test results).

The circuit topology is identical to Knight, but some components and voltage values are different. The test voltages are 30, 80, and 250V, compared to 30, 100, and 250V used by Knight and Heathkit. The neon short and leakage test is more sensitive in Conar; the neon indicator is paralleled by a 680k resistor, meaning the neon will light up for inter-electrode resistance (leakage) under 680kΩ. Heathkit and Knight used 270k sensitivity.

They all used a 30V emission test voltage, so the same test settings can be used for all three brands of testers! Just as Knight and Heathkit, Conar also offered their gear as kits and as prewired units.

Although a few "experts" expressed negative opinions and made some disparaging online remarks about Conar tube testers, they do a decent job of emission testing. Technically, they are in no way inferior to similar Heathkit, KnightKit, and Eico tube checkers.

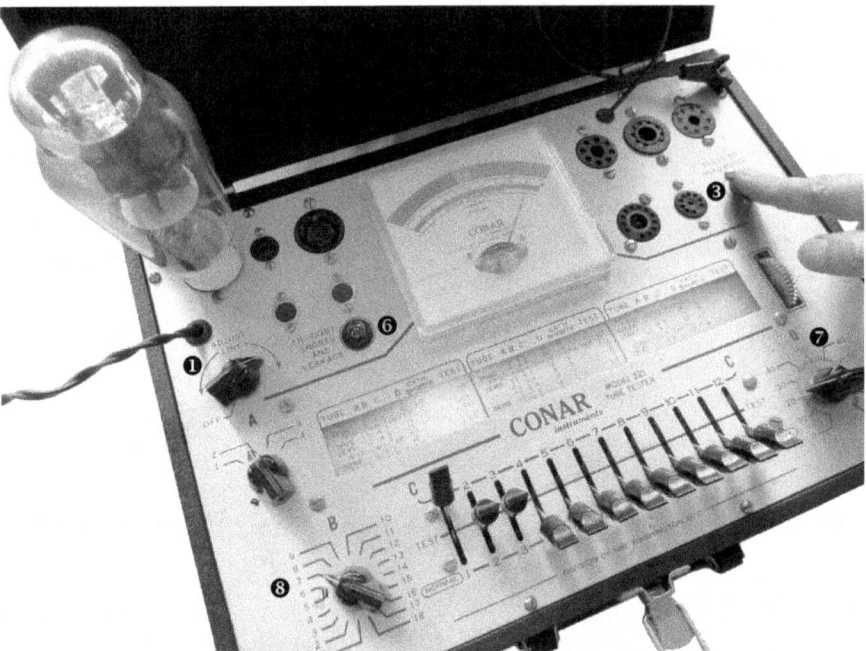

LEFT: Testing a new JJ 300B triode on Conar 221

BELOW: The internal view of Conar 221
1. "Adjust line" rheostat
2. Internal mains fuse
3. "Press to read meter" switch
4. Copper-oxide rectifier (diode)
5. Two globes for chart illumination
6. Neon "short" indicator and associated circuitry
7. "D" rheostat (Shunt)
8. "B" or heater voltage selector switch
9. Power transformer, mounted on the timber case instead of the top panel

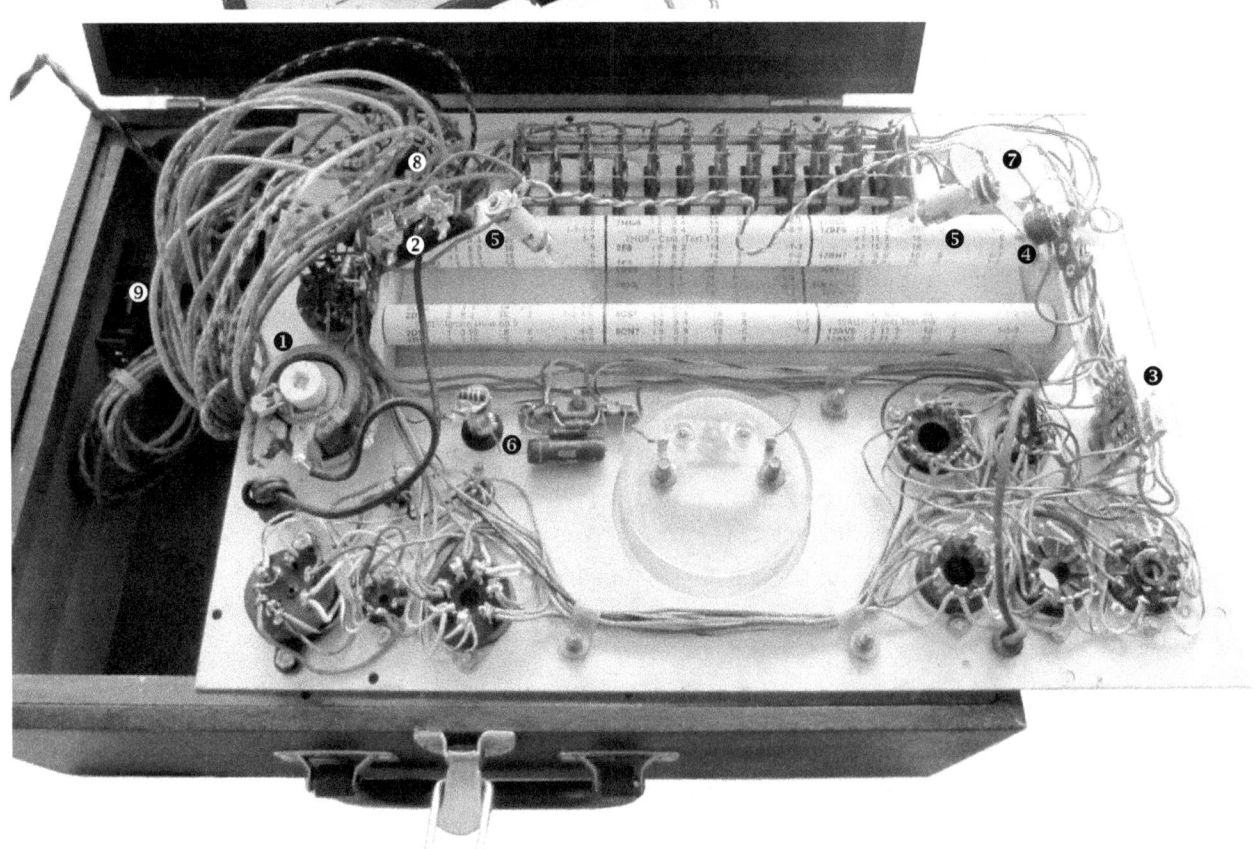

EXPERIMENT: Determining electrode contributions to the emission test readings

The contribution of individual electrodes to the overall reading of emission testers can easily be determined. Triodes can be tested between anode and cathode (always a referent electrode), grid and cathode, and anode & grid strapped together and cathode. Beam tetrodes and pentodes have three electrodes we can play with (anode, screen, and control grid) in seven possible combinations: anode only (screen and control grid left open), screen grid only (anode and control grid left open), and so on. The final combination is all three electrodes strapped together (A+S+G), which is the test regime used in most emission testers.

In this simple experiment, using Conar 221 tester, we will answer a few basic questions. The first hypothesis is "The closer the electrode is to the cathode, the higher the percentage of its current versus total current." So, the control grid will contribute most to the overall (A+S+G) current, the screen will contribute less, and the anode will contribute even less than the screen grid.

EMISSION TESTERS   43

The second question is if the same distribution ratios apply to all similar tubes. For instance, is the percentual contribution of the screen grid the same for 6L6 and 6BQ5 power tubes, or 6550 and EL34? Intuitively, most of us would say no; each tube has a different size and geometry of electrodes and different control grid and screen grid m factors, so the contribution percentages of various electrodes should be different.

And the final question, does such a distribution change for weak (low emission or low Gm) tubes? We didn't have a low reading 6L6 but had a worn-out 6550 and a very weak-testing EL34 tube (30% on Triplett 3444), so we compared them to brand new ones.

	6550 new	6550 weak	EL34 new	EL34 weak
Anode only A	0.5-1.0	0	0	0
Screen grid only S	30	10	8	8
Control grid only G	74	63	71	50
A+S	40	14	15	10
A+G	75	60	71	50
G+S	76	64	73	54
A+S+G	76	66	74	54

Conclusions?

1. The anode contributed nothing when G &S were used for testing or a barely noticeable indication of 1/2 to 1 minor graduation for a new 6550. However, if A+S were used for testing (no control grid used), the anode made a small contribution (14 vs. 10 for a weak 6550 and 40 versus 30 for a new one).

2. The compression problem: the tester indicated a strong 6550 as a 76% tube (too low) and a weak 6550 as 66% (too high), compared to Triplett 3444 tube analyzer's results.

3. The contribution of the screen grid dropped from 30% to only 10% in a weak 6550 tube. So, 30/76 = 39.5% for a new tube, and 10/66 = 15.2%. Thus, the percentual contribution of the screen reduces as the tube ages.

4. The figures are not additive; in other words, using the new 6550 as an example, 0 for A + 30 for SC + 74 for G, a total of 104, does not equal the reading of 76 for A+S+G! For a worn-out 6550, the sum of 0+10+63 = 73 was much closer to the total of 67.

5. Just as with #2, the tester tested a strong EL34 as a 74% tube (too low) and a weak one (30%) as 54% (too high), so it seems its results are artificially "compressed."

PRECISION 640 (NRI 71) & PRECISION 660

Many eBay buyers get fooled by the elegant looks and the "Precision" in the name of these large testers. They look solid and instill confidence. The controls are well laid out, and there are so many of them that one would be forgiven for thinking, "Man, this is one capable tester!"

If you doubt a certain model, you don't need its circuit diagram. Just select a tube you are familiar with, look it up in the tester's data book, and the settings will tell you the whole story.

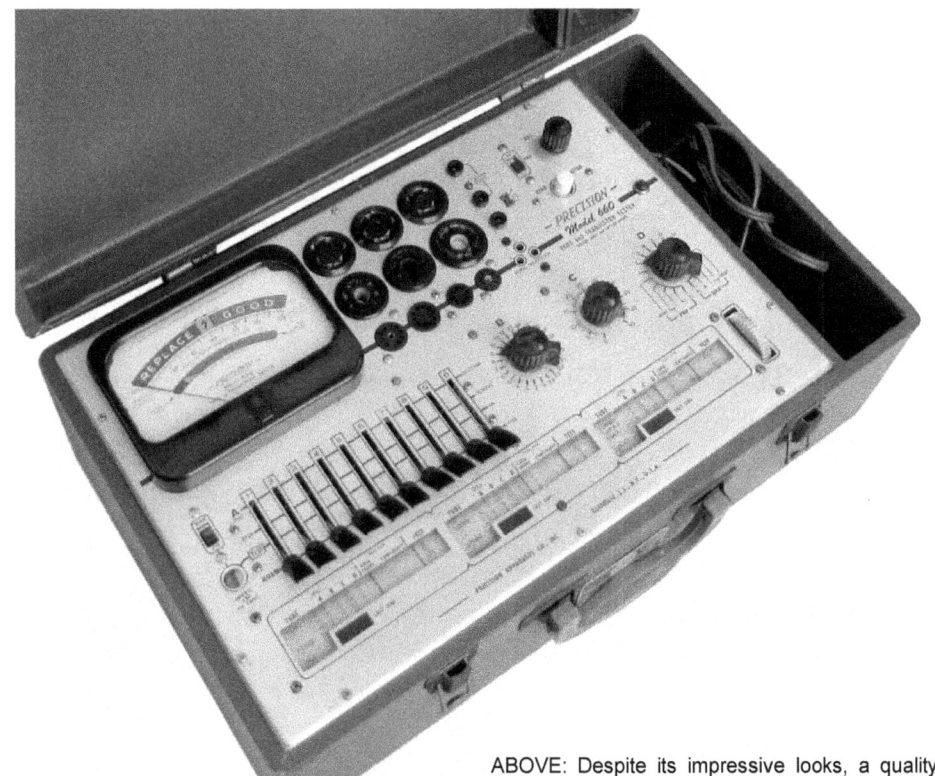

In this case, the 6L6 settings are A=2 (this switch selects the heater pin, pin2 in this case), B=9 (heater voltage, not 9V, just position number 9, 6.3V), D=1 (selects one of many test voltages for anode supply) and "TEST"= 3, 4, 5.

Pins 3 (anode), 4 (screen), and 5 (control grid) are connected together, and 6L6 is tested as a diode, with current flowing between its cathode and all other electrodes strapped together. Thus, it is a simple emission checker.

Precision 640 is functionally identical to model 660, and NRI (National Radio Institute) model 71 is identical to Precision 640.

ABOVE: Despite its impressive looks, a quality analog meter and well laid-out controls, Precision 660 is a basic emission tester of modest capabilities.

TRIPLETT 2413 AND 3414

An elegant and well-built instrument with neatly laid out controls and a large quality meter, Triplett 3414 is one of the better emission testers.

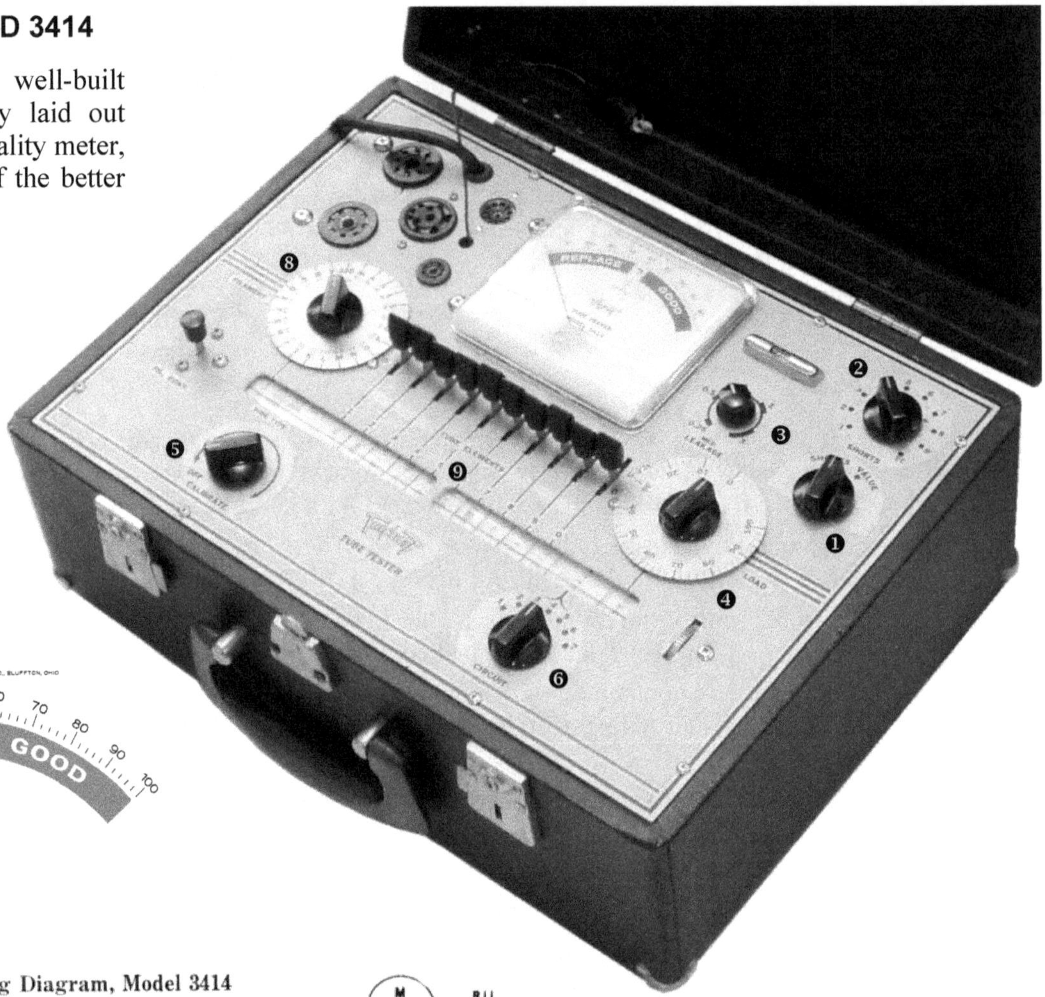

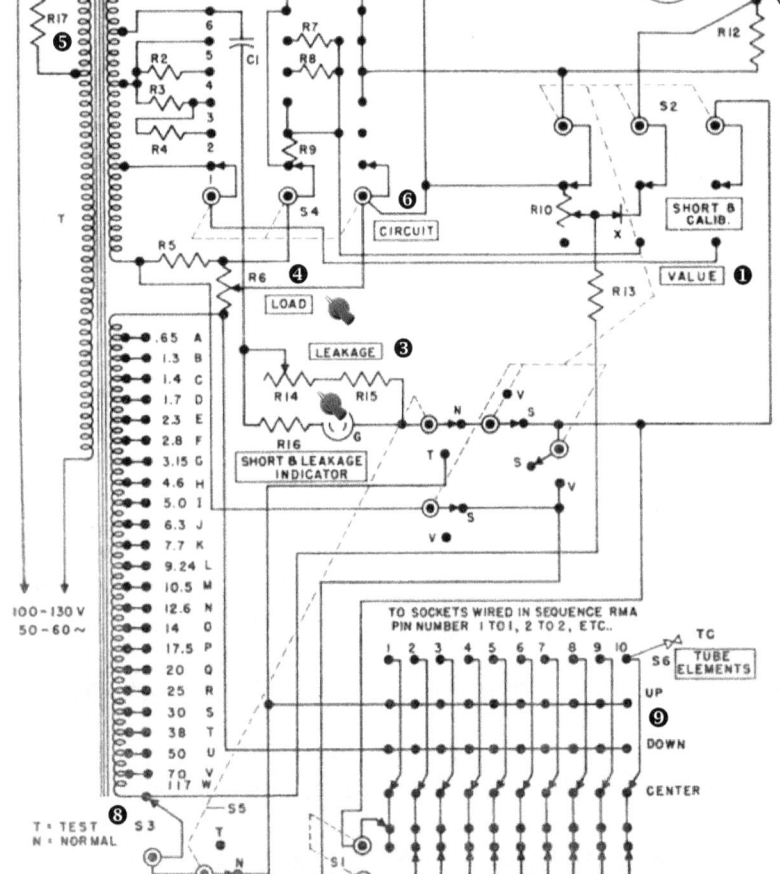

It tests for "LEAKAGE" using an ordinary neon test circuit but with a twist. A variable resistor R14 (5MΩ) was added in parallel with the neon branch, effectively varying the test's sensitivity. Notice the range marked on the tester's control panel from 0.25MΩ (250kΩ) to 3MΩ.

LEFT: The circuit diagram is well laid out and easy to follow. © Triplett
BELOW: A detail of Triplett 3414 controls

EMISSION TESTERS

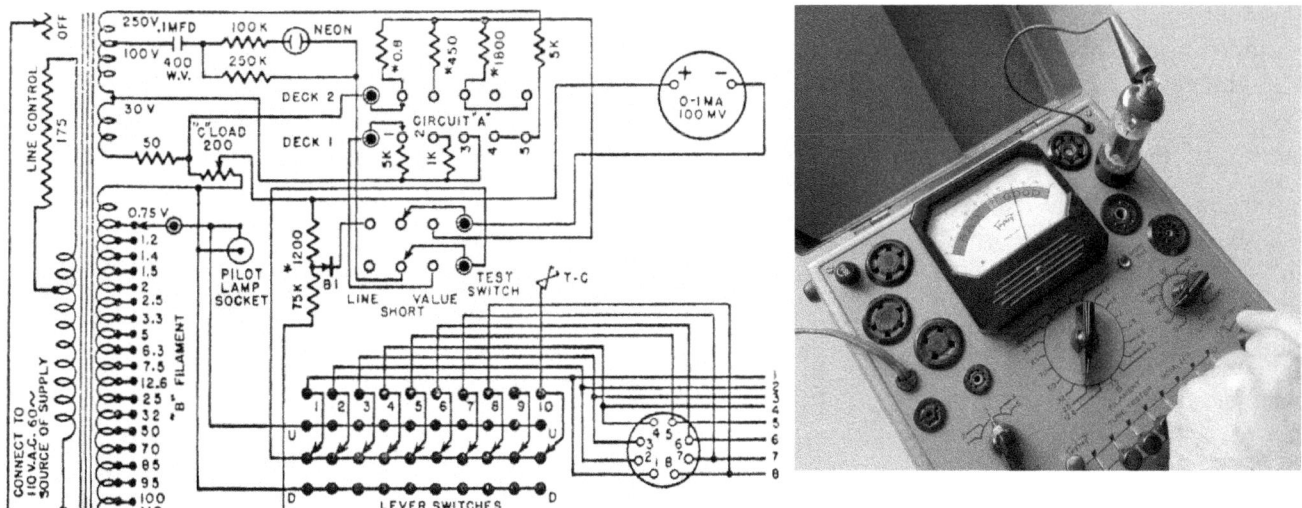

ABOVE RIGHT: Triplett 2413 (3212) is a basic emission tester, but its meter is a large and precise quality instrument, better than meters on most mutual conductance testers!
ABOVE LEFT: Triplett 2413 circuit diagram, © Triplett

Its smaller brother, Triplett 2413, is a more basic tester but still a quality instrument, housed in a steel box and built like a tank. Its "CIRCUIT" or "test regime" switch has five settings while 3414 has seven.

ELETTRA PROVAVALVOLE

Scuola Radio Elettra di Torino was an Italian electronics school, and their "Corso radio per corrispondenza" had students assemble this simple emission tester. It had no indicating instrument, so, as another project, an analog VOM or "multimeter" was also assembled (1) and used to read the tubes' emission currents. Any digital or analog multimeter can be used on the $3V_{DC}$ range for preamp tubes and $10V_{DC}$ range for power tubes.

The third instrument in our set was an RC substitution box, with one variable resistor (2), one switch-selectable bank (3) of resistors (47Ω to 2M2), and a switchable capacitor/resistor bank (4), with values ranging from 15pF to 16µF. It also had a bridge so unknown resistances and capacitances could be measured by comparing them with a known component. Transformer voltage ratios could also be determined, a very useful feature for evaluating unknown power, output, interstage, and other audio transformers.

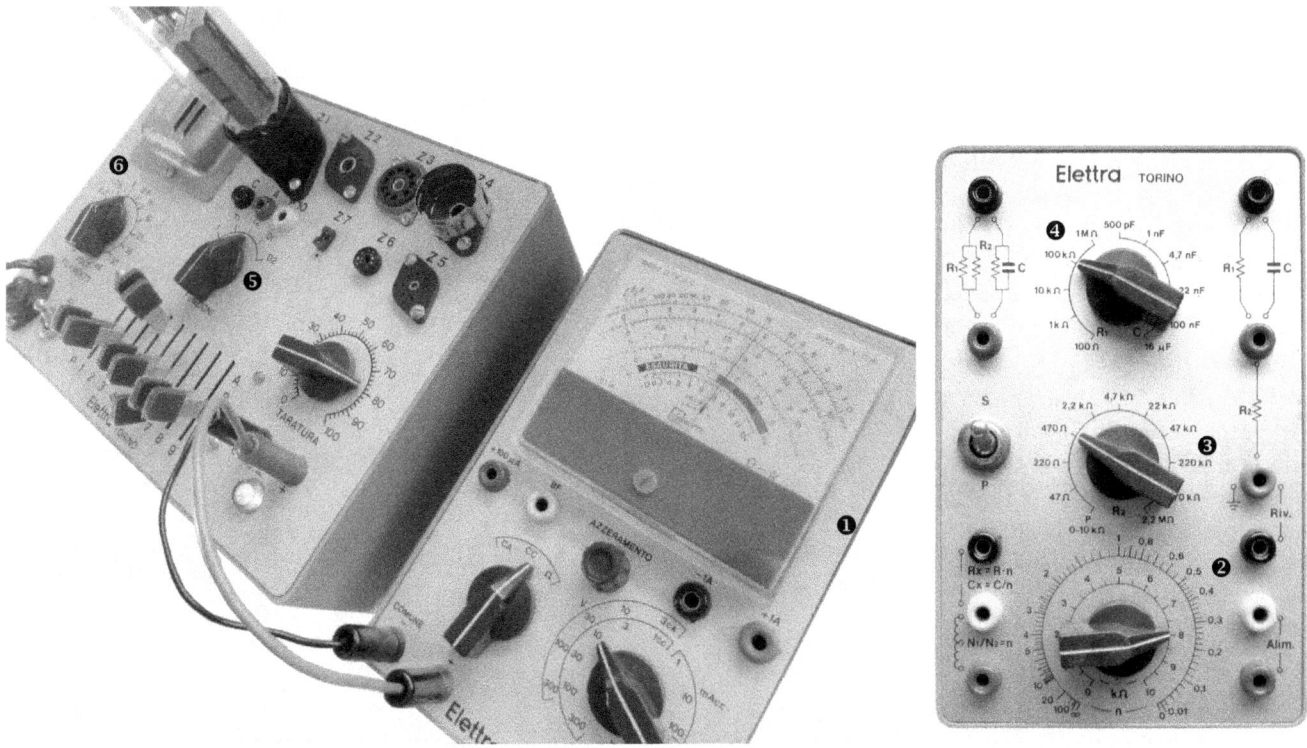

EL34 was tested using the T2 position (5), with $71V_{AC}$ on the anode and screen. The heater voltage in the 6.3V position was only 5.9V unloaded, and even in 7.5V position the voltage dropped from 7.3V with no load to 5.6V with EL34 plugged in. Finally, EL34 was tested in 9V heater position (6), the voltage dropped from 8.7V to 6.05V!

On the power supply side, this emission tester has no fuse and, although it has a 125V tap, it has no 230 or 240V tap, only 220V, the old European standard (1). The new, "harmonized" mains voltage should be $230V_{AC}$.

Insulation or leakage between electrodes is checked in position "I," using $71V_{AC}$, which is half-wave rectified into a pulsating "DC" voltage and fed to the anode bus bar ("A") via the external analog meter and the internal shunt circuit, the 130R resistor and "Taratura" or "Calibration" 1k pot.

The circuit is thus a simple ohmmeter with an external meter indicating the degree of leakage or short circuit.

In two "Diode" positions (D1 and D2), the same $71V_{AC}$ test voltage is used, in one case via the 33k resistor, in the other via the 3k3 resistor.

In "T1" emission-testing mode $48V_{AC}$ voltage is fed via the meter shunt to the anode, and in "T2" mode a higher test voltage ($71V_{AC}$) is used.

The lever switches have four positions, "A" or anode, "B" or disconnected, "C" is the COM or "0V" point, to which TUT's cathode and one heater pin must be connected, and "D" is the heater end, fed from the filament voltage selector switch "Volt filamento".

Testing high Gm tubes can cause oscillation, which can be cured by connecting a capacitor in parallel with the meter and an addition of a simple CLC PI-filter between the anode and cathode. Component values are not critical.

The 48V and 71V test voltages are relatively high for an emission tester, but the lack of line adjustment feature and the incredible sag of all heater and test voltages under load place this checker in the lowest quality category. It's practically a toy!

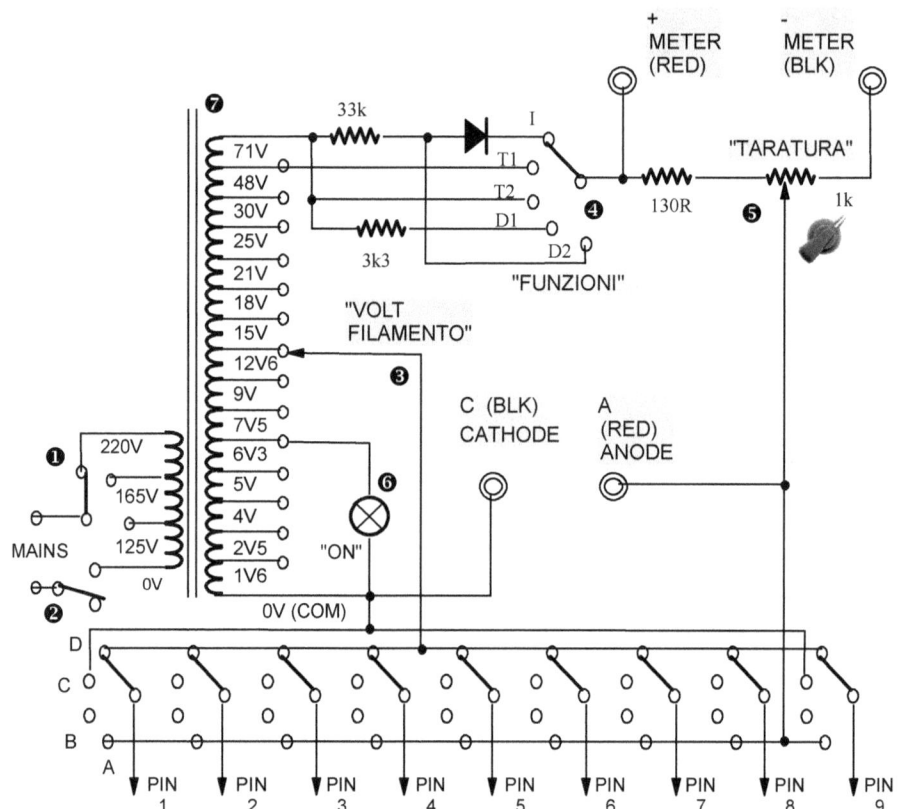

ABOVE: Circuit diagram of Elettra tube tester
BELOW: The internal wiring of the tester

GRID CIRCUIT TESTERS

4

- HOW GRID CIRCUIT TESTERS WORK
- SECO 78, 88, AND 98
- SENCORE "MIGHTY MITE" TESTERS: TC114, TC130, TC-136, TC-142, TC-154
- SENCORE TC162 & TC28 ("THE HYBRIDER")
- B&K 600 & 606 DYNA-QUIK TUBE TESTERS
- B&K 607 & 667
- B&K 625 DYNA TESTER
- B&K 666 DYNA JET TUBE TESTER
- PRECISION APPARATUS COMPANY (PACO) 650 & T-62
- MERCURY 1101, 1101C & 1101CT
- AMERICAN SCIENTIFIC DEVELOPMENT COMPANY TV-20

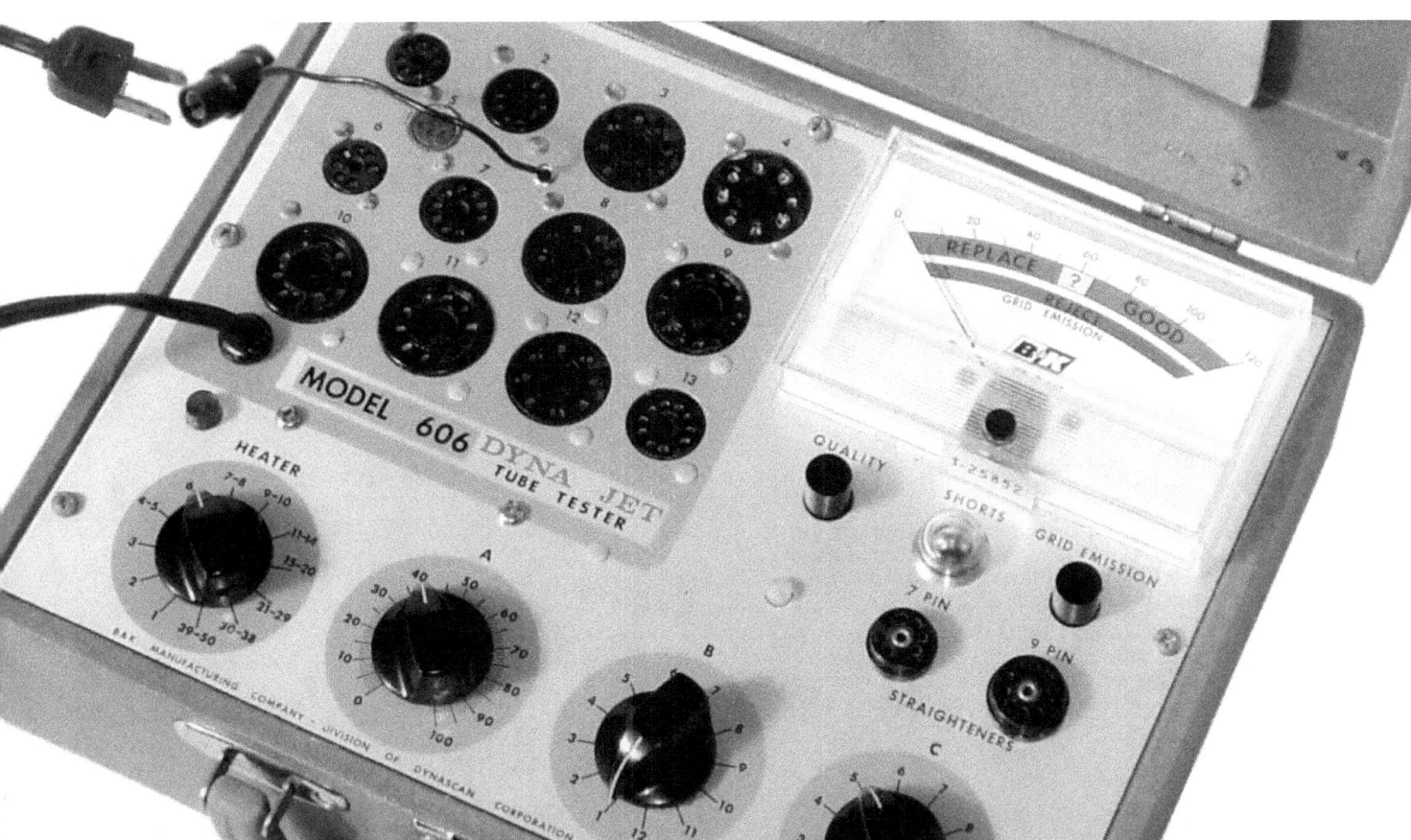

HOW GRID CIRCUIT TESTERS WORK

Grid circuit (GC) testers are a sub-category of emission testers. Instead of the cathode current flowing to the screen grid and anode, grid testers connect the anode and screen grid to the cathode, so the current only flows between the cathode, which emits electrons, and the control grid, which collects them.

No current reaches the screen or the anode, so some testers such as Sico 82 leave those electrodes disconnected. One EL34 tube had its anode internally disconnected from its pin, so it did not work in an amplifier, yet Sico 82 tester passed it as a healthy tube! This is the most glaring inadequacy of grid testers - they only test two electrodes, the cathode, and the control grid! Even then, as with all emission testers, the current controlling action of the grid is not tested in any way; only mutual conductance testers and, to some extent, the dynamic conductance testers do that.

Another problem with grid circuit testers is that very low test voltages and very low currents are involved. Most control grids are would with superfine wire, and even a few mA of current could be enough to melt it (just like a fuse) and destroy a tube under test! This is especially so with voltage amplifying (preamplifier) triodes such as 12AX7, 6DJ8, and alike.

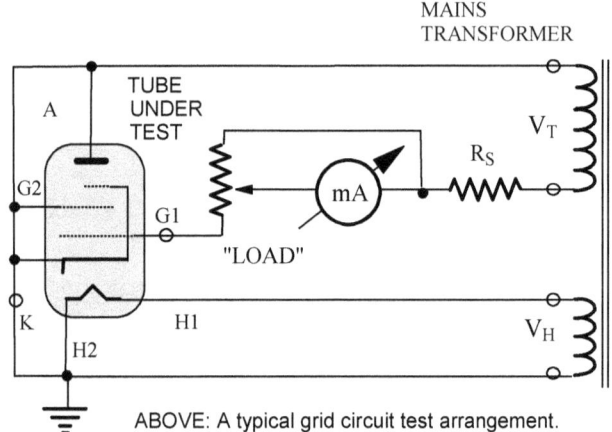

ABOVE: A typical grid circuit test arrangement.

Manufacturers liked GC testers because they could save even more money on transformers, the most expensive part of any tester. Instead of 50 or 60 VA-rated power transformers in ordinary emission testers, tiny transformers of less than 20 VA could be used with these "instruments," which would now be lighter and smaller, so further cost savings would be achieved.

SECO 78, 88 AND 98

Seco Electronics, Inc. was a small business based in Minneapolis. Their line of budget tube testers consisted of three grid testers with a simple internal vacuum tube voltmeter, models 78, 88, and 98, and Model 107, a Gm tester, later revised and expanded as 107B and 107C. In 1962, model 88 was selling for US$69.95.

While their testers aren't any worse than B&K and similar budget brands, Seco's technical documentation is atrocious, arguably the worst of all.

Seco testers' schematics are primitive hand-drawn sketches, missing most important details and often very different from "as built" testers, all in all, next to useless for troubleshooting and repair purposes.

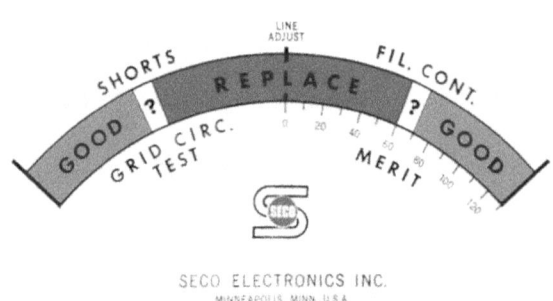

ABOVE: The left half of the scale was used for the gas/grid leakage and shorts tests. The right half of the scale indicated the tube's "merit," or cathode emission picked up as a tiny grid current, just as with all grid testers. That scale is so small and compressed that it borders on useless.

RIGHT: Sico 78/88/98 are cute little tube checkers but far from serious instruments. What Seco calls "grid circuit test" is actually "grid current," "gas," or "leakage" test, while the emission current between cathode and control grid is called "merit" test.

GRID CIRCUIT TESTERS

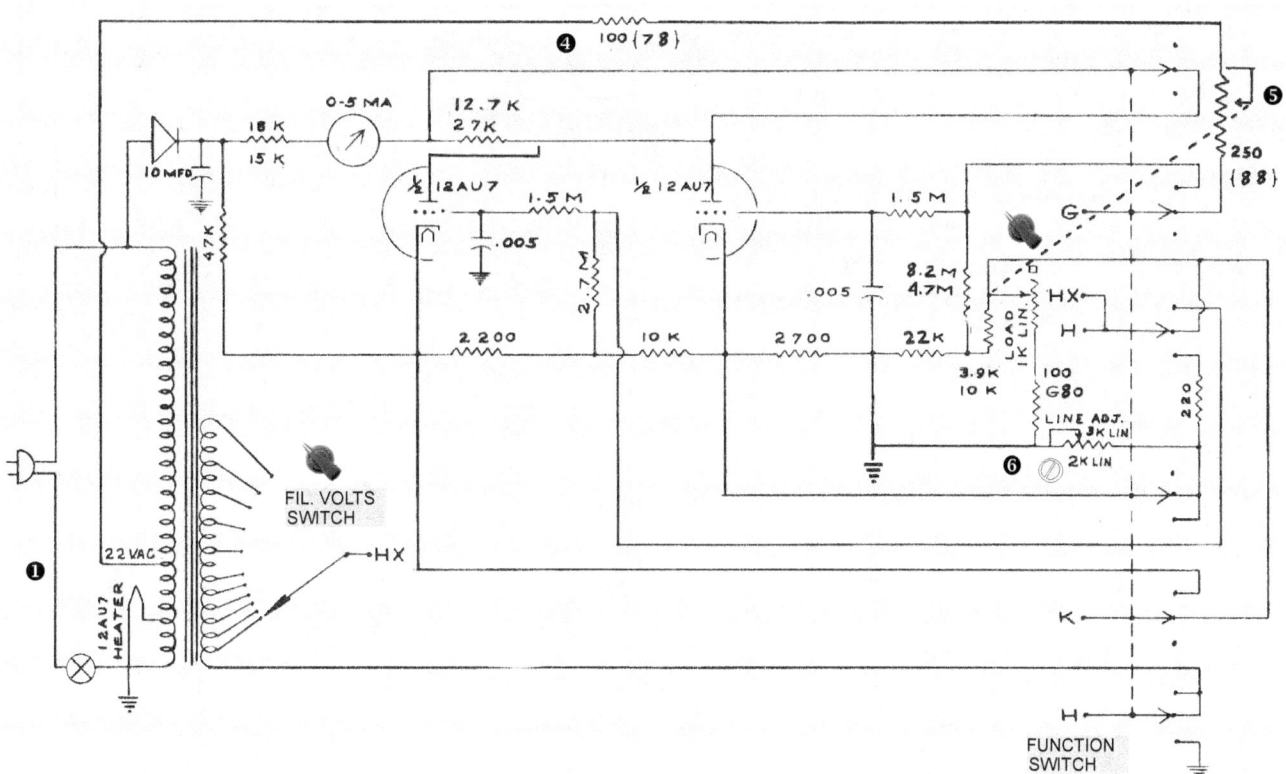

ABOVE: Circuit diagram of Seco 78 & 88 tube testers. If there are two figures next to a component, the upper value is for model 88 and the bottom value is for model 78. If a component is present in one of the models only, the model's name is in brackets.

The on-off switch (1), the "L" switch (2), and the "R" switch (3) were not even drawn the schematics. Model 78 (photo below) used a fixed 100R resistor in the GRID test line (4), where model 88 included a "Load" potentiometer (5). The "Line adjust" is actually a zeroing feature for the VTVM and does not change heater or test voltages which will fluctuate as much as the mains voltage does.

Instead of a proper drawing, Seco included a table for switch L's contact assignments in the tester's manual and a similar one for switch "R." According to the manual, "the lever switch 'L' is a multi-rotor switch arranged for opening circuits to selected pins.

DETAIL OF LEVER SWITCH "L" MODEL 88 SECO TUBE TESTER

POSITION OF SWITCH "L"	9 PIN (4 & 5 HTRS.)	7 PIN (3 & 4 HTRS.)	OCTAL	LOCTAL	9 PIN SPEC SOCKET
NORMAL	All socket pins except htrs. connected to circuit thru ROTARY switch "R".	All socket pins except heater pins. Rest of pins in circuit thru switch "R".	Pins 7 & 8 are heater pins. Rest of pins in circuit thru switch "R".	Pins 1 & 8 are heater pins. Rest of pins in circuit thru switch "R".	This socket wired for 1X2 etc. Also tests 12B4 & simil. Set LEVER SW. "L" as shown in set-up chart.
X	All pins in (same as above) except #9 is out of circuit.	All pins in circuit except #7.	Pins 2 & 7 are the heaters. The rest of pins in circuit.		Pin #1 disconnected.
Y	All pins in the circuit except #6.	All pins in the circuit except #6.	Pins 2 & 8 are the heater pins for 5Y4 etc. Pin #5 open.		Pin #7 disconnected.
Z	All pins in the circuit except #3.	All pins in the circuit except #5.	Pins 2 & 7 are heater pins. #3 is out of circuit.	Pins 1 & 8 are heaters. Pin #4 is out of circuit.	Pin #3 disconnected.

All modern TV and radio heater type tubes (including rectifiers) that have the following heater pin combinations may be tested on the Model 88 Seco Tube Tester:

9 PIN	7 PIN	OCTAL	LOCTAL	SPECIAL SOCKET FOR
4-5	3-4	2-7	1-8	1X2
3-4-5	3-4-5	7-8	1-4-8	12B4
4-5-6	3-4-6	2-8		etc.
4-5-9	3-4-7	2-3-7		
		2-7-8		
		2-5-7		

MODEL 88
Grid Circuit & Compound Shorts Test

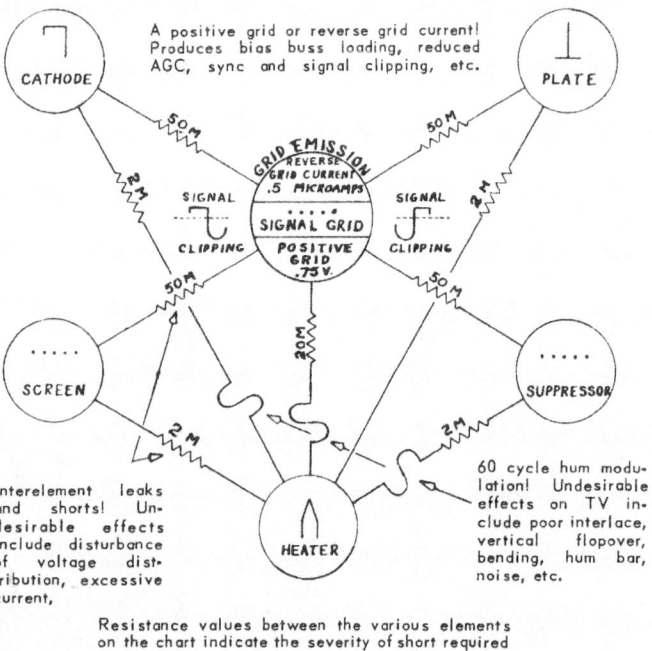

Interelement leaks and shorts! Undesirable effects include disturbance of voltage distribution, excessive current,

60 cycle hum modulation! Undesirable effects on TV include poor interlace, vertical flopover, bending, hum bar, noise, etc.

Resistance values between the various elements on the chart indicate the severity of short required to deflect the meter to the question mark on left of meter.

It also transposes the heater voltage to different pins. Study of the chart will reveal that 14 heater pin combinations are available in addition to the 9-pin special socket."

Apparently, switch "R" is a "dual ganged switch. One section is an isolating type while the other is a selecting type." I understand what that means, but I've spent years fixing and modifying tube testers. The real question is would an average user of such testers be any wiser after studying this table? I doubt!

ELEC. DETAIL OF SWITCH "R"

TUBE TYPE / SWITCH R POS	A (GREEN)	B (ORANGE)	C (BLUE)	D "Y" OPENS (RED)	E "X" OPENS (BROWN)	F (GRAY)	G "Z" OPENS (WHITE)	H (PURPLE)	J (YELLOW)	K (CLEAR)	TUBE TYPE
7 PIN	1	2		6	7		5	TC			7 PIN
NOVAR	1	7	9	6	2	8	3	TC			NOVAR
10 PIN	1	2	7	6	9	8	3	10			10 PIN
OCTAL	1	2	4	5	8	6	3	TC			OCTAL
COMP.	11	2	3	4	5	6	7	8	9	10	COMP.
NUVIS.	4			2				8	TC		NUVIS.
LOCTAL	3	2		5	7	6	4				LOCTAL
SPEC. 10 PIN				7	1		3	10			SPEC. 10 PIN
SPEC. NOVAR				5	9		7	TC			SPEC. NOVAR
SPEC. 7 PIN		2		4	6		3	TC			SPEC. 7 PIN
TUBE TYPE / SWITCH R POS	A	B	C	D	E	F	G	H	J	K	TUBE TYPE / SWITCH R POS

SENCORE "MIGHTY MITE" TESTERS: TC114, TC130, TC-136, TC-142, TC-154

Sencore "Mighty Mite" series of tube testers was one of the most popular in their day; tens of thousands were sold to DIY constructors and radio/TV repairmen over the three decades (the 1950s-1970s). TC-114, TC130, TC-136, TC-142, and TC-154 all share the same test settings and, apart from some cosmetic and minor internal differences, are the same tester.

The setting for 6L6 is 6C6E, meaning A: position 6, B: position C, C: position 6, and D: position E!

Position 6 of "A" means a heater voltage of 7.7V. Although 6L6 needs 6.3V, since it is a power tube, its heater draws almost 1A of current, the voltage sag of small power transformers inside these budget testers is so large that 7V7 with no load will drop to 6V5 or even lower under load.

One pole of switch "D" picks the control grid to separate it from other electrodes; this is pin 5 in 6L6, and since pin 1 is position A, pin 2 position B, etc., pin 5 is position E of the switch "D."

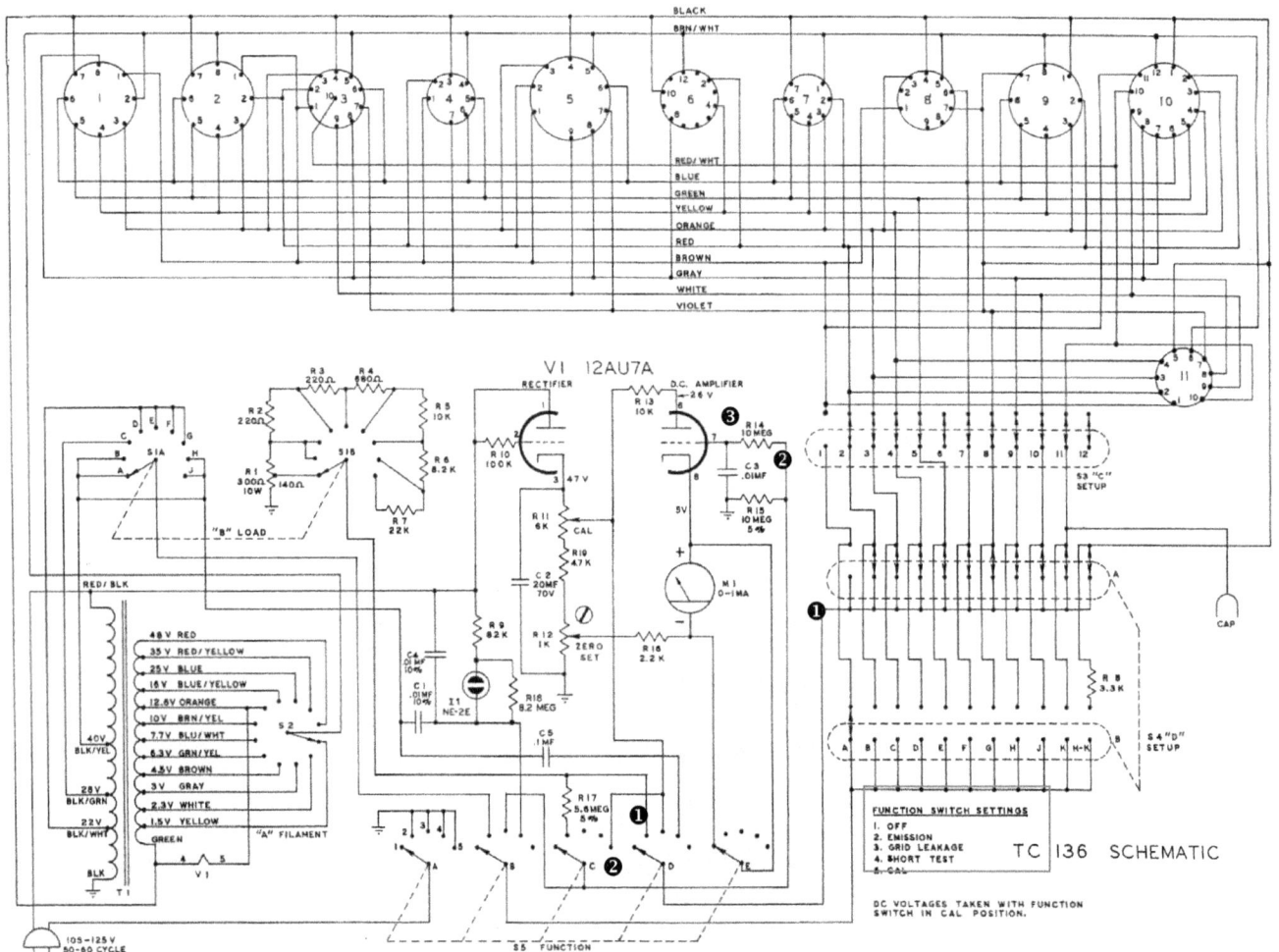

TC-136 circuit diagram, © Sencore

GRID CIRCUIT TESTERS

The other pole straps together all other electrodes of the tube-under-test since those must be connected to the load resistor and the metering circuit. Follow point (1) from the "D" switch, through one contact of the "Function" switch, via the 5M6 resistor, the other contact of the "Function" switch (2), and then onto the 10M resistor.

Those two resistors form a voltage divider, followed by the 10M-10nF RC filter (3). The half-wave rectified voltage pulses on the cathode are thus smoothed into a DC voltage proportional to the DC cathode current. This voltage is then fed into the control grid (pin 7) of one 12AU7 triode and amplified. The analog meter is in the cathode circuit of the triode, so the metering circuit is a fairly simple single-stage amplifier.

After checking for shorts and measuring cathode emission, the next test is usually the gas/grid emission test. There are three faults detected in one test since all result in the same symptom - grid current flows if the tube is gassy, if there is grid leakage, and when the grid is contaminated, it emits electrons (acts as a cathode).

Testers employing this method measure the voltage drop across the large value resistor connected between the grid and ground (cathode), created by the flow of negative grid current. This voltage is then amplified and indicated on the analog meter.

Understanding tester's scale: Sencore "Mighty Mite" example

With zero volts on 12AU7's grid, the analog meter in its cathode will read very low, about 0.1mA. As the DC voltage is increased, the cathode current will rise in an almost linear fashion. A tube with normal emission will have about 0.78 mA flowing, and the meter will indicate 100% on the "emission-quality" scale. At 50% of normal, the meter's indicator will fall into the "Bad" area. A DC voltage of +5V is required for the 100% indication (0.78mA).

At that point, the 12AU7 triode is almost at zero bias, so any further DC grid voltage increase will cause very little change in the meter's indication (saturation region), meaning the cathode current will increase very slowly. To reach the maximum or full scale reading of 130%, a whopping +40V on the grid is needed.

While the increases between 0 and 100% are relatively linear, the increase from 100 to 130% is very compressed, i.e., nonlinear. While anything nonlinear is usually complicated and "bad" in electronics, this is actually a useful feature. Even if a shorted tube is tested or a load selector is in the wrong position, the meter will not get damaged. Also, the readings for "extra strong" tubes will not shoot off the scale!

A typical 0-100% scale of a simple emissions checker (SICO). A new or strong tube will test around 80-85%!

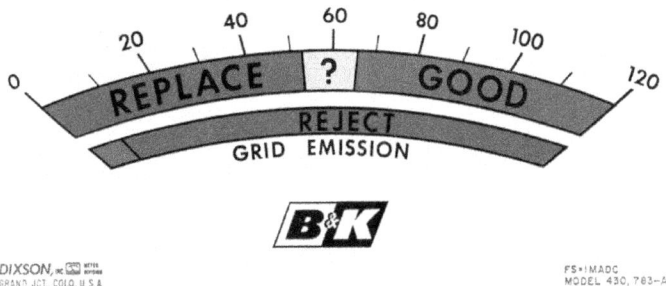

Mighty Mite testers' scales are 0-130%, similar testers by B&K (models 600-606-607-667) use 0-120% scales. A new tube will test around 95-100%.

Designers of vintage testers using 0-100% scales had to allow for those extra-strong tubes, so their "Shunt," "Sensitivity," or "Load" resistor settings were determined so even the strongest tube could not take the reading past 100% (full scale). This means that an average reading tube would test at around 75-85%, which confused users, who intuitively expected a new tube to test at 95-100%. As a result, they would question the quality of such tubes or the honesty of sellers who sold them such tubes instead of understanding the flawed design of their primitive tube checkers. Thus, a 0-120 or 0-130% scale is a much better option.

However, due to heavy compression in that last segment (100-130%), a tube tested at 110% will not have a 10% higher current than a tube tested at 100%, but almost 50% higher! An average user would automatically assume that any scale with equal graduations is linear, i.e., fully and equally proportional.

Calibration

The manuals for older models, TC-114 to TC-142, don't even mention calibration; only the TC-154 manual does. TC 154 has two internal calibration trimmer potentiometers, R11 is "Grid leakage calibration," and R17 is "Emission cal.". The older models have a fixed resistor for R17 and only one adjustment, R11, in the metering circuit.

Instead of using a diode and resistors and connecting them into the tester's socket #1 for calibration, making a calibration module (next page) to ensure proper contact between various components is better.

Making a calibration module for Sencore "Mighty Mite" testers

DIY PROJECT

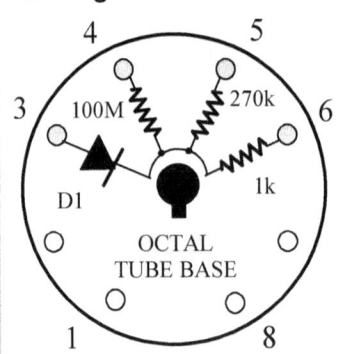

Connect the three resistors and a silicon diode (as per diagram) inside an empty octal tube base (available for sale on eBay). If you don't have a 100M resistor (a very high value that is never used in tube amps or testers) string 4M7, 10M or similar resistors in series.

The 100M resistor is for grid leakage calibration, the silicon diode simulates the tube under test, and the two lower value resistors, 1k and 270k, are for shorts and emission calibration.

SENCORE TC162 & TC28 ("THE HYBRIDER")

Despite its modern looks and a solid-state circuitry, "The Hybrider" is still only a grid circuit tester. It's a Mighty Mite VII or TC-162 tester with transistor testing capability. However, it is still the best of all Sencore testers, except the transconductance models MU140 and MU150. These Gm testers also test for grid current only in their emission testing mode.

The metering circuit

The meter circuit is a resistive bridge with a single JFET transistor in one of its arms, a so-called 1/4 bridge, since only one of its four branches contains an active element.

Instead of the second FET transistor (as a differential amplifier), a fixed resistor is used (R136). The other two branches of the bridge are R129 and the lower portion (below the wiper) of the R10 ("Meter zero" control pot).

Diode CR101 protects the meter from overload. The three trimmer potentiometers (*Emission Cal.*, *Leakage Cal.* and *Grid Leakage*) are internal calibration controls; the only external control is the meter zero which balances the bridge.

The switching arrangement is not shown. The same metering bridge is used for other measurements, not just emission (tube merit).

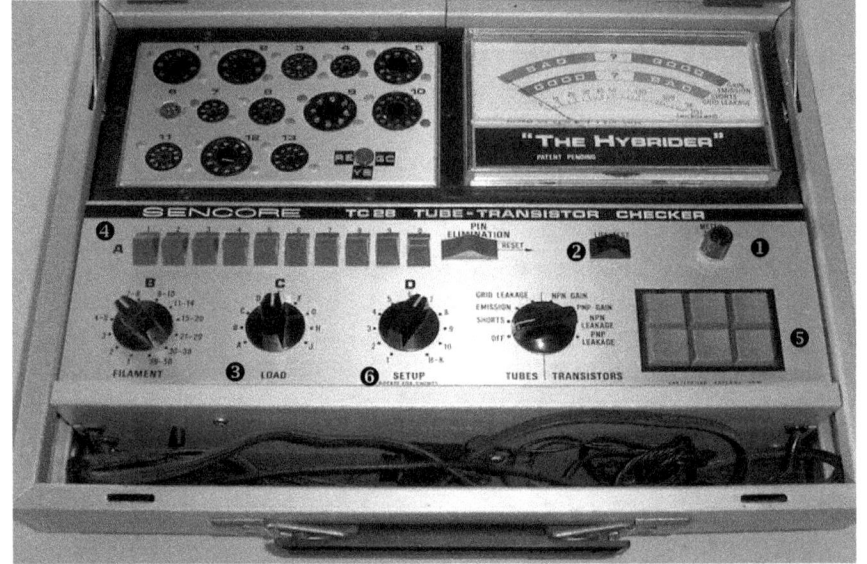

ABOVE: 1) Meter zero 2) Life test 3) "Load" (Meter shunt) 4) Pin "elimination" (disconnect) slide switches 5) Transistor test pushbuttons 6) "Setup" (grid pin pickup switch)

BELOW: The meter driver circuit of Sencore TC162 and TC28 ("The Hybrider") testers

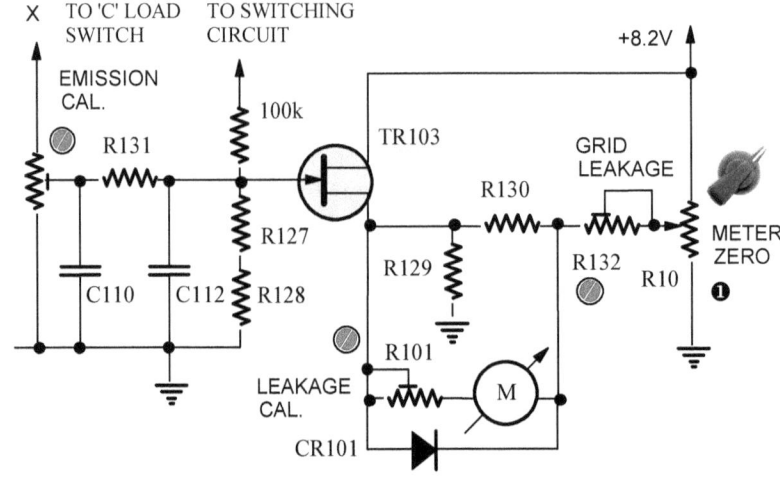

Notice that *Leakage Cal.* trimmer is in series with the meter, affecting the emission calibration result. For that reason, leakage calibration must be carried out before the emission calibration!

GRID CIRCUIT TESTERS

The cathode emission test circuit

The C or "Load" switch (3) selects one of the 3 test voltages (22, 28, or 40 V_{AC}) and different load resistors connected between the bundled electrodes (all accept control grid) and ground, a choice of nine test circuits.

The test AC voltage is fed to the control grid. The tube acts as a rectifier, and the amplitude of the pulsating voltage at its cathode (and other electrodes bundled with it) is proportional to its emissive capabilities.

These cathode pulses are fed to the metering circuit already discussed, are filtered ("smoothed out") by the CRC filter at the input of the metering circuit (C110-R131-C112), and fed to the gate of the JFET amplifier.

Life test

Most testers with "life test" capability have power transformers with a tap on their primary windings. The mains voltage would be brought to such 80% or 90% primary tap during other tests and the whole primary winding during the life test. That means all secondary voltages would be proportionally reduced during the life test, heater voltages, anode/screen voltages, and bias voltages.

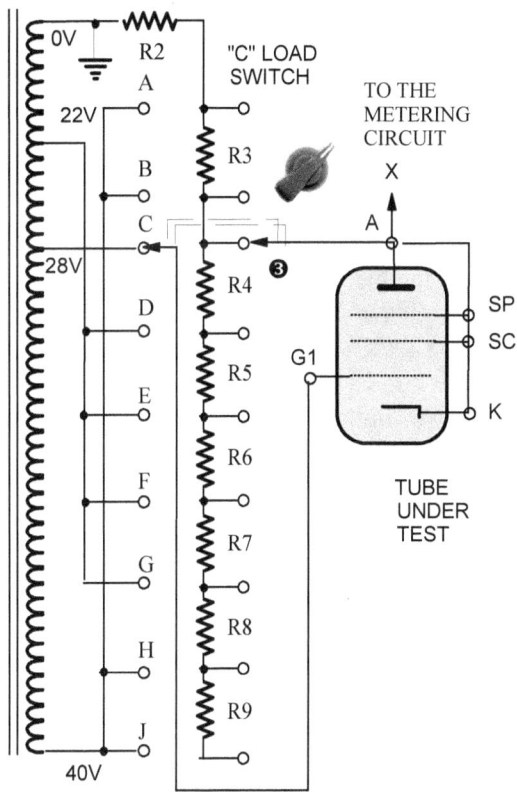

The life test on TC162 and TC28 models is much simpler and cheaper for the manufacturer to implement. The slide switch (2) under the meter (previous page) connects a 1Ω resistor into the heater circuit, in series with the heater filament of the tube-under-test.

The heater current will produce a voltage drop across that resistor; the heater will receive a lower voltage, and the tube's emission will drop.

For 12AX7/AT7/AU7 duo-triodes tested at 6.3V, their 300mA heater current will reduce the heater voltage by 0.3V. However, the 900mA of 6L6 heater current will drop the voltage by a whopping 0.9V or 0.9/6.3= 14.3%! This seems way too much; even a new tube's emission will drop when its heater voltage is reduced from 6.3V to 5.4V.

Thus, another drawback of this dubious method of life testing is that tubes that draw a high heater current, such as 6L6 and other power tubes, would result in proportionally higher heater voltage reduction, so even perfectly healthy tubes with plenty of life left are more likely to fail this "test"!

ABOVE: The cathode emission circuit connections of Sencore TC162 and TC28 ("The Hybrider") testers

BELOW: The internal view of TC28

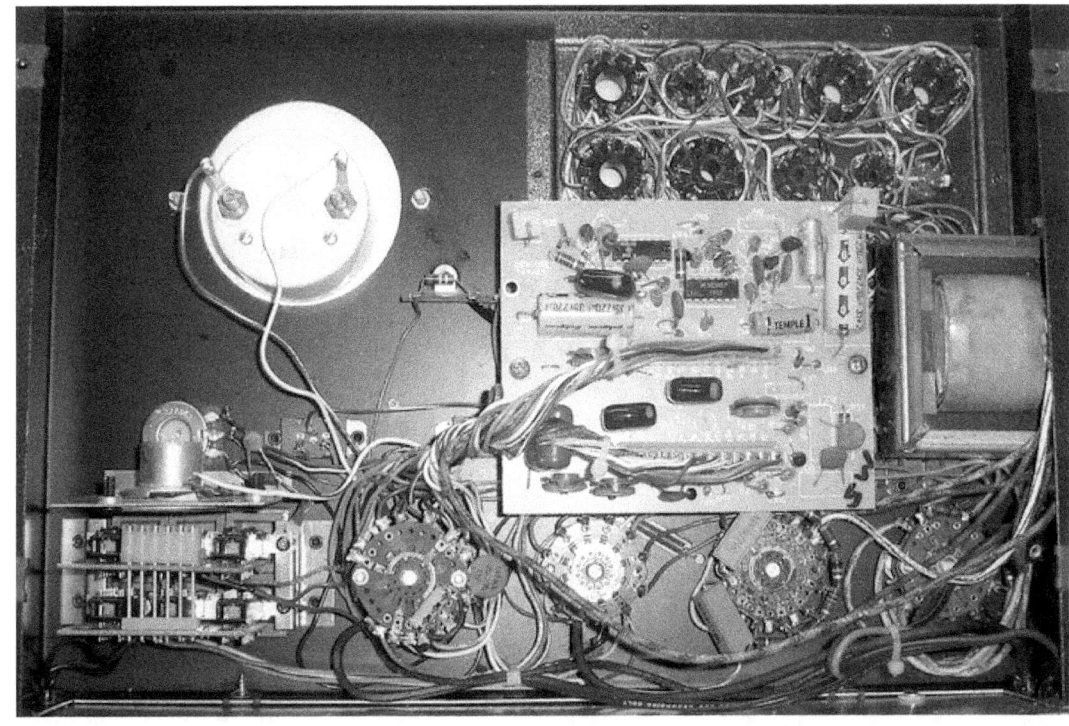

B&K 600 & 606 DYNA-QUIK TUBE TESTERS

Models 600 and 606 were B&K's versions of Sencore's Mighty Mites. Instead of 12AU7, B&K chose 6BN8 (duo-diode + triode) tube for its amplifier circuit. Only one of the two vacuum diodes was used, with triode working as a DC amplifier, just as in Sencore's circuit.

While Sencore offered a choice of three test voltages, B&K only has a single 60V_{AC} voltage.

The setting for 6L6 is 37-1-5, meaning A:37 (load or shunt), B: position 1, C: position 5!

Switch "C", marked (2) on the diagram and photo, picks up the control grid of the tube-under-test, and switch "B" straps all other electrodes (except heater pins, of course) together (3) and connects them to the load resistor and metering circuit, just as in Mighty Mite testers.

Switch "4" is marked "Quality" on the control panel, while switch SW-5 is the "Grid Emission" test push button. Both are of a momentary push-button type (meaning the non-latching switch is only closed while the button is depressed.

The control "A" is the shunt rheostat, marked (6) on the diagram, in parallel with the meter.

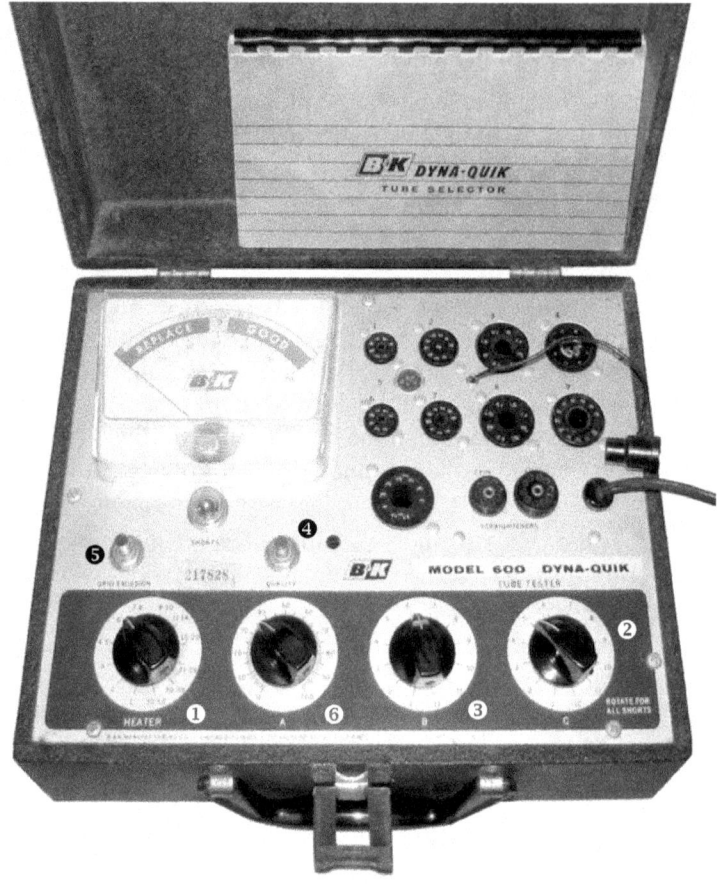

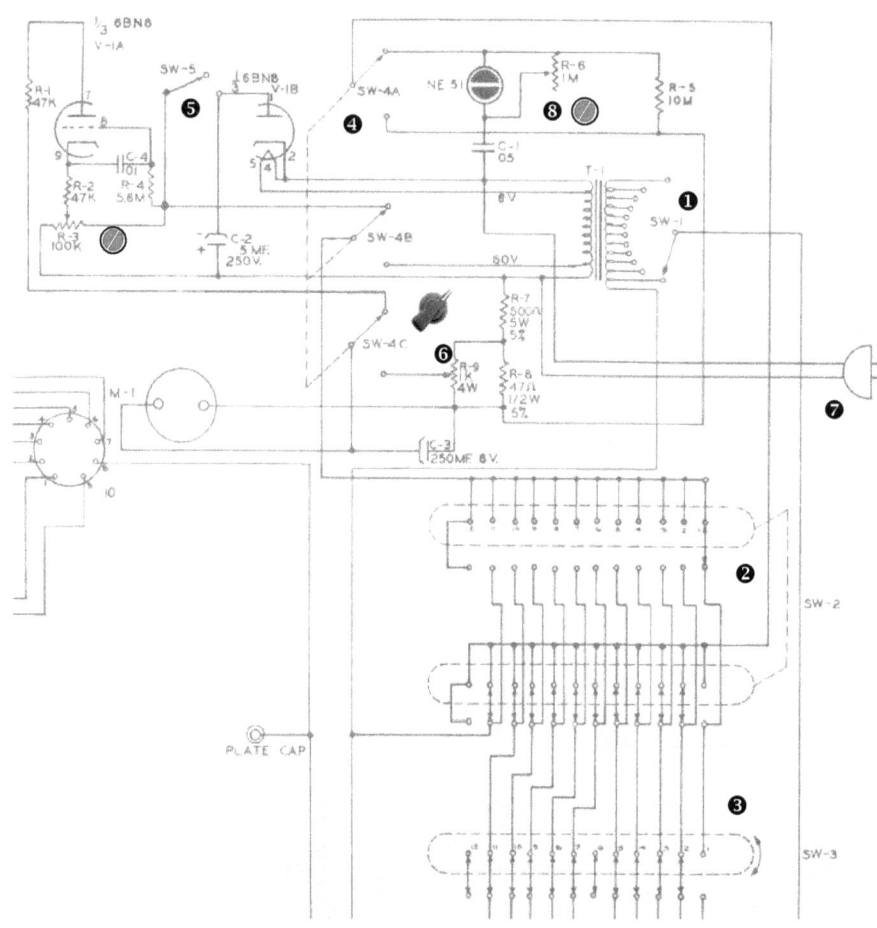

ABOVE: partial circuit diagram of B&K 600 © B&K

The 6BN8 diode rectifies the 117V line voltage, and the 5mF elco (C-2) smoothes it into a DC voltage for gas/grid emission testing. When SW-5 is closed, all electrodes except the grid receive that negative DC voltage.

Only heater voltages (1) were galvanically isolated from the mains; all other voltages (6V for 6BN8's heater and 60V for tube testing) were taken from the primary taps of the mains transformer. This arrangement is unsafe and now illegal. An isolation transformer must be used with these testers, either externally or added internally.

Again, there was no fuse or "Power On" indicator, either (7)!

The 1M shunt trimmer potentiometer (9) enables the user to make the short test more or less sensitive. It is a nice feature, but it should have been brought to the front panel and its scale marked so the user could see at what leakage resistance the neon indicator ignited.

GRID CIRCUIT TESTERS

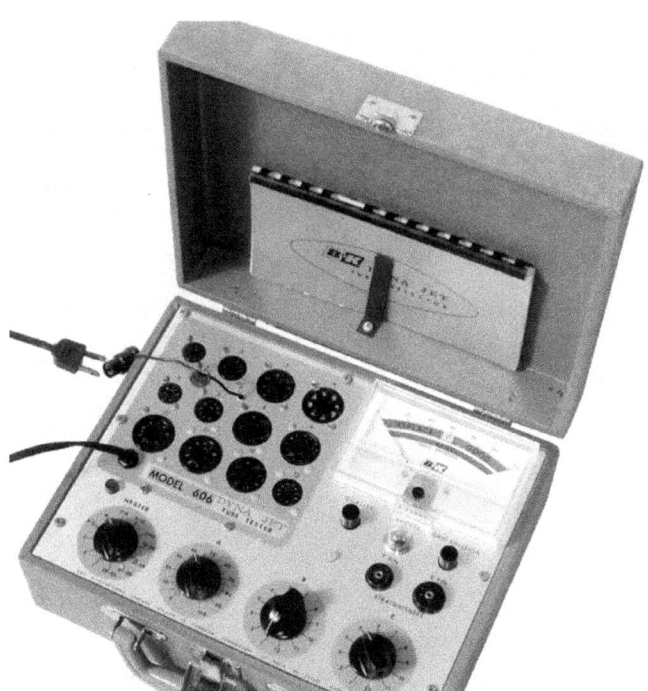

This way, such half-measure is of little practical value. Just as it had done with the cosmetic black-to-blue transition from model 700 to 707, with 606 B&K abandoned the black case and control panel of model 600 and modernized the looks with pale blue case covering and control panel accents of model 606.

The meter was also different, better looking, and of a slimline profile, although still missing the required minor graduations. Unfortunately, the internal design and functionality of the instrument were not improved or enhanced in any way.

LEFT: B&K 606 is a cute and compact quick checker

B&K 625 DYNA TESTER

Although not identical, B&K 625 is very similar to B&K 600 (6BM8 internal tube). A basic VOM section (Volt-Ohm Meter) has three DC and three AC voltage ranges (10, 100, and 1,000V). The internal resistance is low, only 1,000 ohms/volt (10kΩ on 10V range, 100kΩ on 100V range, and 1MΩ on the 1kV range).

The CRT test panel is another addition; it checks for and removes shorts between cathode ray tubes' electrodes and measures cathode emission.

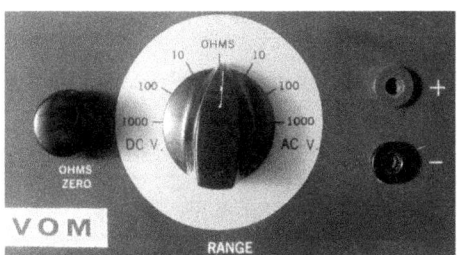

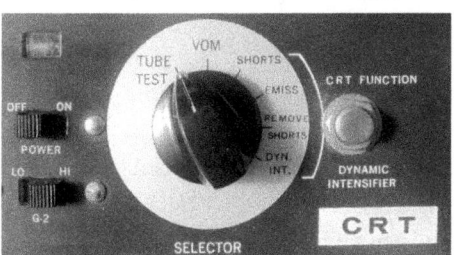

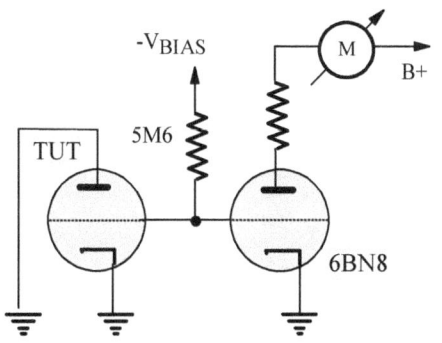

ABOVE: The principle behind B&K 600 - 606 - 675 grid emission/gas test

During the grid emission/gas test, all electrodes except the control grid remain bundled together, just as in the "Quality" or cathode emission test, but this time they are grounded. A negative high voltage is brought to the control grid via a 5M6 resistor.

Any grid current flowing will create a DC voltage drop across such a high resistance, which will be amplified by the 6BN8 triode amplifier and indicated on the meter in its cathode circuit.

B&K 607 AND 667

After a single test voltage and tube-based circuitry of the earlier models 600 and 606, B&K released their modernized versions using a JFET DC amplifier, just as in Sencore TC28 and TC162. Who copied who is unclear, but instead of 22, 28, and $40V_{AC}$ with Sencore, B&K chose 22, 40, and 50 V_{AC} for test voltages in their 607 and 667 tube testers.

While there are minor differences in the circuits, B&K 607/667 and Sencore TC-162 can be considered the same tester from the functional perspective. Sencore's TC162 has "Meter zero" and "Emission calibration" controls, two internal adjustments absent from B&K 607/667 testers. B&K 607 and 667 feature "Shorts sensitivity" adjustment while TC162 does not. TC162 has 14 sockets, compared to B&K's 11, Loktal socket being one of them.

The emission current flows between the cathode and control grid; the voltage drop (between K and GND) on this resistor is amplified by a solid-state DC amplifier (a single JFET), which drives an analog meter.

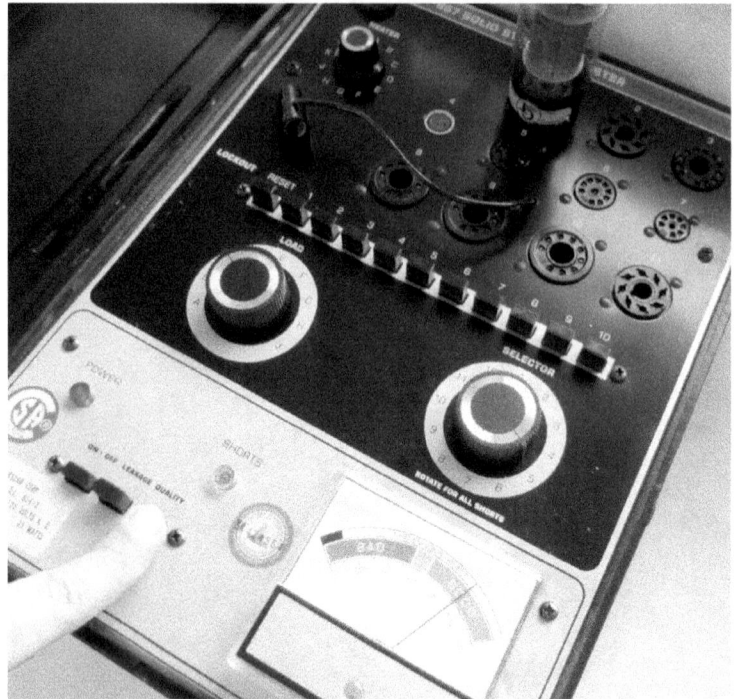

ABOVE: Just as with predecessors, models 600 and 606, B&K 607 and 667's meter scales lack minor graduations, another annoying omission that confirms the status of these "testers" as mere quick "checkers."

Four distracting paper stickers (the largest was removed before the photo was taken) were placed around the control panel, making the otherwise clean and "streamlined" panel look messy.

BELOW: The principle behind B&K's cathode emission (left) and grid emission tests (right)

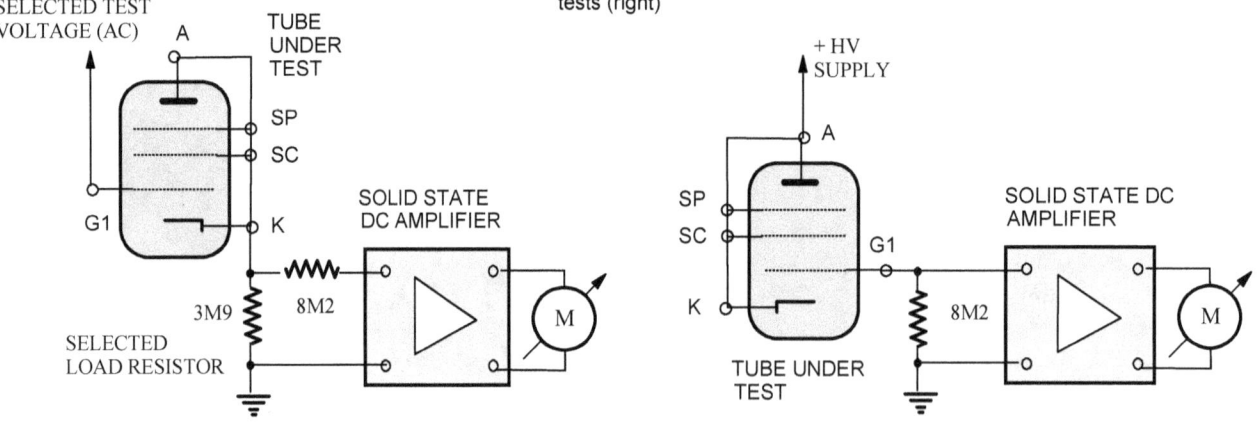

Again, B&K 607/667 testers use the mains voltage for the "Shorts" test, so an isolation transformers use or installation is required for safety reasons. At least Sencore TC28 and TC162 have a 3-pin plug, and their chassis is earthed. All voltages are taken from power transformers' secondaries, so there are no safety issues.

To measure grid-cathode current, we broke the line from the selector switch to the grid and inserted a 10W resistor to measure the voltage drop across it. 12AX7 and EF86 control grids were subjected to a 5mA current, way too high; its prolonged flow could deform the grid structure, overheat and ultimately melt the thin grid wire, destroying the tube. 6CG7's grid passed 7mA, and 6BQ7 is tested with 9mA of grid current. Power tubes' were even more stressed out, grid current levels ranging from 25mA for 6L6 and 30 mA for EL84 to unbelievably high 58mA for EL34!

B&K 666 DYNA JET TUBE TESTER

After changing the meter, the timber case, and the color scheme to cosmetically update model 600 into 606, B&K did the same here, by "modernizing" model 606 into a devilishly sounding model 666, probably after winning a contract to supply hundreds of those to the purgatory.

The meter is the same as on model 606, as is the tube test circuit (6BN8), but the timber case was replaced by a plastic one, just as those used for models 607 and 667. This tester is relatively rare, not often seen on the resale market, so it seems that not many were produced before the focus shifted onto solid-state models 607 and 667.

GRID CIRCUIT TESTERS

> ***Warning: Tube damage or destruction likely!***
>
> The B&K manual claims, "Quality is checked in a test circuit that determines the full capability of cathode emission under current loads simulating actual operating conditions." I haven't seen a hi-fi tube amplifier that connects anode and cathode and then amplifies grid current?
>
> During the test of EL34, the measured grid current was 58 mA, way too high for the tube's fragile grid, so don't keep that test button pressed for too long, or you may destroy the tube.

PRECISION APPARATUS COMPANY (PACO) 650 & T-62

PACO 650 and T-62 may look different, but the difference is purely cosmetic. The controls are identical, even positioned in the same manner. Indeed, under the hood, it is the same grid circuit analyzer and megohmmeter.

The photo below shows a true RMS digital multimeter measuring the test voltage on "Megohmmeter" test terminals (76.4V_{DC}). Even a quality instrument such as this Escort meter is not an ideal voltmeter (which would have an infinite input impedance and would not "load" the measured circuit *at all*), but has a declared Z_{IN}=10MΩ.

Indeed, notice the "Megaohms" scale (1) indicating around 11MΩ! The potentiometer "A" (2) serves as a zeroing control for MΩ tests and as a "Shunt" or "Load" control for "Cathode emission" tests.

A small trimmer pot (3) accessible from the top (just under the meter) is marked "VTVM ADJ." Does this mean there is a vacuum tube voltmeter inside? It certainly does.

A 12AU7 duo-triode is configured as a differential amplifier (see circuit diagram on the next page and the in-principle circuit below) with trimmer R16 in its anode circuit, acting as anode resistor for both triodes. By moving its slider one way or the other, one anode resistance increases, and the other increases, thus balancing the voltage gains of the two halves of the differential amplifier and zeroing the meter.

The meter M is connected as a load between the two cathodes through the "Cal. control" trimmer resistor R15.

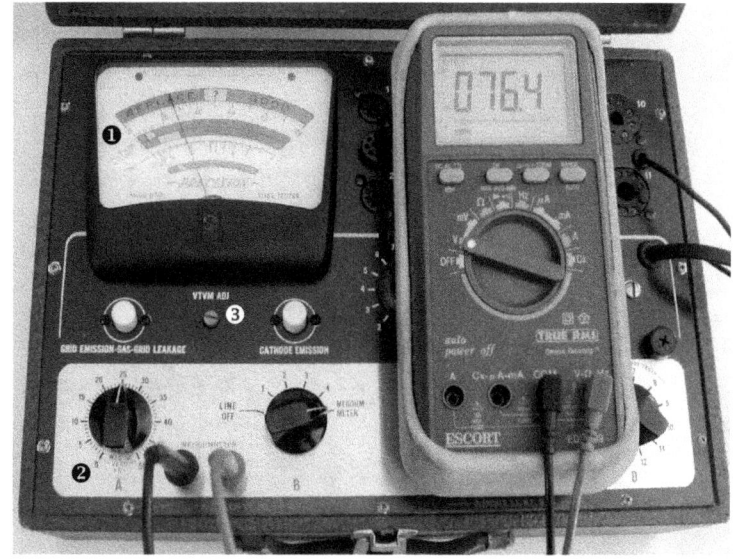

ABOVE: 76V_{DC} is the test voltage for leakage tests, as demonstrated here. The 10 Mohm input impedance of the digital meter was enough to show as leakage on the tester's sensitive leakage test!

BELOW: Testing a 300B tube for emission (through a 4-pin adapter) in the octal socket of T-62

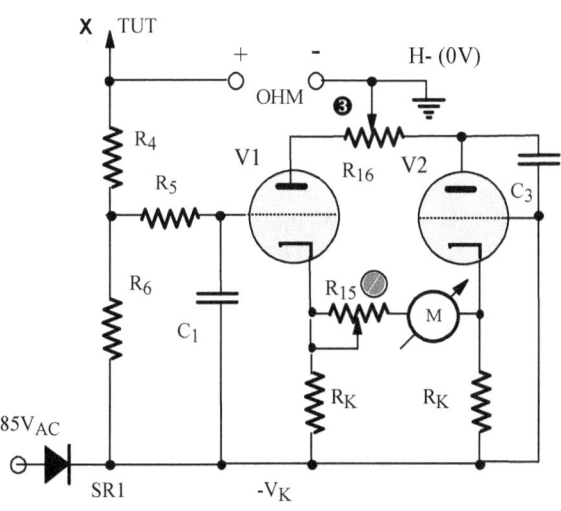

ABOVE: The vacuum tube voltmeter (VTVM) at the heart of PACO 650 is a differential amplifier

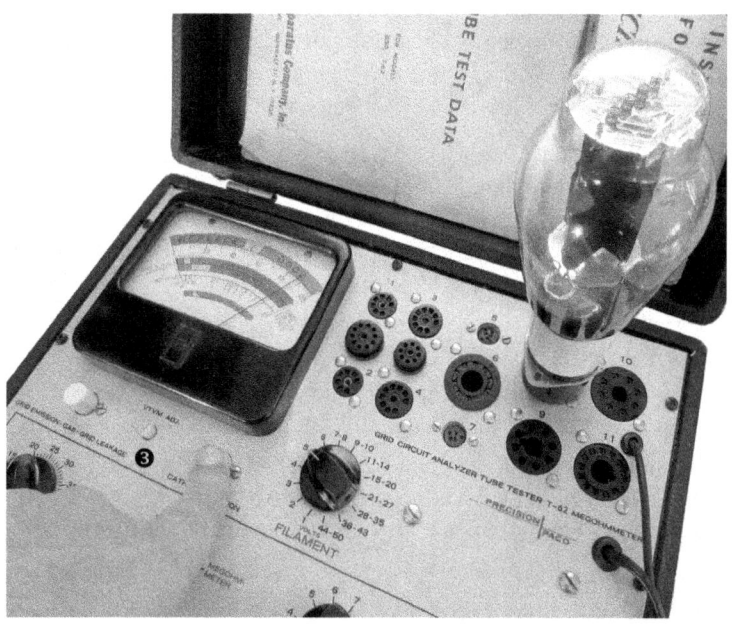

Triode V1 is the "active" half; the signal is brought to its grid through the voltage divider R4-R5 and a low pass filter formed by R5 and C1 (shunting high frequencies away from the grid). The grid of V2 is grounded for AC signals through capacitor C3.

Normally a positive anode voltage $+V_{BB}$ is on the anodes, and cathode resistors R_K are at the GND or 0V potential. Here, the anode resistor R16 is at GND or 0V, while the cathodes are at a negative DC voltage provided by the rectifier SR1, which rectifies one of the power transformer secondary test voltages $85V_{AC}$, which is then filtered and smoothed out by a 10µF capacitor C3. This makes no difference; as long as anodes are more positive than cathodes, a tube will work as it should. Voltages (potential differences) are a relative term!

A differential amplifier VTVM is superior to single tube or FET circuits used in other grid testers. The meter reading is not affected since supply voltage fluctuations, drift, and aging affect both triodes equally.

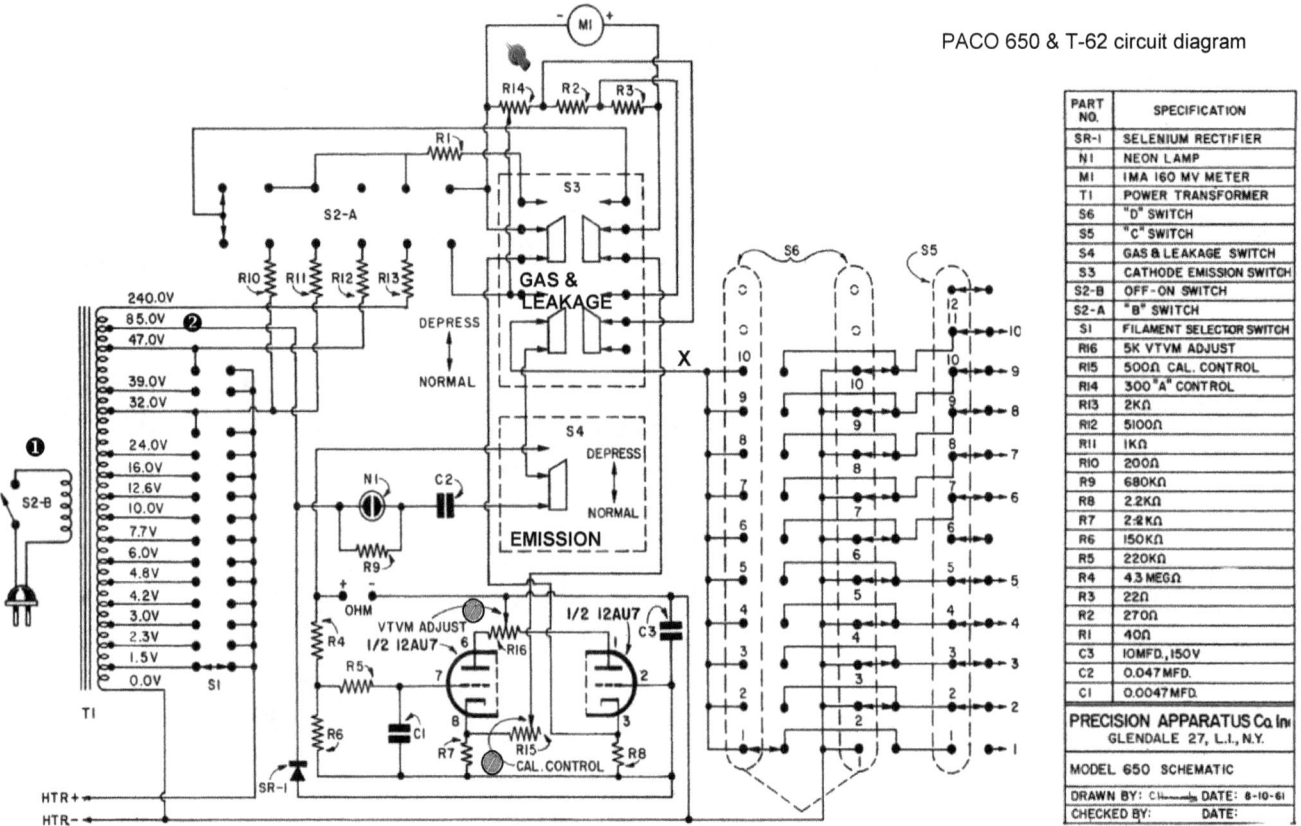

PACO 650 & T-62 circuit diagram

If used in a differential amplifier, meaning in identical DC and AC conditions, both triodes inside one tube will age at the same rate, so such a gradual and long-term drift will have no impact on accuracy.

As for the shortcomings, the idiotic practice of tester manufacturers to save money on a poxy fuse and fuse holder is an irritant here again (1), with no primary or secondary fuse protection at all! Strangely enough, the tester has a 240V secondary tap but has no 117V heater voltage (2), so it cannot test many older tubes with 117V heaters.

The MΩ-meter can test high voltage electrolytic capacitors for leakage. Using $76V_{DC}$ is not as conclusive as 300V or 400V leakage tests on condenser checkers such as Heathkit C3, Sencore LC53, and Eico 950B but is still superior to 9V tests by digital LCR meters.

As for "Gas and grid leakage," notice different scales for voltage amplifying tubes and power tubes (3), reminding the user of the fact that power tubes generally have a higher heater-cathode leakage and higher levels of gas!

Precision testers generally featured high-quality moving coil meters, almost as good as Weston's or Simpson's lovelies! 650 and T-62 are very enjoyable testers to use.

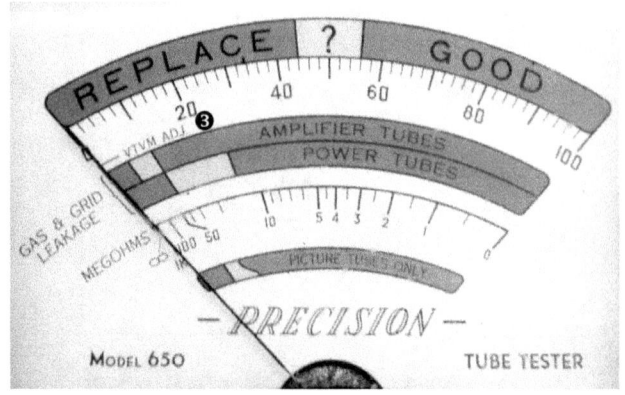

ABOVE: PACO 650 ohmmeter (leakage) scale goes above 100MΩ, something very few other testers manage to achieve!

GRID CIRCUIT TESTERS

MERCURY 1101, 1101C & 1101CT

USA-made Mercury 1101 is a typical example of a low spec tube checker: an auto-transformer is used for heating and plate voltages (meaning there is no galvanic isolation between the tester and the live mains), all tubes are tested at $25V_{AC}$ ("Normal") or $110V_{AC}$ ("Special"). Add to that a standard neon short/leakage circuit and switchable meter sensitivity (four load resistors). There's no line adjustment of any kind.

The only unusual feature is the bridge circuit for the meter. Selector S-5 picks the control grid, while S-4 selects the cathode, meaning this is a simple grid circuit tester.

In terms of testing for emission, Mercury 1101C and 1101CT are functionally identical to 1101 but feature a better gas/grid emission test, using a JFET DC amplifier (metering circuit), similar to B&K 607/667 and Mighty Mite testers by Sencore.

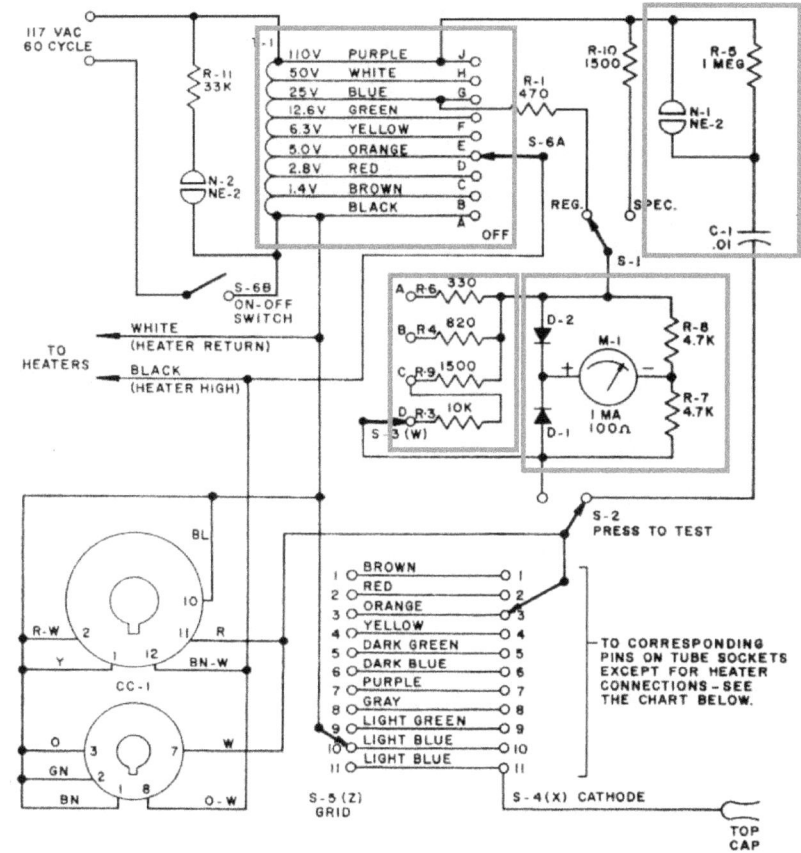

ABOVE RIGHT: Mercury 1101 is a simple grid circuit checker. Eico model 635 is electrically and functionally identical. While sharing the same simple emission circuit, models 1101C or CT also include a gas/grid emission JFET test circuit (BELOW).

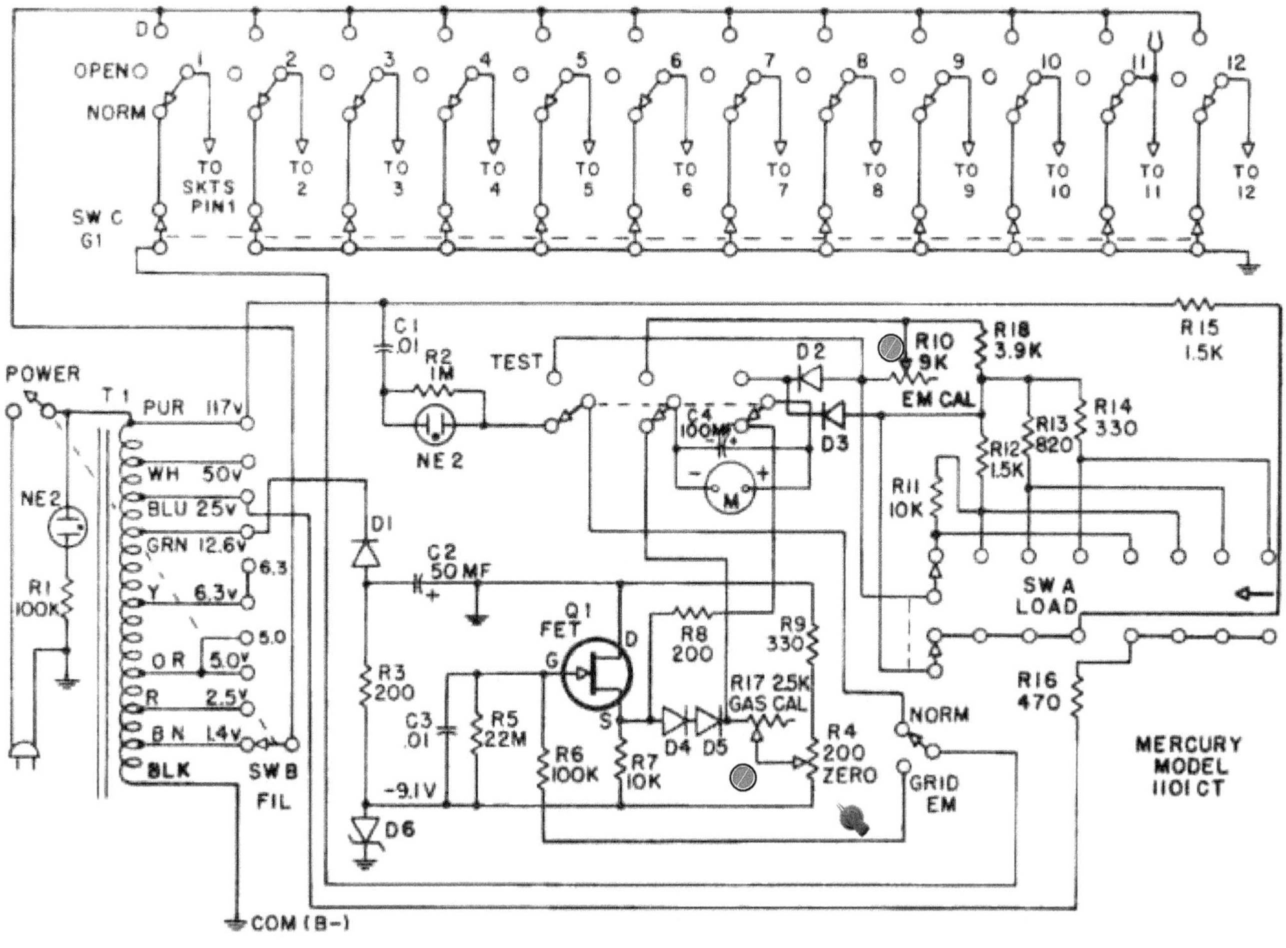

AMERICAN SCIENTIFIC DEVELOPMENT COMPANY TV-20

Although it didn't test picture tubes, TV-20 was a brainchild of Jacob Anthes and marketed as a special purpose "Television Tube Tester." A large and solid timber case, upholstered in deep burgundy leatherette, with a large and good quality meter. Twenty prewired sockets, so only the test mode called "Switch" and "Load" had to be set.

When you add the name such as "American Scientific Development Company" to all those visual attractions, you'd think TV-20 would be one competent emission tester. Wrong!

One pin of the power plug is directly connected to the "cathodes" busbar (1), so 110-120V_{AC} will end up on cathodes if the plug is reversed.

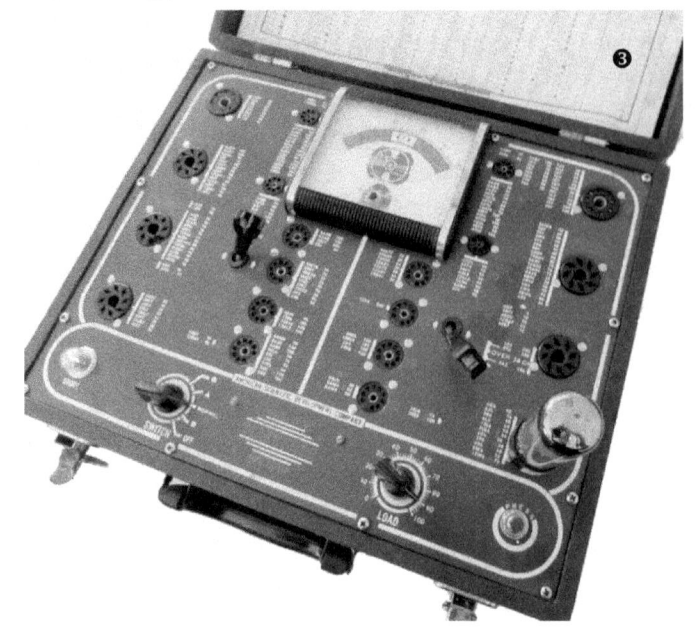

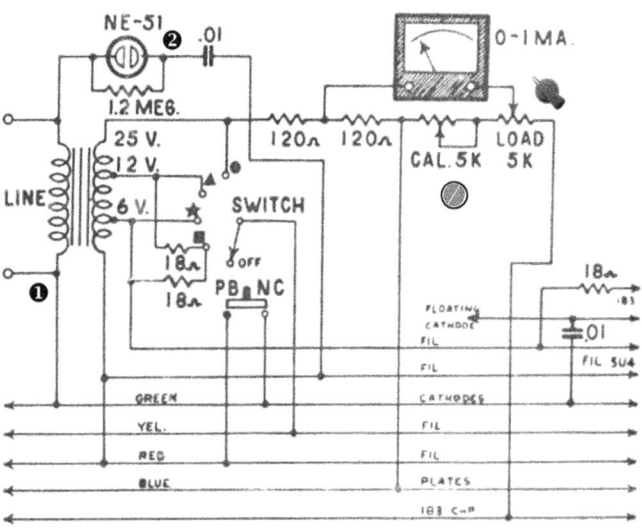

ABOVE: The circuit diagram was printed on tester's tube setup chart (3).
BELOW: The "Quik-Check" tester by "Reliable Electronics Corporation" seems to be if not identical then very similar. Notice the 5th position of the switch imaginatively called "Switch", marked "+", most likely a heater voltage higher than 25V on TV-20. It had 22 sockets without any tube names printed on the top panel, so a tube test chart was needed.

The "SWITCH" is actually a heater voltage selector, 3V/6V/12V/25V.

The neon short test circuit of 1M2 sensitivity (2) is permanently wired between the "FIL" busbar and "CATHODES" (via the primary winding of the mains transformer). The control grids and screen grids of pentodes and beam power tubes are strapped to one heater terminal, and anodes and cathodes of all amplifying tubes are connected to the source of test voltage. Thus, a short or leakage between heater cathode, grid to the cathode, or screen to anode will close the circuit, and the neon indicator will light up.

This tester was designed for TV servicemen, with speed as the primary design goal, so its accuracy is questionable. For instance, the leakage sensitivity was made so high to make it easier to convince customers to buy as many new tubes as possible (to replace some functioning tubes).

If you study vintage adverts for tube testers, especially those by B&K, you will notice how they pitch their gear to radio & TV repairmen, talking about additional profit from an increased number of sold tubes! The meter indicates the current between the cathode and control grid, so TV-20 is a grid circuit tester.

High power rectifiers that require a 5V heater supply (such as 5U4) were tested with 6V on their heaters. Due to its tiny transformer with poor regulation, the 5V/3A (15 Watts) heater loads the transformer so much that its 6V secondary tap's voltage drops to under 5V!

Notice the 10n capacitor and "floating cathode" busbar. While some el-cheapo testers test duo-triodes by connecting both in parallel, this is the only design I have seen that connects them in series. The capacitor is needed to detect H-K leakage in the upper triode.

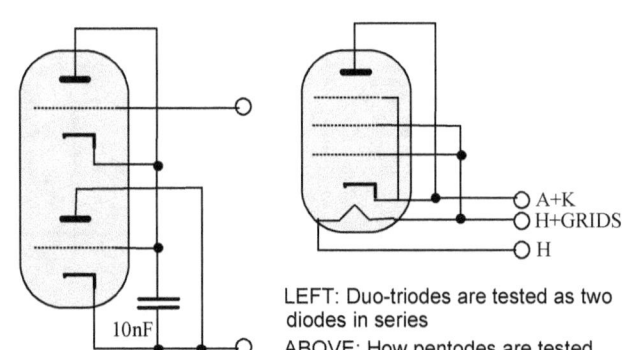

LEFT: Duo-triodes are tested as two diodes in series
ABOVE: How pentodes are tested

DYNAMIC CONDUCTANCE TESTERS

5

- THE OPERATING PRINCIPLE BEHIND DYNAMIC CONDUCTANCE TESTERS
- SICO MODEL 85
- SICO TV-12
- SYLVANIA 139 & 140
- SYLVANIA 219 & 220
- JACKSON TUBE TESTERS
- EICO 666 & 667 (SIMPSON 1000)

THE OPERATING PRINCIPLE BEHIND DYNAMIC CONDUCTANCE TESTERS

The three electrodes (control grid G1, screen grid G2, and anode A) get an AC voltage of a chosen amplitude and phase (through selector switches). The tube acts as a rectifier (rectifies those three AC voltages) and as an amplifier at the same time.

This circuit doesn't only test for cathode emission (the DC anode current is proportional to emission) but also includes the effect of control grid voltage on the plate current. Therefore, Sylvania's meter reads "Composite Transconductance and Emission."

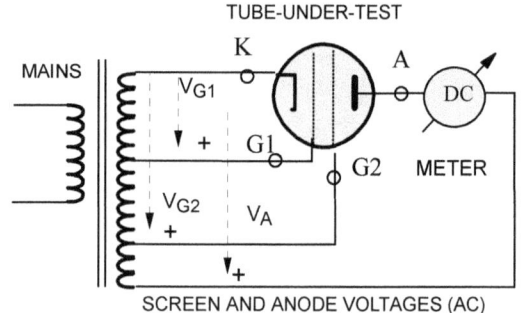

ABOVE: The operating principle behind dynamic conductance testers

Precision touted this "electronamic" principle (a combination of 'electronic' and 'dynamic') by using this graphical explanation. Since mains frequency alternating voltages are applied to all electrodes, the operating point of a tube under test is dynamically swept along a curved path 50 or 60 times a second. Plate voltage changes from zero to some maximum value and back, and the plate (anode) current rises and falls accordingly.

Anode curves for a power pentode are shown, but the principle works for triodes, too.

However, notice that all three AC voltages are in phase, so the grid AC voltage V_{G1}, after being rectified by the control grid, biases the tube positively. Normal screen & anode voltages of 300-400V would result in very high anode & screen currents and the destruction of the tube under test, so these testers use much lower voltages to keep the currents lower not to exceed the anode power dissipation.

Notice also that Precision's graph (right) shows the operating point swinging path through the negatively biased pentode. Yet, the swing is above the $E_G=0$ curve in the positively biased region in these testers, so this graph is grossly misleading.

Sylvania also adopted the same principle in their testers, models 139/140, 219/220, and 620. Jackson, another reputable tube tester brand of its day, chose dynamic conductance testing for their very popular testers, such as the 648 and 658 series.

Today, Simpson 1000 and its copies Eico 666 and 667 are the most commonly found vintage dynamic emission testers.

Sylvania 219 and Eico 666 test the same tube (6L6) with almost identical screen voltage, but Sylvania's plate and control grid voltages are lower than Eico's. As a result, the plate current is much higher on Eico 666. The anode dissipation during Sylvania's test is only 4.7 Watts, while Eico subjects 6L6's anode to dissipation of 199V*49mA = 9.75 Watts (table on the right)!

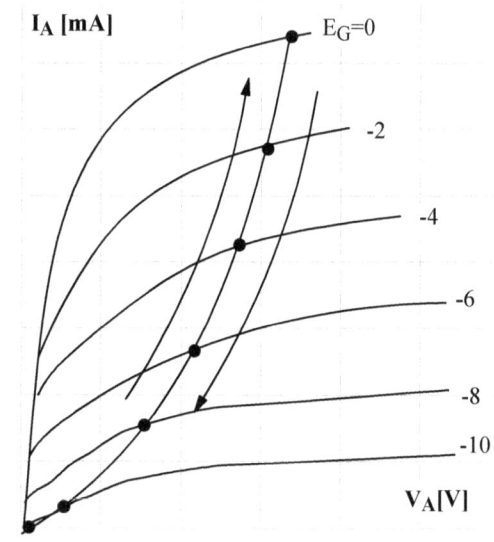

ABOVE: The misleading illustration by Precision Apparatus Company of the "electronamic" sweep of a power pentode.

BELOW: The AC voltages on various electrodes and DC anode current measured with a True-RMS multimeter (6L6 tube tested on two common dynamic conductance testers).

AC volts at 6L6's	Sylvania 219	EICO 666
ANODE:	134	199
SCREEN:	95	100
CONTROL GRID:	4.8	17.5
I_A (mA DC)	35 mA	49 mA

FTC complaint against "Precision Apparatus Company" and their use of term "Precision Dynamic Mutual Conductance Tube Testers".

FURTHER READING

Lodged in Mar. 1942 and finally ruled upon in Nov. 1951 (almost ten years later!), in this court case, the Federal Trade Commission (of USA) determined that Precision is prohibited from claiming that theirs were "Dynamic Mutual Conductance Tube Testers." The 16-page ruling makes an interesting read. It does not say who lodged the complaint, probably one or more of Precision's American competitors making Gm testers at the time, and those were Hickok and Weston.

Even though both sides (the Commission/examiner and the "respondents") used external expert witnesses to prop up their claims, some of the "findings" and claims are not technically correct! The case can be downloaded from online sources.

DYNAMIC CONDUCTANCE TESTERS

SICO MODEL 85

Sico 85 and TV12 are also relatively common budget testers. Since their circuits are much simple and easier to reproduce and comprehend than the very convoluted and confusing Sylvania or Jackson schematics, let's start with Sico.

A bigger and better "dynamic" brother of the pretty useless grid circuit tester, Model 82, Model 85 is a dynamic emission tester. Two different AC voltages are brought to tube electrodes, $5V_{AC}$ to control grid and $50V_{AC}$ to anode and screen grid. The tube rectifies both AC signals, grid, and anode/screen, into a half-wave rectified pulsating waveform, and a certain amount of DC anode current flows through the tube and the analog (moving coil) meter.

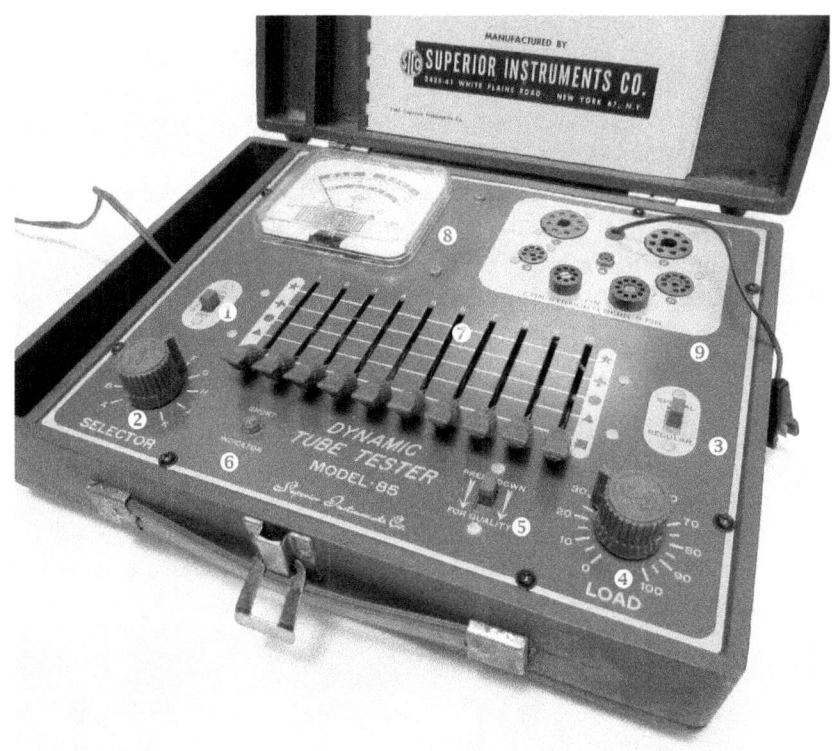

Circuit diagram

The first thing one notices is the bare-bones nature of this el-cheapo tester. There is no fuse and no "Power on" indicator on the circuit diagram below (1).

The tester is not galvanically isolated from the mains; an autotransformer (2) was used to save on manufacturing costs. This may have been OK in 1959 but is illegal today. Again, the first thing that needs to be done is the installation of an isolation transformer. However, the tester's case is so slim and small that there is no room for an internal isolation transformer.

ABOVE: 1) "On-off" switch
2) Heater voltage "Selector" switch
3) "Regular-Special" test regime switch
4) "Load" - a shunt resistor across the meter
5) "Test" switch
6) Neon "Shorts" indicator
7) Bank of 10 lever switches, one for each tube pin
8) Analog meter
9) Five tube sockets and two pin streighteners

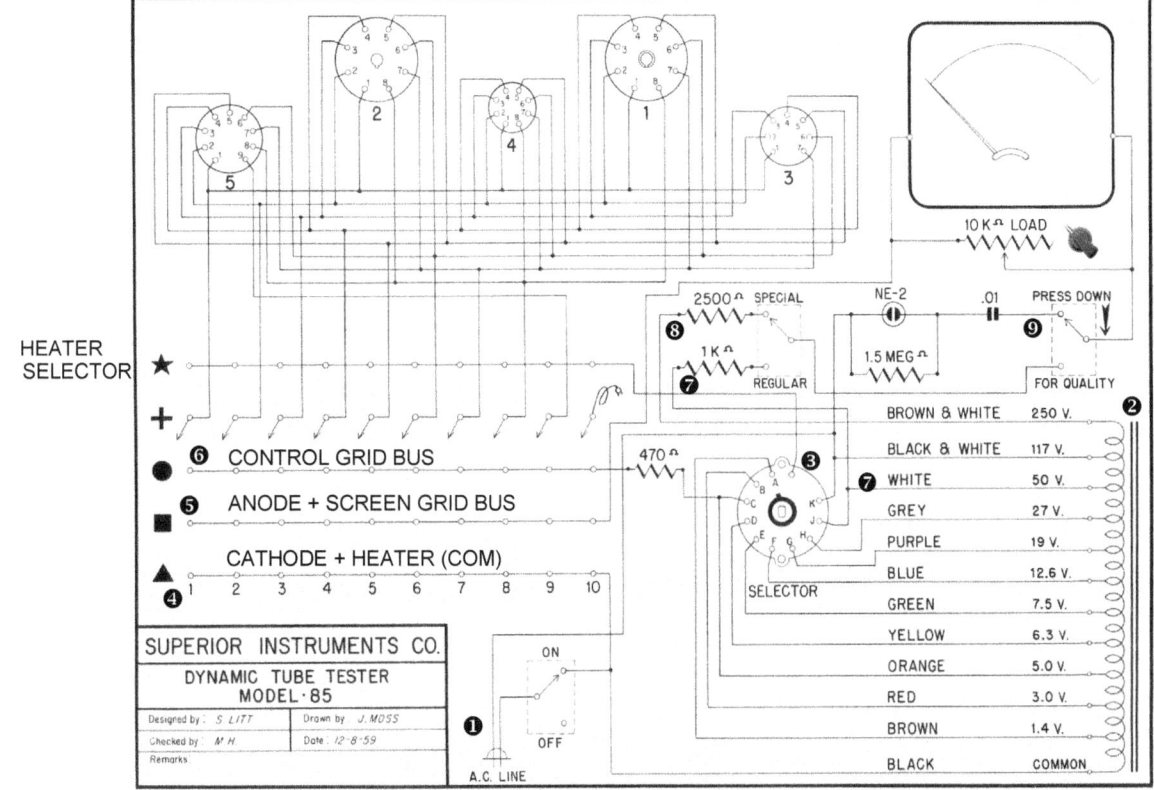

The "Selector" switch connects one of ten transformer taps (from 1.4V to 117V) as a heater voltage to the "Star" busbar, which will be connected by one of the lever switches to one of the tube's heater pins. The other end of the heater must stay connected to the "Common" end of the transformer (4), the "Square."

There is an error on the original factory circuit diagram; the square (5) and triangle (6) symbols need to be swapped around.

In the "Regular" test mode (7), once the "Test" switch is pressed (9), an AC voltage of 50V is brought to the analog meter through a 1k resistor, the other side of the meter is directly connected to the anode busbar ("Triangle"). In the "Special" test regime for some tubes, $250V_{AC}$ is used (8) via a 2k5 resistor.

Finally, before the *Test* button is pressed, the shorts circuit is enabled, connected directly between the $117V_{AC}$ line and the meter, and then to the "Triangle" busbar. The shorts are detected by moving each lever (except those placed in the heater "Star" or "+" disconnected position) to the anode position ("Triangle"). A short or leakage under the neon limit of $1.5M\Omega$ would cause the neon lamp to glow.

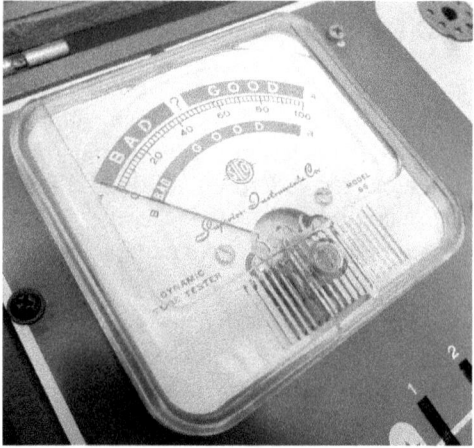

Sico's analog meters are among the smallest and cheapest of all tube testers. Their mechanical damping is almost nonexistent, taking quite a few seconds of oscillation before they settle down.

Test regime

The easiest way to figure out how a specific tube tester works is to take a tube you are familiar with, measure DC and AC voltages on its significant pins (anode, screen, control grid, cathode) in the test regime, and correlate that with tester's circuit diagram.

Model 85 tests 6L6 by applying 3 V_{AC} to the control grid and 50 V_{AC} to the anode and screen (although the measured voltage was only 35.5 V_{AC}). The plate current was 8.6 mA_{DC} (see the Philips multimeter below). With 6L6/6V6, and similar octal tubes pins 3&4 (anode and screen) go to "triangle" (see the test data below), which must be the anode voltage, not ground or "common" as on the diagram, while pin 8 (cathode) plus all unused pins stay at the "common" or default setting, which must be "Square"!

TUBE	SELECTOR	★	+	SOCKET	INDICATOR SHOULD LIGHT	●	▲	LOAD	FUNCTION	READ SCALE
6K6	D	2		2	7	5	3, 4	45	R	A
6L6	D	2		2	7	5	3, 4	25	R	A
6V6	D	2		2	7	5	3, 4	30	R	A

LEFT: Test setup for three common octal power tubes

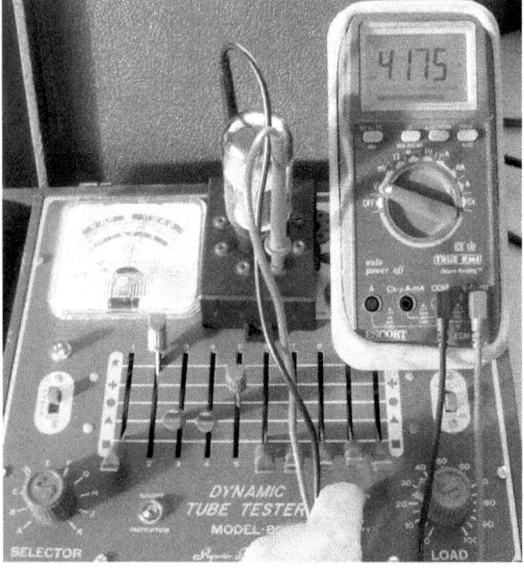 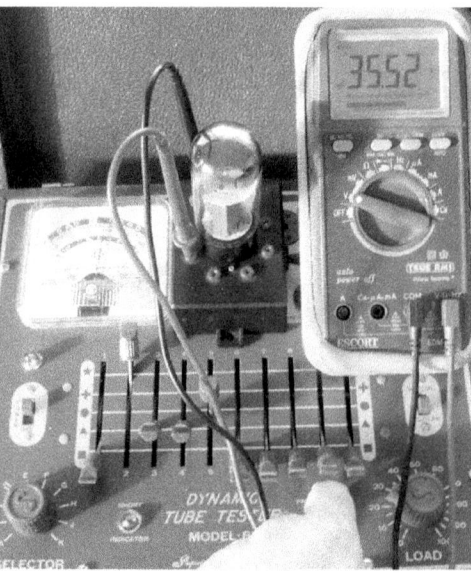

FAR LEFT: The AC voltage between pins 8 (cathode) and 5 (control grid) was 4.175 V.

LEFT: the AC voltage between pins 2 (heater) and 4 (screen) was 35.5V.

> "...the indication on the meter is a composite of transconductance and emission. The reading on a power tube is predominantly influenced by the emission capability of the tube's cathode element, while the reading on a high gain, low power tube is predominantly influenced by its mutual conductance." Sylvania's 620 tube tester manual

QUOTE

DYNAMIC CONDUCTANCE TESTERS

SICO TV-12

Sico brazenly claimed that their model TV-12 was a "Transconductance" tube tester, but its looks and the wiring are very similar to Sico model 85 just discussed.

Instead of silly and meaningless graphic symbols being used, this time, the same five positions of the lever switches are marked F (heater or Filament selector), N (Not connected), "G" (control Grid), "P" (Plate) and "K" for cathode and common heater point. So, even the order of positions is identical.

The manual states: "TV-12 Transconductance Tube Tester was designed to test tubes under dynamic conditions closely resembling the operating conditions of the tube. It performs this function by measuring the plate current of the tube under test while under the direct influence of an in-phase grid voltage applied to the tube."

In most cases, tube circuits work with a negative DC grid bias, and here the in-phase AC voltage on the grid biases tubes under test positively. Claiming that such a regime "closely resembles" the tube's operating conditions inside equipment is a total lie!

> ### Are they serious: SICO'S "explanation" of how TV-12 "measures" transconductance
>
> The Model TV-12 is a dynamic trans-conductance type of tube tester. Although an understanding of the term trans-conductance is not at all necessary for the average Radio and TV Serviceman, the detailed explanation given below may be of some benefit to Radio Engineers who employ instruments in research and development.
>
> The trans-conductance of a tube is the change in plate current divided by the grid voltage causing that change. The resultant figure is "milliamperes — volts" and when that is multiplied by 1000 the result is referred to as "conductance." (MICROMHOS)
>
> In the Model TV-12 this is accomplished by placing a 5 volt in-phase voltage on the grid of the tube under test and observing the plate current meter. Since a D.C. meter is used in the plate circuit, the reading will be an average value falling between the plate current flowing when +5 volts is applied to the grid and zero volts is applied to the grid. The tube is in a non-conducting state for all values of grid voltage swing between zero grid volts and −5 grid volts. The meter then reads the average change in plate current divided by the change in grid voltage or $\Delta Ip / \Delta Eg$. The result is the trans-conductance of the tube under test.

ABOVE: An excerpt from TV-12 manual   BELOW: Sico TV-12 circuit diagram

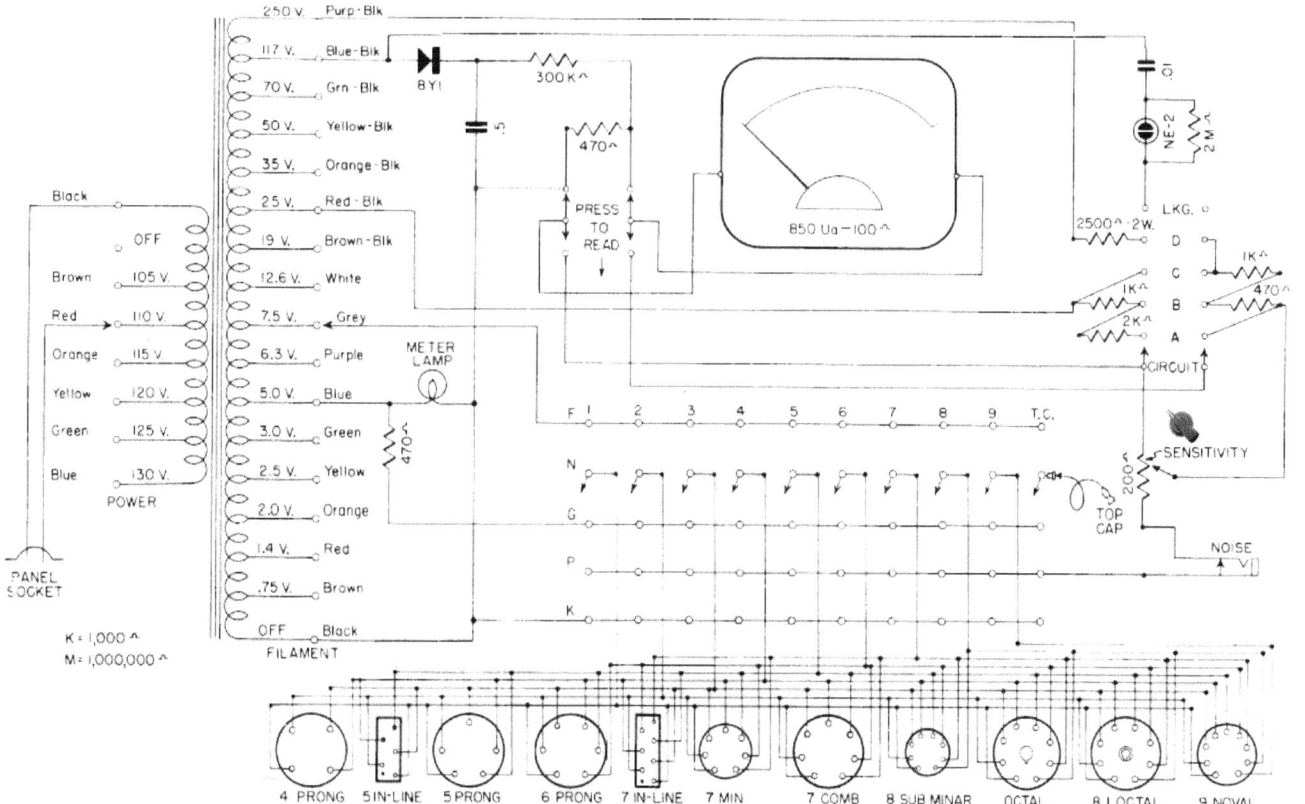

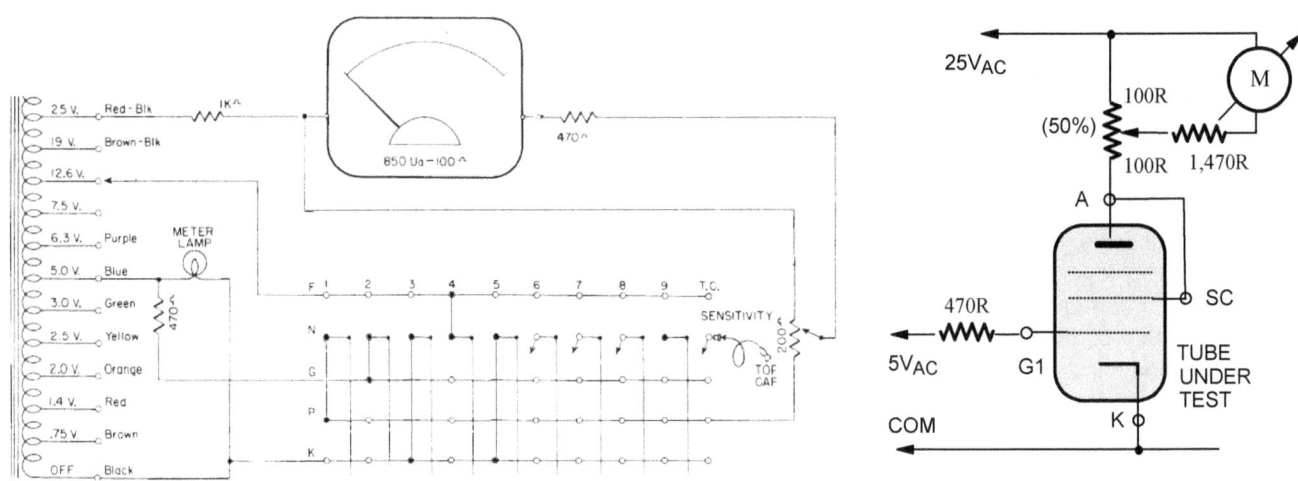

ABOVE LEFT: Testing one half of 12AU7 duo-triode on TV-12 (Circuit "B") ABOVE RIGHT: "Measuring" mutual conductance with TV-12.

Calculating mutual conductance from anode current measurements

Sico states that to determine Gm, all tubes must be tested with "Circuit" at "C" ($25V_{AC}$ on anode & screen grid) and with "Sensitivity" at 50%. In that case, the anode current is 1/10 of the figure displayed on the meter. We checked this claim (about current only), and it is correct. In the photo below, with a reading of "102", the 300B under test is indeed passing 10.2mA.

The RMS voltage on the anode/screen was 22.7V, so Sico claims that the internal resistance of the triode is the peak voltage on the anode divided by its anode current, or R_I= 22.7*1.41/0.0102= 32/0.0102 = 3,137Ω. Sico now instructs users to find the tube's mju from its datasheet (for 300B, μ=3.9) and to calculate its Gm as Gm=μ/R_I, in this case Gm = 3.9/3,137 = 1.24 mA/V. Normally, Gm for 300B ranges from 5mA/V to 5.3mA/V, but that is at normal operating conditions, with anode voltages of 350V and 300V, respectively.

For 6L6, the anode current was 10.5mA, so according to Sico, its R_I= 22.7*1.41/0.0105= 32/0.0102 = 3,048Ω. The data sheet specifies that when triode connected its μ=8, so Gm = 8/3,048 = 2.62 mA/V.

Sico then contradicts itself in the framed explanation on the previous page, quoted from the TV-12 manual. Now users are instructed to calculate Gm by dividing meter indication ("the average change in plate current") by "the change in grid voltage".

With $4.87V_{AC}$ on the grid, for 300B Gm=10.2mA/4.87V= 2.09mA/V and for 6L6 Gm=10.5/4.87=2.16 mA/V.

The final arbitration was done on our trusted Triplett 3444 Tube Analyzer. Setting "9" of the "C" switch uses 30V on the screen and anode, the closest to Sico's test voltage.

With 300B, we adjusted the bias until the anode current was also 10.2mA (around -3.0V) and measured Gm as 1.8 mA/V.

With 6L6, even with zero bias, the anode current was only 9.0mA, but that was close enough; Gm was 2.2mA/V, almost identical to the TV-12 result using the second Gm calculation method (2.16mA/V).

The 300B result was also close, 2.2mA/V, versus Sico's 2.09mA/V.

ABOVE: Testing 300B triode on TV-12 as per the Gm instructions: Control C at "50" and circuit at "C", meaning a $25V_{RMS}$ test voltage is applied to the screen and anode.

Replacing the "shorts" neon circuit with an analog megohmmeter

IMPROVEMENT PROJECT

We've already discussed this improvement in the chapter on emission testers. However, an additional switch was needed and substantial rewiring. For Sico TV-12, this modification is much simpler and easier.

The neon and its resistor and capacitor are disconnected from terminal (1) of the "Circuit" switch. A 180k resistor is soldered to the terminal (2) of the same switch (which was not used before), the other end of that fixed resistor goes to the "Zero adjust" control pot, the anode of the rectifier diode (1N4007), and the negative of the 22µF elco (3). The positive terminal of that elco goes to the common point of the whole tester (GND or COM), which is the cathode "K," point (4).

Once the "Press to read" switch (5) is pushed, the meter is in the ohmmeter circuit, which closes through the fixed ends of the sensitivity rheostat (6) and the normally-closed noise jack to the "P" busbar (7). Since leakage is tested by connected pins one by one to "P" (all other electrodes are at "K"), if there is any leakage, the circuit between "P" and "K" will close, and the current will flow through the meter.

The "Meter Lamp" on the original drawing simply did not exist on the actual tester, so we installed a 117V neon "Power on" indicator (8).

Each tester is different, so the values for the fixed and variable resistor in the MΩ-meter circuit (180k and 50k here) must be found by experiment.

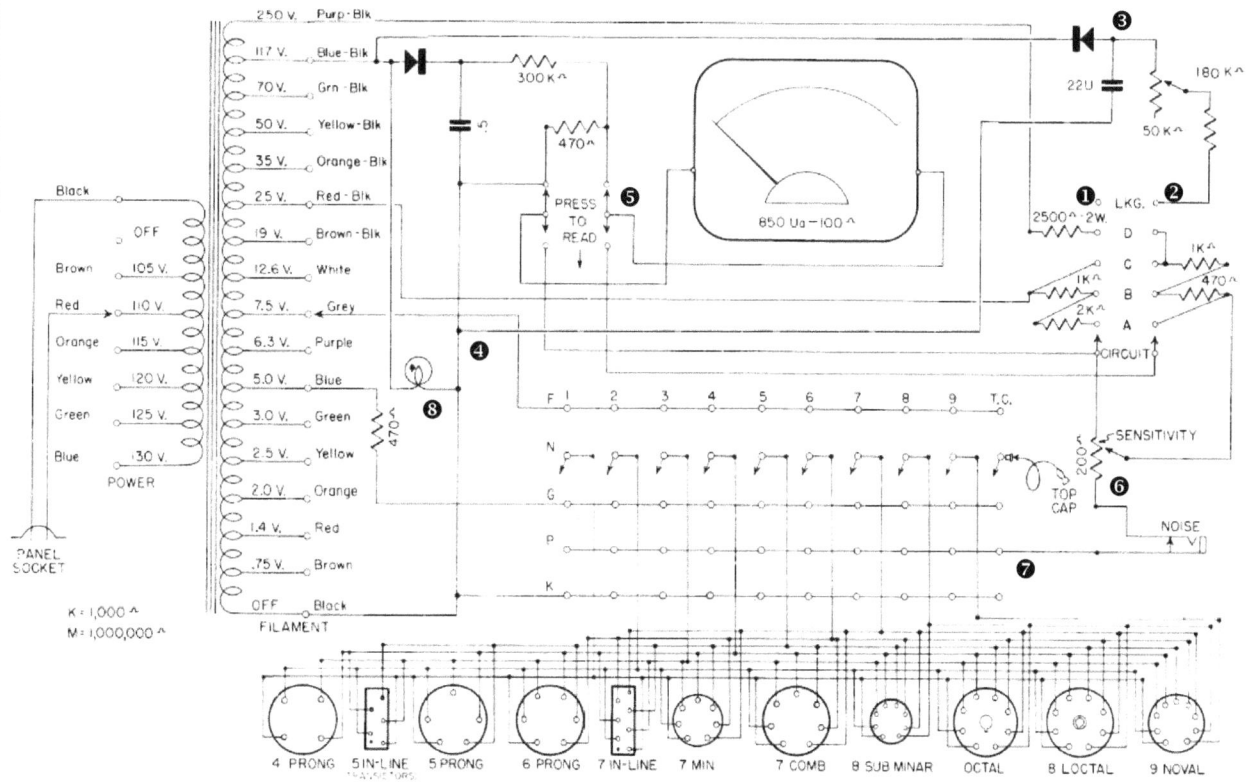

RIGHT: Since the indication of the MΩ-meter (leakage test) depends on the setting of the "Power" control switch (105-110-115-120-125-130V), the "Zero adjust" potentiometer (9) had to be externally mounted. We reused the neon shorts indicator hole just underneath the "Power" switch.

FAR RIGHT: The 1.2MΩ indication is 8 minor divisions from "infinity" or open circuit ("0" on the tester's scale). The 2.4MΩ indication was at 4 minor divisions.

SYLVANIA 139 & 140

Released in Sept. 1946, Sylvania 139 (the benchtop model with nice art deco looks and timber sides) and 140 (the portable model in a steel case) initially sold for US$69.50. Those were replaced by models 219/220 in 1950 and model 620 in 1955. All are very similar, both electrically and functionally. Three in-phase AC voltages are fed to the tube under test, 100V anode secondary winding (1), 110V screen supply (2), and 5V control grid supply (3).

Our unit, bought on eBay, had a 9-pin socket installed for the then newly-released tubes such as 12AX7/12AT7 & 12AU7, but some earlier units did not, and were shipped with a covered cutout for a spare socket.

This is how the factory wired the Noval 9-pin socket on newer model 140 units:

Pin 1 (red) to pin 2 of the 4-pin socket - Pin 2 (green) to pin 3 of the 4-pin socket - Pin 3 (pink) to pin 6 of the 7-pin mini socket - Pin 4 (black) to pin 4 of the 4-pin socket - Pin 5 (pink) to pin 1 of the 4-pin socket - Pin 6 (black/red) to pin 4 of the 6-pin socket - Pin 7 (yellow) to pin 5 of the 6-pin socket - Pin 8 (black) to terminal #8 of switch "B" (top cap) - Pin 9 (beige) to pin 5 of the 7-pin mini socket.

We updated the schematics below to reflect this modernization; the 9-pin socket is first from the left.

Switch "B" selects "Filament high" pin (4) while passing all others through, and switch "D" selects the other end of the filament (heater) and connects it to the COM point to which all four secondaries are tied (5).

In short-test mode, switch "E" isolates pins one-by-one while shorting all others together. The "test" mode switches anode voltage to the required pin (6). Before reaching switch "E," the anode voltage passes through and is controlled by the bank of test switches T-U-V-W-X-Y-Z (7).

In the "C" position (UP), circuit switches 0-9 provide through-connection for the negative heater voltage (8) and also enable isolation of any tube pin(s) from the test circuit. In the "F" position (DOWN), its contacts switch the screen and grid voltages to the required pins via two vertical "bus bars" next to (9).

Control "G" is the meter shunt potentiometer. The only internal control is R-105, 400R cathode resistor of the internal 1LE3 triode, which changes its grid bias and its anode current, which in turn passes through its anode resistor R-114 (68k), enters the (-) pole of the 0-1mA meter, exits the (+) pole of the meter and reaches the +110V_{AC} secondary via series resistor R-110 (1k5). This means that R-105 is a meter calibration resistor for "Line Control."

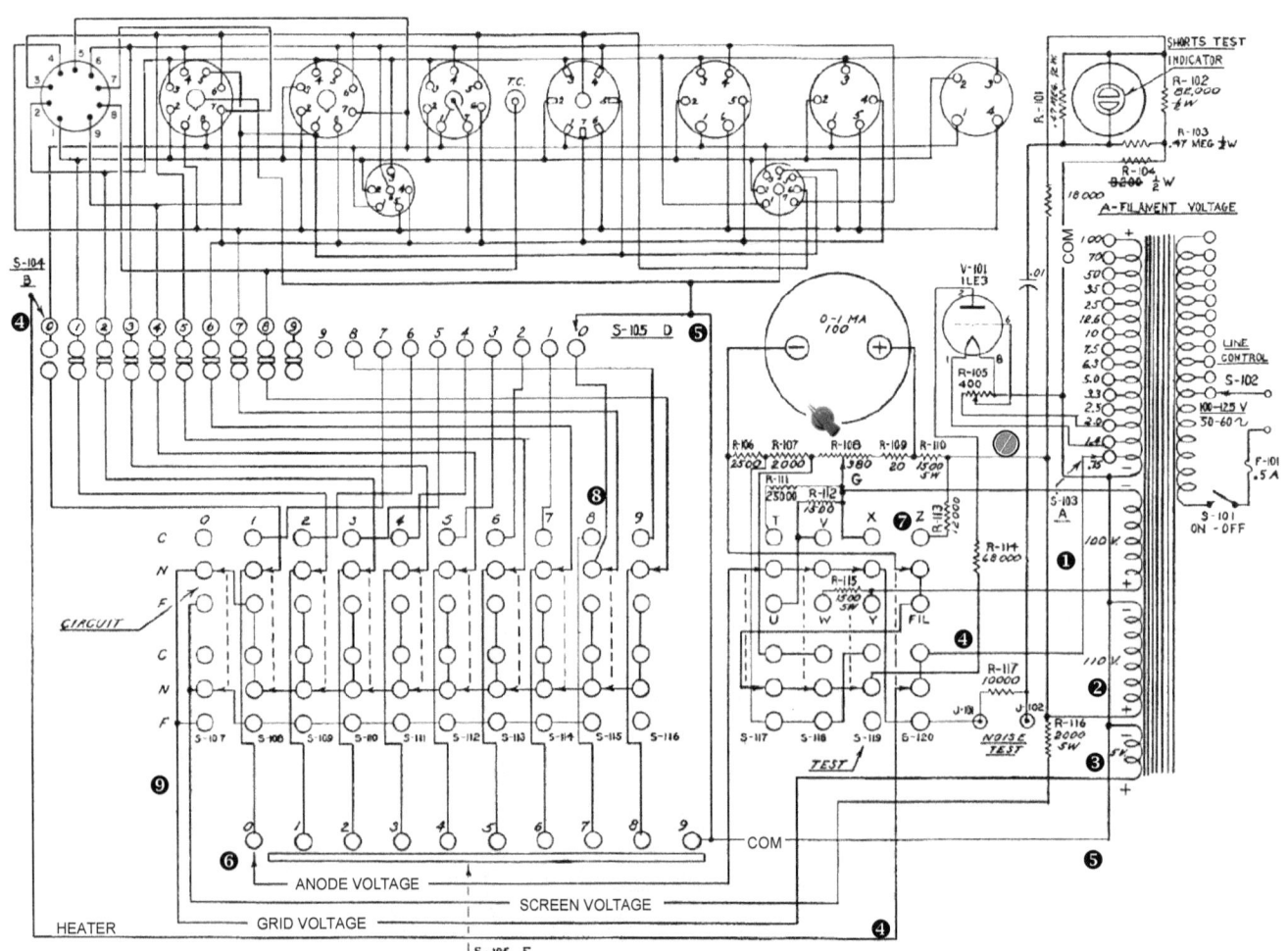

Warning: Check if 139/140 tester you are considering buying has the nine pin (Noval) socket installed (1). If it hasn't (unless you are prepared to install and wire one yourself), stay away from it!

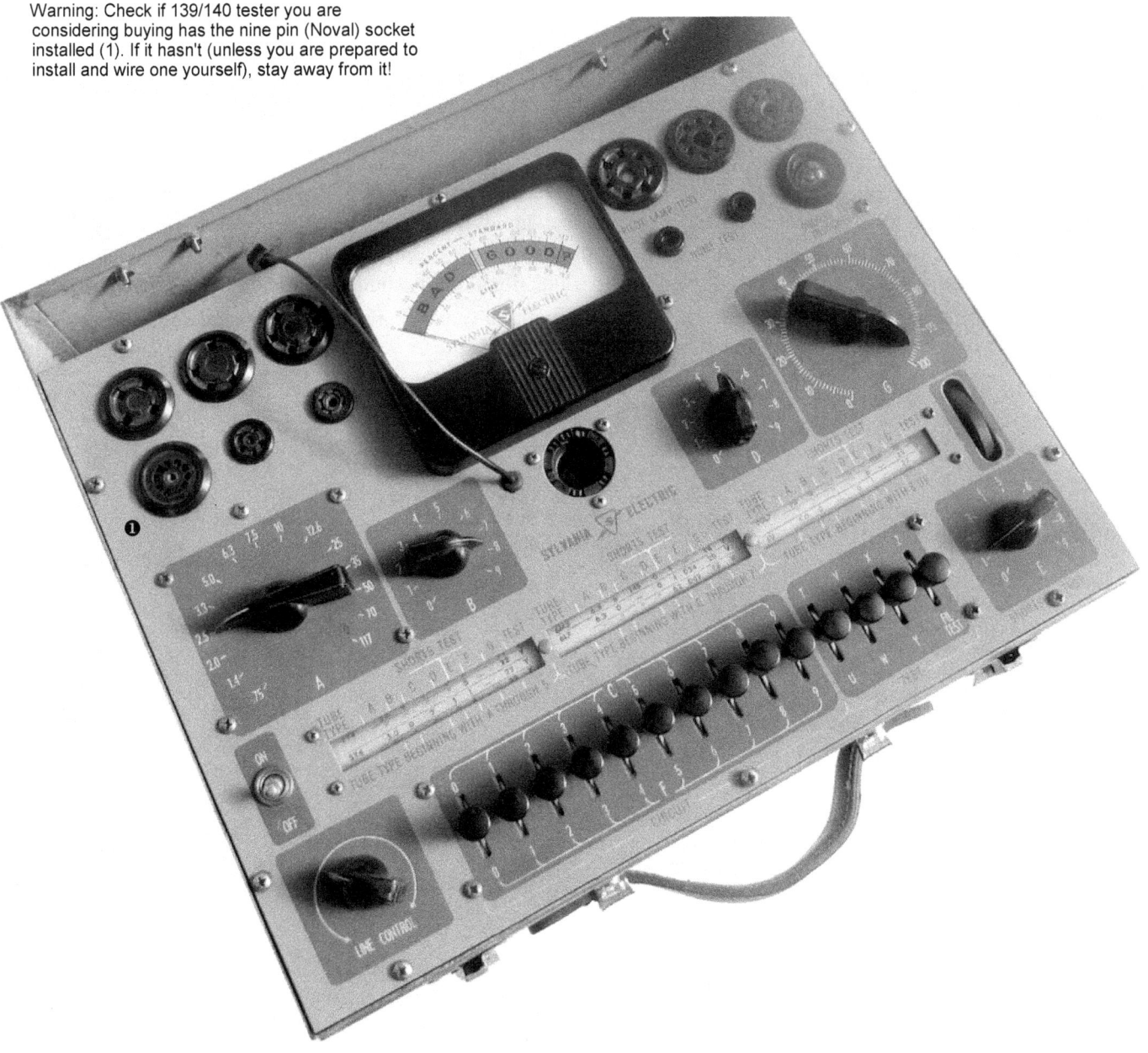

Should we adjust "line control" with or without load?

Quite a few testers use this stepped method of "Line Control," so let's elaborate a bit further on it. There are two ways of approaching this issue.

The "no-load" approach requires an AC multimeter connected across any of the heater's secondary taps and COM, for instance, the 6.3V tap. The "Line control" switch is adjusted to a position that results in a multimeter reading closest to 6.3V. The tester's meter should indicate 67% of the scale or right on the "LINE" mark. If it doesn't, tweak R-105 until it does.

The "high load" approach involves plugging in a power tube with a 6.3V heater, such as 6L6 or EL34, and following the same steps as with the "no-load" approach. Of course, make sure that switches are set, so the tube's heater is energized and glowing. This time the heater is loaded with a 0.9A or higher current so that this adjustment will compensate for its natural "sag" under load. The heater voltages will rise with this method while tubes with lower heater current draws are tested. Instead of 6.3V with 6L6, there may be 6.6V with preamp tubes such as 6DJ8 (ECC88), for instance.

With the previous "no-load" approach, high heater load (power tubes) would result in heater voltages sagging significantly. In contrast, preamp tubes (lower heater current) would be tested with slightly reduced heater voltages. In other words, all tubes would be tested with lower heater volts than prescribed, some with slightly lower, others with significantly lower voltage levels.

Sylvania 219 & 220

If it weren't for its hard-to-read and follow, and thus almost incomprehensible circuit diagram and its weird switching controls, Sylvania 219 would be one of my Top-5 testers. A desktop version of the portable model 220, it looks really elegant and reminds one of the bygone art deco era. The sides are solid hardwood and steel panels are gently curved, with not a single sharp edge anywhere!

Inside, it is neatly and solidly built, a quality instrument by all means. There is no transformer buzz, nothing rattles or squeals, the switches are smooth and solid, the roll chart is one of the least troublesome we've used, in short, this tester (and all Sylvania testers) is an exemplar of solidity and quality.

Models 219 and 220 replaced models 139 and 140 in 1950, so newer tubes such as 6BQ5, EL34 and 5DJ8 may not be listed in its tube setup book. Model 620 followed in 1955, functionally it is almost identical to 219/220.

Some controls are intuitive and obvious: the On-Off switch with line control (1), the heater voltage selector marked "A", the heater (+) or active pin selector marked "B" (2) and meter shunt or sensitivity, marked "D" (7).

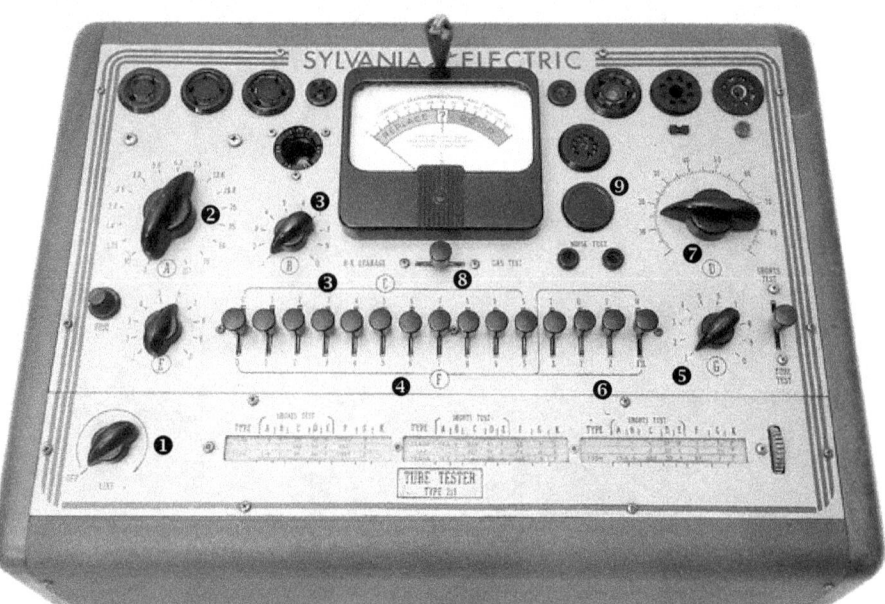

1) On-off switch + "Line Adjust"
2) Heater + selector switch "A"
3) Heater - (return) selector switch "B" (with any pins 1-9 in "C" position)
4) G1 and G2 pin selector in position "F"
5) Pin selector during "Shorts Test" + anode pin selector in "Tube Test" mode "G"
6) Anode loads, meter shunts and anode/screen voltage selection circuits
7) "Load" or "D" - a shunt resistor across the meter
8) "HK Leakage" - "Gas" test switch
9) Spare cutout for a future tube socket

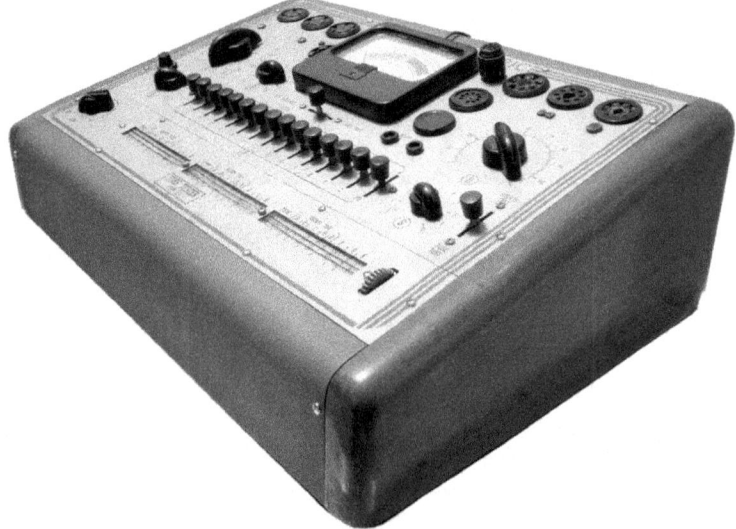

The rest can be perplexing, however, as always, a few minutes spent studying some typical tube setup examples clarifies the logic behind designer's choice all those decades ago.

Any of the lever switches 0-9 turned up to position "C" connect corresponding socket pins to switch "E" (3), which is heater (-) or return.

Switch "G" (5) serves a dual purpose. It selects tube pins (one-by-one, as the user flicks it through all its positions 1-10) and shorts together all others during "Shorts test".

In "Tube Test" or "merit" mode Switch "G" selects the anode pin.

The most confusing aspect is the "F' position. Levers 1-9, when in "F" position, connect G1 and G2 to corresponding pins on tube sockets. The lowest numbered pin will connect G1 (control grid), while the higher numbered pin will route screen voltage to that pin. However, in some tubes screen is connected to a lower pin number than the control grid, for instance in 6L6 screen (G2) is pin 4 and G1 is pin 5. In that case lever switch zero "0" must be in position "F", its job is to reverse the connection.

It seems all this weird switching is a consequence of Sylvania trying to save money by not having 4- or 5-position lever switch bank, and trying to do with only 3-positions (only two "active" positions, up & down)!

DYNAMIC CONDUCTANCE TESTERS

The last column on the roll chart is marked "K" for the cathode; however, it isn't marked anywhere else on the control panel. It is used only during the HK-leakage test.

With control "G" at the pin listed under "K," once the test switch is flicked into "Shorts test," the heater-to-cathode leakage is being tested. If the meter's needle flicks to the right, into the green or "Good" segment, the manual says HK leakage is "not excessive." What that means, heaven knows. There is no graduated leakage scale, but even this is better than the primitive neon shorts checks.

The same applies during "Shorts" testing, as switch "G" is flicked through all positions, one-by-one pins are selected, and the leakage between that pin and all other electrodes (bundled together by the 2nd wafer of the same switch) is displayed on the meter. Green means no short (high insulation resistance), red means a short or some leakage (low resistance).

	A	B	C	D	E	F	G	K
	HEATER VOLTS	H +	TO "E"	SHUNT	H -		ANODE	CATHODE
6BQ5	6.3	4	5	17	5	2-9-Z	7	3
6L6	6.3	2	7	19 (11 for EL34)	7	0-4-5-Z	3	8
6DJ8 T1	6.3	4	5 - 8	42	5	2-T	1	3
6DJ8 T2	6.3	4	3 - 5	42	5	7-T	6	8
12AU7 T1	12.6	4	5 - 8 - 9	35	5	2-Z	1	3
12AU7 T2	12.6	4	3 - 5 - 9	35	5	7-Z	6	8

ABOVE: A few examples of control setup configurations for some common audio tubes

12AU7 and 6L6/EL34 examples

A few test setup examples will clarify the operation of the switching circuits. 12AU7 is tested at 12.6V heater voltage, so pins 4 and 5 are used as H+ and H-. Pin 5 must be under "C" for both triodes, T1, and T2. Ppins 8 and 9 are also switched to "C" in triode 1, meaning they are disconnected since "E" is at 5, and only that pin is "active." Likewise, with T2, pins 3 and 9 are disconnected. Pin 9 is disconnected since it is the CT (center tap) of the heater, and as such, it is not used here. During each triode's test, the cathodes of both triodes are disconnected.

Pin 1 is the anode of T1, so "G" is at 1, and pin 2 is its grid, so pin 2 is listed under "F," and circuit Z selects the chosen combination of grid and anode voltages to test this tube. 6DJ8 does not have heater CT at pin 9, so pin 9 is not listed under its "C" setting. Its test voltages are "T," not "Z," as for 12AU7.

The heater for octal power tubes is at pins 2 and 7, so "B"=2 and "C" and "E" are set at 7. The anode is pin 3, so "G"=3. The screen is pin 4, and the control grid is pin 5, so they are both under "F." However, since the G1 pin number is higher than the screen's pin number, the circuit must be reversed. That is achieved by also flicking switch "0"(zero) to position "F," which is why 0 is listed under "F." The test voltages are the same as for the 12AU7 triode, combination "Z."

Grid current ("gas") and heater-cathode leakage tests

Most dynamic testers (Sico 85, Sico TV-12, Eico 666/667) don't have a dedicated "Gas test" button. The issue of testing for gas in those testers' manuals is not even mentioned! However, Sylvania 219/220 has such a button. When thrown to the "HK Leakage" position, the switch disconnects all electrodes from the shorts test except the heater and the cathode. To avoid conductance errors, the direction of the applied shorts voltage is reversed during this test.

In the "Gas Test" position, the switch opens the grid circuit of the tube, and since most of the dynamic current during the "merit" test flows through the grid and not through the faraway anode or screen, the meter reading (tube's composite current) should drop. If the current does not drop or even goes up with open G1, the tube has a problem - gas, grid poisoning (grid emission), or grid leakage!

On most units (but not all!), the meters even had a green arrow, pointing CCW and saying "GOOD," and a red arrow, pointing CW and saying "BAD," just in case the user didn't read the manual properly.

> **Simpson's claim of correlation between plate conductance and mutual conductance**
>
> "The percentage of Plate Conductance indicated on the meters of several Tube Testers Model 1000 are made to agree with the percentage of rated Mutual Conductance measured on each sample tube with a laboratory type Mutual Conductance Bridge. This correlation assures you of proper indications for evaluating all tubes according to standard rating systems when you test them on your Simpson Tube Tester Model 1000."
>
> Plate Conductance Tube Tester Model 1000 Operator's Manual, Simpson Electric Co., 1958

JACKSON TUBE TESTERS

Paul Jackson incorporated his Dayton, Ohio business in 1933. Model 548A was released at the time of his death in 1957, followed by 648R two years later. Models 648S, 658, and 598, the last three Jackson tube testers, date around 1962. Model 658 was then priced at a whopping US$189.95, while 648R was more affordable at US$129.95.

Mercury 1000, a mutual conductance tester, was only US$79.95 at the time, and even the much superior Triplett 3444 was only US$349, so less than double the price of 658. While Sico, Simpson, Eico, Precision, and other makers of similar dynamic conductance testers used a single power transformer, better Jackson models such as 658 use two, one to supply heater power, the other as a source of the anode, screen, and control grid voltages.

Apart from that, it is hard to see why Jackson testers were (and still are by a section of the tube tester user community) held in such high regard.

Dynamic emission testers by Jackson had sold reasonably well for over four decades. As a result, such vintage units are often offered for sale on eBay and other online sources.

Apart from a few models that used lever switches, most used a complex and temperamental push-button switching assembly. We had two such testers, 648R and 648S, and both arrived from the USA with (mechanically) faulty switching units. One we managed to repair, the other was taken apart for parts.

Even when they worked, those rectangular PB switches were wobbly, looked untidy, and it wasn't easy to see which ones were pressed.

Their markings, printed on the buttons and in frequent contact with users' fingers, would rub off very quickly. The only Jackson tester we didn't mind using was 598 with its lever switches.

A particularly annoying issue is Jackson's poor quality drawings, which are very difficult to follow and comprehend, making the repair of their testers a time-consuming and frustrating task.

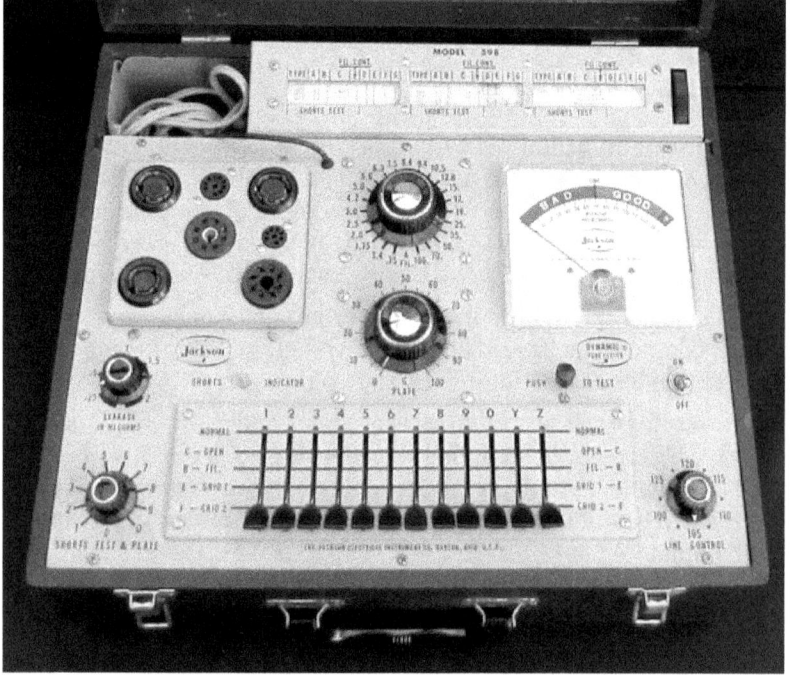

ABOVE: Jackson 598 is an elegant, well laid out tester. However, despite its very large size, there are only seven test sockets, and it is still an emission tester first and foremost.

LEFT: The scale lacks minor graduations, so matching tubes on Jackson testers is a dubious activity. The "Relative micromhos" marking was pushing the boundaries of misleading advertising, making users think the tester measures mutual conductance; it doesn't!

Jackson 658 diagram and switches

We have seen the same or similar drawing symbols on Simpson 330 tester. Still, Jackson switches reach a much higher level of complexity and cause utter confusion in the minds of unsuspected troubleshooters, so let's use this opportunity to briefly point a few issues out.

First, Jackson engineers used rotary switches yet drew them using "linear" symbols (next page). One could argue that it may be easier to mentally picture sliding contacts than rotating ones, so while this may be a faux pass from the consistency aspect, it may not affect the user-friendliness of such an approach.

More seriously, it is unclear (not marked on drawings) if the contacts should move up or down. Switch S1 is the "Line Input Control" and is easy to comprehend, so let's start there.

Its single contact picks on tap of the primary winding. The common contact (1) must always be connected to the sliding bar, and the raised segment must touch only one tap (showed connecting tap "D").

Assuming the contact slides UP on the drawing, in the next position ("D") would be disconnected. The common contact CC would connect to itself, and no primary tap would be connected in that or any subsequent switch positions! So, clearly, this would not work - the contact "bar" must slide DOWN!

However, if you mentally or visually slide the contact "bar" down one "notch," so the raised (thicker) section touches tap "E," the common contact CC is not touching the bar any longer.

Again, the circuit would be broken, and there would be no power feeding the primary. So, something is wrong again, and that is the wrong way the common contact CC was drawn. We have corrected that part of the drawing below.

Normal logic would tell you that if one switch was drawn a certain way, all other similar switches would be too, so let's move to the next example. Switch S6 is a "Plate V." (anode voltage) selector switch, and on the control panel, its positions are marked "Q-R-S-T-U-VR1-VR2-VR3".

No positions are marked on the drawing. The contacts were drawn in the first position and most likely in the CCW position, which would mean position "Q," whatever that means. The schematic says "250V". The switch has three wafers, each with two separate contacts, so it is a 6-pole switch, but the control panel shows eight positions while the drawing indicates only 5!

Our aim here is not to clarify and fix their mistakes (frankly, I don't know the answer to this riddle and have no intention of wasting my valuable time on it) but to try to explain the logic (if any!) behind vintage drawings.

We have a common or "pickup" terminal or contact (3), always connected to the bar. In position "1," it connects to the secondary tap (4), in position "2" to the next tap down, and so on. The other contact of S6C has one common terminal (5) and two pickup terminals, (6) & (7).

In position "1," all three are connected together. In position "2", (7) is disconnected, so only (5) and (6) are connected. In the last position, #5, all three are strapped together again due to the upper raised segment.

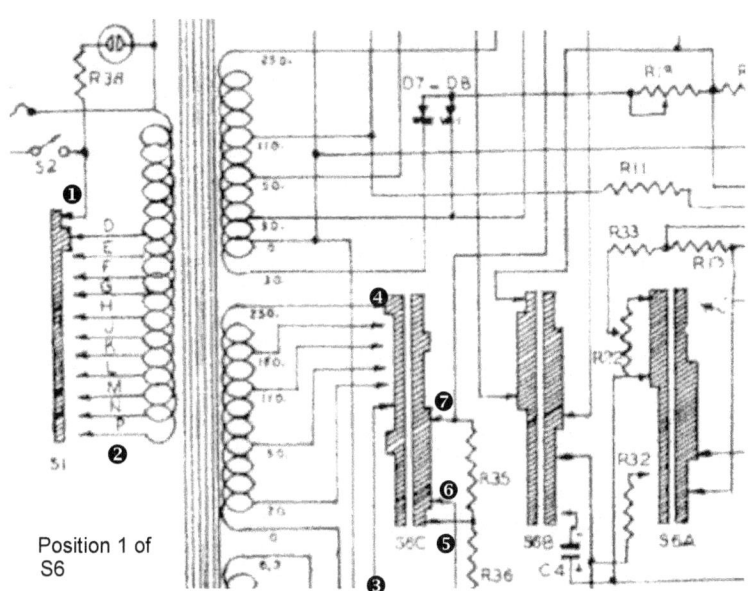

Position 1 of S6

ABOVE: The way switch S1 is drawn on the original drawing would not work. In the corrected drawing (BELOW) the contact moves down and connects taps D-E-F-G- ... etc.. to common contact CC!
RIGHT: The contact "C" of switch S6 shown in positions 2 and 5

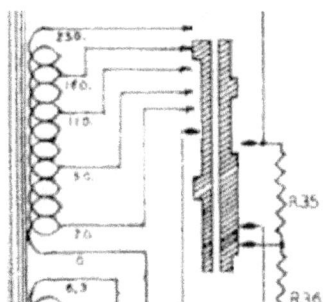

Position 2 of S6

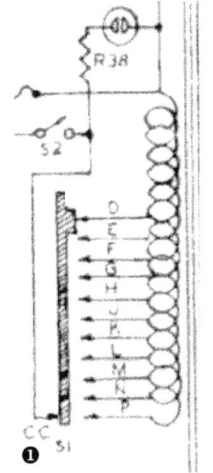

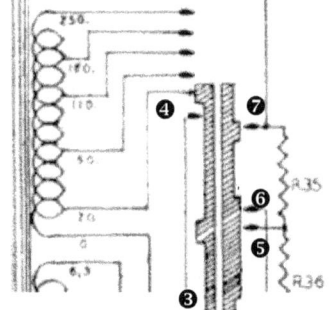

Position 5 of S6

EICO 666 & 667 (SIMPSON 1000)

When released in 1955, Simpson 1000 cost US$135.-, a significant sum. The production stopped in 1962. The earlier version had a roll chart at the bottom of the top panel, close to the handle, which was letter moved towards the top, just underneath the analog meter. Apart from such minor changes, the internal circuit remained the same.

The devilishly named Eico 666 seems to be an almost exact copy of this tester. Even test settings are identical, so if you have a problem with your Eico not testing a particular tube well or at all (there are numerous errors on Eico roll charts, some with catastrophic consequences!) consult Simpson 1000 test setup instead.

Since Eico sold tens of thousands of model 666 and almost as many of model 667, meaning they are two of the most commonly offered tube testers for sale and certainly much more common than Simpson 1000, we will analyze them in detail here.

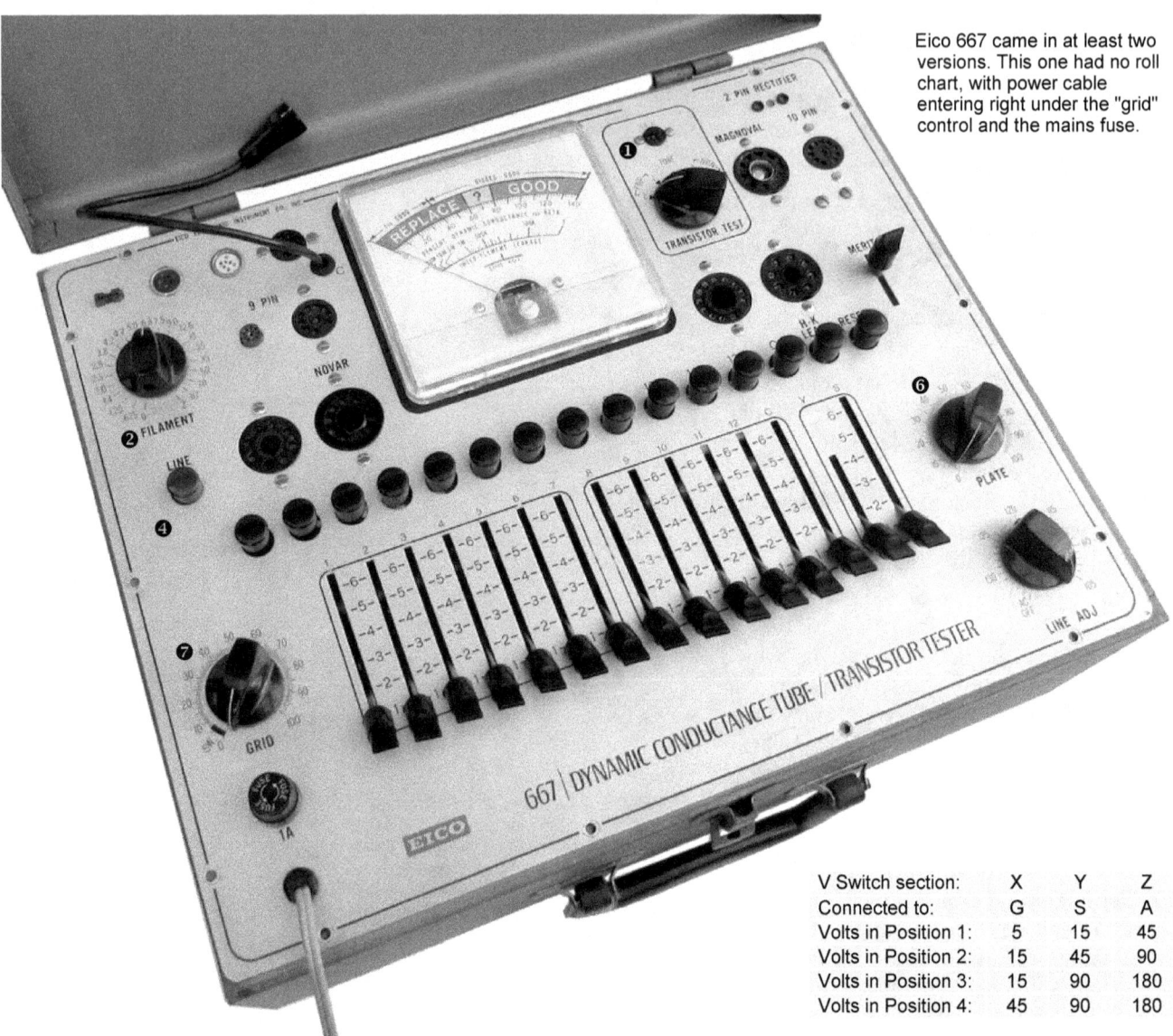

Eico 667 came in at least two versions. This one had no roll chart, with power cable entering right under the "grid" control and the mains fuse.

V Switch section:	X	Y	Z
Connected to:	G	S	A
Volts in Position 1:	5	15	45
Volts in Position 2:	15	45	90
Volts in Position 3:	15	90	180
Volts in Position 4:	45	90	180

The older model, 666, has ten lever switches, while the newer variant, 667, featured thirteen, so it could test 12-pin Compactrons and other tubes with more than 10 pins (9 pins plus top cap). However, model 667 lost a couple of older sockets, including the useful 4-pin socket, so it cannot test tubes such as 2A3 or 300B, and gained a few later socket types. Otherwise, 666 and 667 are functionally identical testers. So, if you want to restore old tube radios or if you are a directly-heated triode fanatic and want to test 2A3, 300B, and similar triodes, 666 is a better tester for you.

Functionality and circuit analysis

Transistor testing circuitry is very simple (1); a half-wave rectified $7.5V_{AC}$ voltage is filtered and applied to various transistor terminals, depending on its type, PNP or NPN. The Line Adjustment rheostat incorporates the ON-OFF switch. When the Line button is pressed (4), the "Line Calibration" potentiometer is connected to the meter to center it at the mark. The circuit closes through a diode and "leakage calibration" trimmer pot (3).

A separate transformer secondary winding has taps for 20 heater voltages selected by S3 (2).

The bank of dual switches operated by pushbuttons selects pins to be tested for shorts and, finally, during the tube merit test, selects the anode.

The heart of the circuit is the three secondaries (5) and the triple "V" switch, which selects one of four combinations of five AC voltages to test a tube. Its bottom pole connects the wiper or point X to one of three grid voltages: 5V, 15V, or 45V, and then passes that voltage to the Grid bus when the Merit switch is pressed.

Likewise, contact Y selects the Screen voltage (15V, 45V, or 90V) fed to the Screen bus. These "buses" are four vertical lines in the frame marked (8), connected to switches S15 to S27. The Z contact selects the anode or plate voltage (45V-90V-180V-180V).

Single-pole 6-position switch "S" selects the value of the shunt resistor placed across the meter and "Plate" rheostat (6). It is a coarse meter sensitivity control, and "Plate" is the fine adjustment.

DYNAMIC CONDUCTANCE TESTERS

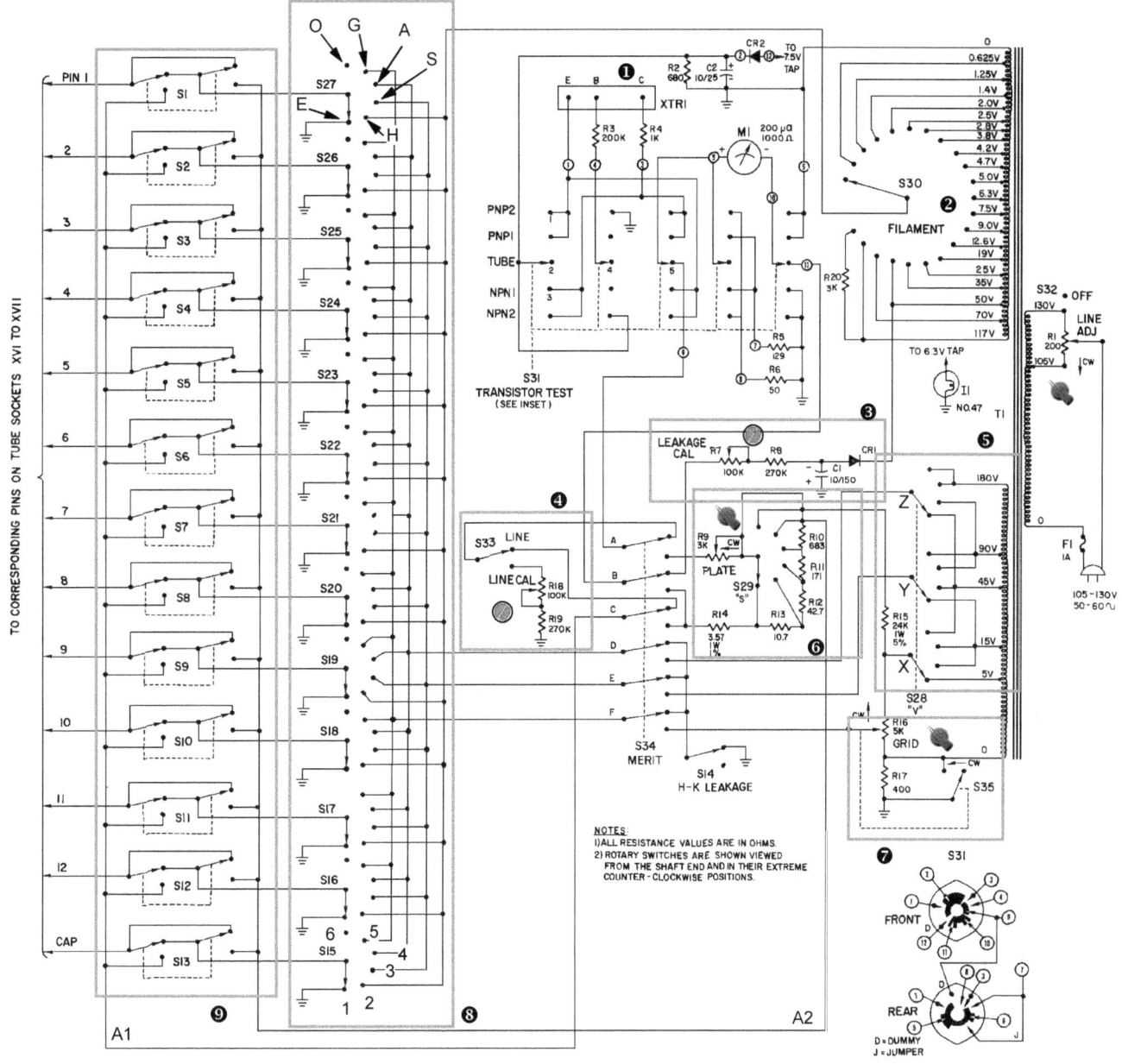

ABOVE: The original detailed circuit diagram of Eico 667 tube tester with major functional blocks framed

Position "5" is the highest sensitivity setting for low current (preamp) tubes, and position "1" (the lowest sensitivity) is used for testing power tubes.

The bank of lever switches S15-S27 (8) connects the right tube pins to the proper points. Each pin is served by one switch (pin 1 by switch S27, pin 2 by S26, and so on. Each pin can be connected to one of the 6 points. They are marked 1 to 6 on the tester's control panel, and their meaning is as follows.

6 is O - open circuit, 5 is G - control grid (G1), 4 is A - anode, 3 is S - screen grid (G2), 2 is H - heater (one side), and E - earth or ground or COM (also the other side of the heater connection). You can remember it as "OGASHE".

Only one heater connection is needed since the other side of the heater winding is permanently wired to the common terminal (marked 0 in the top right-hand corner), so the other pin connected to the heater in the tube would be connected to E (earth). Notice that all switches are drawn in the E (earthed) position.

Once the grid tap is selected (5, 15, or 45V_{AC}), the "grid" potentiometer acts as the final adjustment of the exact grid AC voltage specified in the tube charts.

"Merit" test

This diagram (next page), omitting all but one of the S15-27 switches (S26, feeding Pin 2 on a 6L6 tube), and all but one of the S1-S13 pushbuttons (S3, isolating pin 3), is shown with all switches in the active state (Merit switch pressed). Only one of the pushbuttons S1 to S13 is pressed, in this case, S3, because pin 3 is the anode on 6L6.

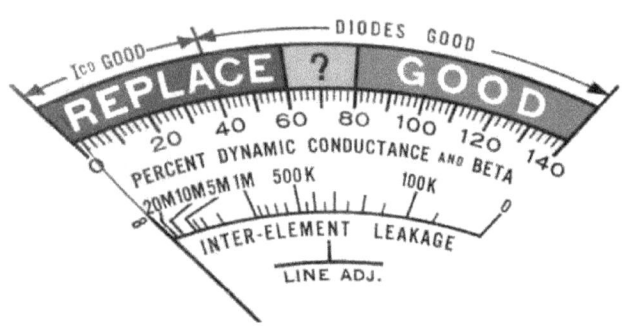
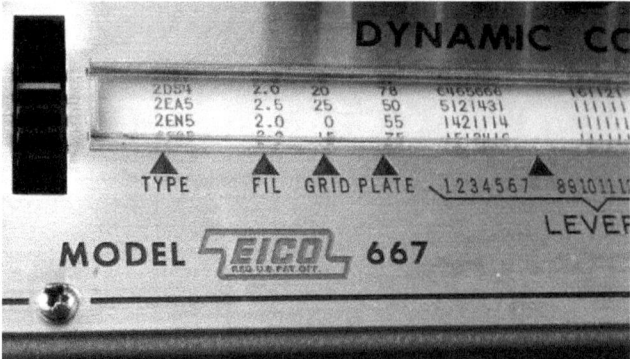

Eico 667 meter faceplate (ABOVE) and details of the roll chart on some units (RIGHT)

S3 breaks the anode circuit and inserts the metering circuit between points A1 and A2. I have also marked these points on the main schematics (previous page). The anode break switch (S3 in this example) connects A1 to the anode of the tube-under-test and A2 via the anode busbar and the uppermost contact of the "V" switch to the selected tapping on the mains transformer's secondary (in this case to 180 V_{AC}).

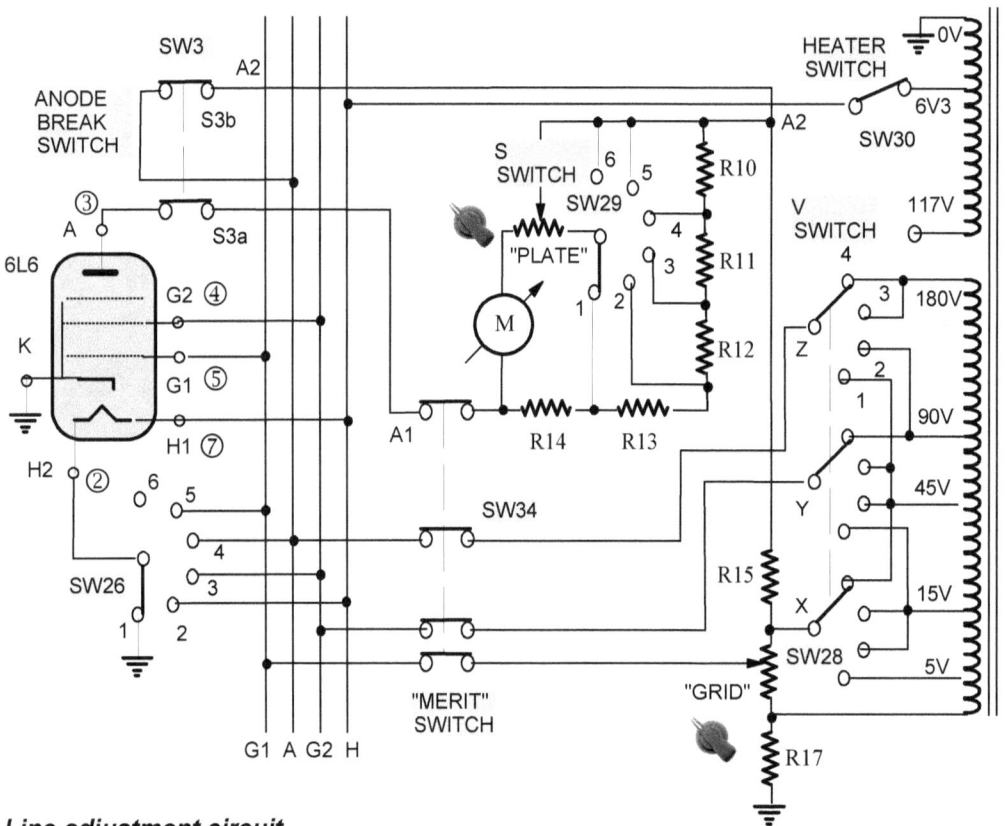

LEFT: The simplified diagram of Eico 667 tube tester while testing a 6L6 tube.

Line adjustment circuit

Although Eico 666/667's operating principle is relatively simple, the myriad of switches makes test circuits difficult to trace and troubleshoot. Let's take the simplest function, "Line adjustment." Quite a few eBay sellers report an issue such as "The needle moves upon power-up, but "line" cannot be adjusted to the middle mark." A faulty line adjustment rheostat can cause this. Still, if the meter's indication varies with that rheostat's adjustment, the problem is most likely in the indication circuit, which is shown here.

50V_{AC} is half-wave rectified and filtered by C1, so the first test would be to measure the DC voltage on C1 (around -70V_{DC}). This DC voltage then feeds a series metering circuit comprised of two fixed resistors, two calibration trimmer pots (not just "Line CAL." but also "Leakage CAL." as well), and no less than five switch contacts!

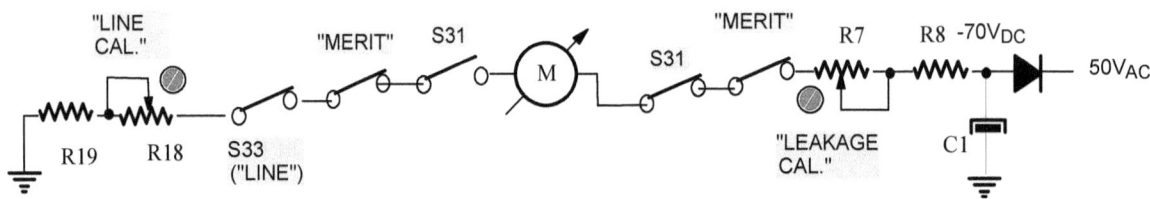

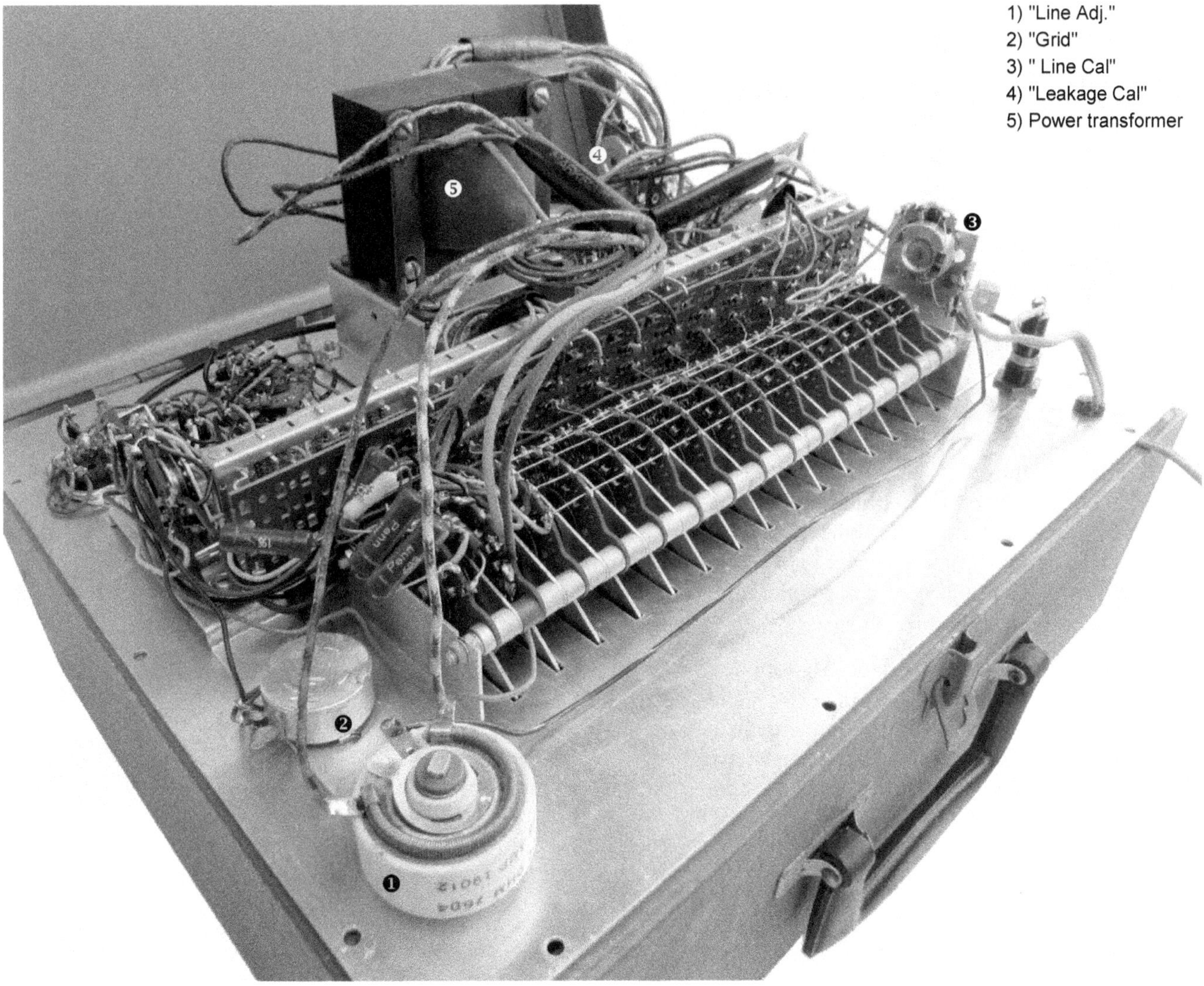

DYNAMIC CONDUCTANCE TESTERS

1) "Line Adj."
2) "Grid"
3) " Line Cal"
4) "Leakage Cal"
5) Power transformer

Adding a fuse or a globe in the grid control line

IMPROVEMENT

A common fault on Eico 666/667 testers is a burned "Grid" control potentiometer. This is caused by a tube with a short between its control grid and screen or anode or by the incorrect setting of the tester's controls. Finding a replacement rheostat is very difficult unless another tester is kept for spare parts.

This protection measure uses an approach used in Hickok-type testers. Simply unsolder the wire from the pot's slider (wiper) or center lug and insert a miniature bayonet lamp socket with a type 47 incandescent globe (rated at 6.3V and 150mA).

If all is well, there will be no current drawn by the grid, and the globe will be off. If a short circuit is present, the high voltage will be present between the globe's ends, it will flash momentarily, and its thin filament wire will burn out, just like a fuse, before the thicker wire in the "Grid" rheostat had the time to overheat and burn out.

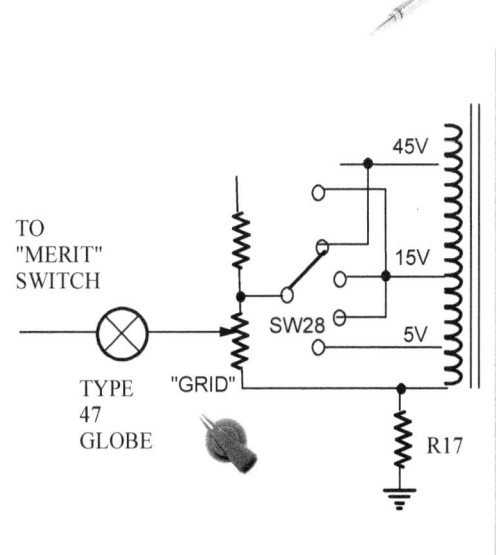

BY THE SAME AUTHOR:

Sound Improvement Secrets For Audiophiles: Get Better Sound Without Spending Big

Publisher: Career Professionals
Year published: 2021
Language: English
Paperback: 328 pages
ISBN: 978-0648298205

Avoid the hit-and-miss approach and stop wasting money on overpriced high-end products in the blind hope of sonic improvement. Achieve the ultimate audio synergy and get more enjoyment from your audio system by making it as good sounding as possible.

"Sound Improvement Secrets for Audiophiles" will teach you how things work, why some circuits, designs, and technologies sound the way they do, and how to make them sound even better through simple modifications and improvements.

It is like having an audio and acoustic consultant by your side to guide you through optimizing and voicing your audio system and your listening room.

While relatively technical and in-depth, this practical manual goes way beyond "a dozen quick tips" and the simplistic advice you read elsewhere. Instead, the focus is on dozens of DIY projects, case studies, and examples of commercial audio components – turntables, preamplifiers, amplifiers, loudspeakers, power supplies, and acoustic treatments.

With over 400 photographs, diagrams, and illustrations, "Sound Improvement Secrets for Audiophiles" makes it easy for you to understand and comprehend complex technical concepts and issues.

The author does not shy away from many controversial and hotly debated topics. Tubes vs transistors, objectivists vs subjectivists, measurements vs listening, and digital vs analog: all of these are discussed in detail.

The money invested in this book would not even buy you a budget-priced pair of cables: it will prove to be one of the best financial investments you ever make. Even if you implement only a few improvements from the hundreds described within its pages – you will never look back!

BOOK CONTENTS:

1. WHY YOU SHOULD READ THIS BOOK AND HOW YOU WILL BENEFIT FROM IT
2. BEFORE YOU BUY AN AUDIO SYSTEM OR COMPONENT - THINGS TO DO & MISTAKES TO AVOID
3. WHAT DO WE LISTEN FOR AND WHAT DO WE ACTUALLY HEAR?
4. CLEANING UP THE POWER SUPPLY TO REDUCE NOISE, HUM, AND INTERFERENCE
5. CABLES, FUSES, CONTACTS, AND CONNECTIONS
6. UPGRADING & FINE-TUNING THE SOURCES: OPEN REEL RECORDERS, TURNTABLES, PHONO STAGES AND CD PLAYERS
7. AUDIO AMPLIFIERS - HOW THEY WORK AND HOW TO IMPROVE THEIR SOUND
8. HEADPHONES AND HEADPHONE AMPLIFIERS
9. LOUDSPEAKER TYPES, TESTS, AND IMPROVEMENTS
10. COMPONENT MATCHING AND AUDIO SYSTEM INTEGRATION ISSUES
11. LOUDSPEAKER POSITIONING
12. OPTIMIZING THE ACOUSTIC PERFORMANCE OF YOUR LISTENING ROOM
13. ACOUSTIC TREATMENTS
14. MINIMIZING UNWANTED VIBRATIONS & OSCILLATIONS
15. TROUBLESHOOTING YOUR AUDIO SYSTEM

PROPORTIONAL MUTUAL CONDUCTANCE TESTERS

6

- HOW PROPORTIONAL MUTUAL CONDUCTANCE TESTERS WORK
- WESTON 798
- TRIPLETT 3423
- TAYLOR 45D
- AVO VALVE CHARACTERISTIC METER (MK III)
- METRIX 310CTR
- SIMPSON 330

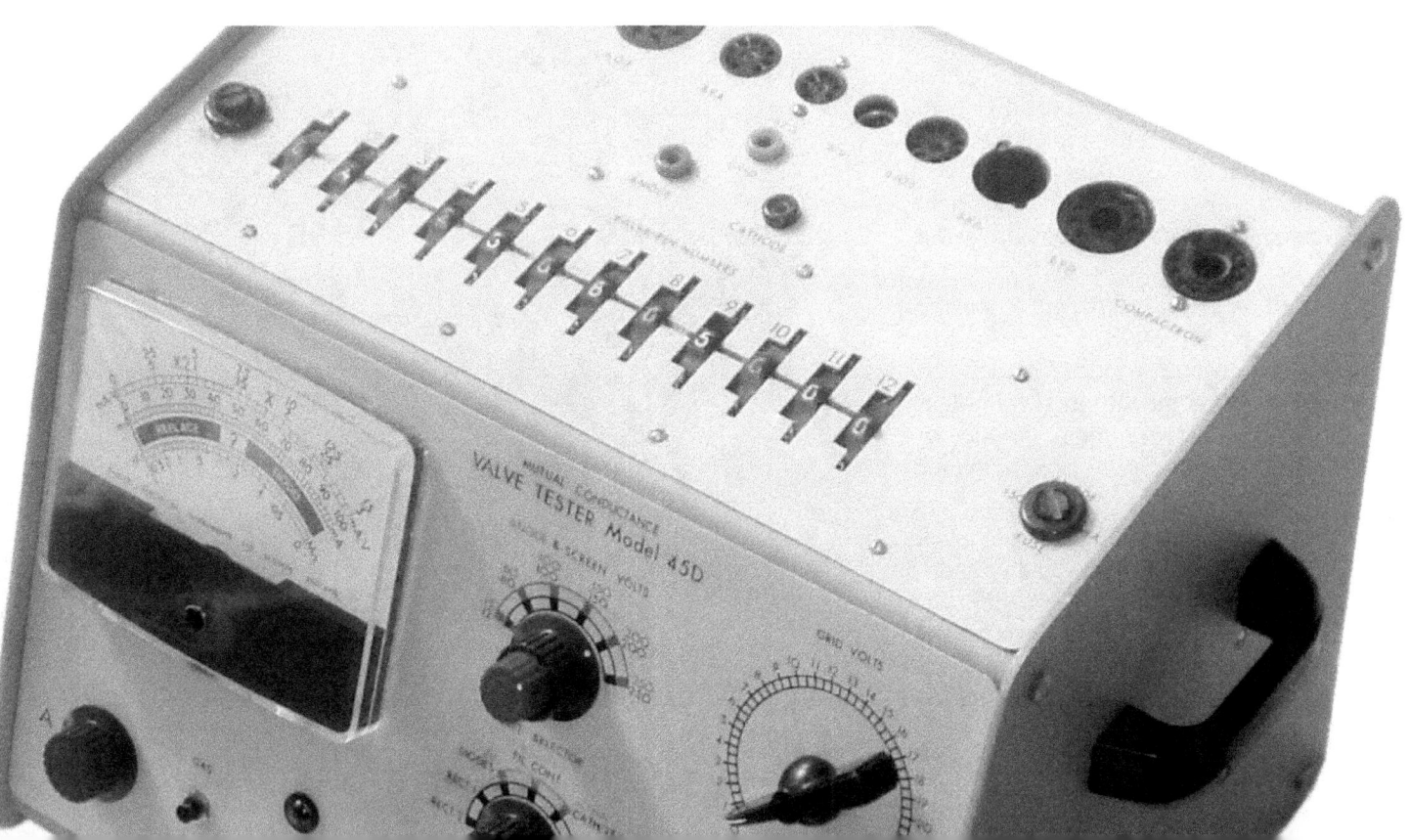

HOW PROPORTIONAL MUTUAL CONDUCTANCE TESTERS WORK

The test circuit is similar to dynamic plate conductance testers; AC anode and screen voltages are used exclusively, even the bias voltage is AC, but in this case, it is always out of phase with the rest of the secondary voltages, so the tube-under-test is negatively biased. All AC voltages in dynamic plate conductance testers are in phase, so tubes are positively biased.

They are called "proportional" Gm testers because proportional AC voltages are used (proportionally higher, screen voltage is lower than the anode voltage, for instance) and not real voltages used in tube amps. Thus, since anode voltages are lower, screen and bias voltages are lower as well, "in proportion."

The TUT acts as its own rectifier, the control grid rectifies its grid bias voltage, the screen grid rectifies its voltage, and the anode rectifies its voltage.

Some of these testers use an AC grid signal source and AC metering in the anode circuit. When a small AC signal is applied to the grid, either from a secondary winding of the power transformer or from a dedicated higher frequency oscillator (a better option), only the AC component of anode current is measured.

The meter is calibrated to read mA/V or micromhos directly. Other testers belonging to this family don't measure the AC anode current and Gm directly. Instead, they leave it to the user to perform a static Gm test or use the "backing off" method of zeroing the DC anode current through the meter.

In essence, the AC voltages on all electrodes are common to all approaches, the only differences being in the metering approach.

WESTON 798

While B&K uses a bridge that keeps the mains frequency grid signal (somewhat) constant, Weston uses a dedicated transformer as part of the 5kHz oscillator circuit. Using the mains frequency grid signal (Hickok, B&K, and all lower-spec testers) is considered an inferior approach because the mains frequency hum (50 or 60Hz) affects the meter reading.

The internal view of Weston 798 shows a neat layout (next page), which makes it easy to identify parts and repair or replace them. Unfortunately, we could not source a circuit diagram online, and none came with our tester bought in the USA.

While it can test older tubes such as 45, 2A3, 300B, and similar directly-heater triodes, model 798 lacks more modern sockets. There isn't much spare room on the top panel for adding such sockets, but it is possible.

We added a fuse holder (1) and anode current measurement binding posts (2) and replaced some elcos (3) and film capacitors (4). A step-down autotransformer (240 to 117V) was installed next to the power transformer (5).

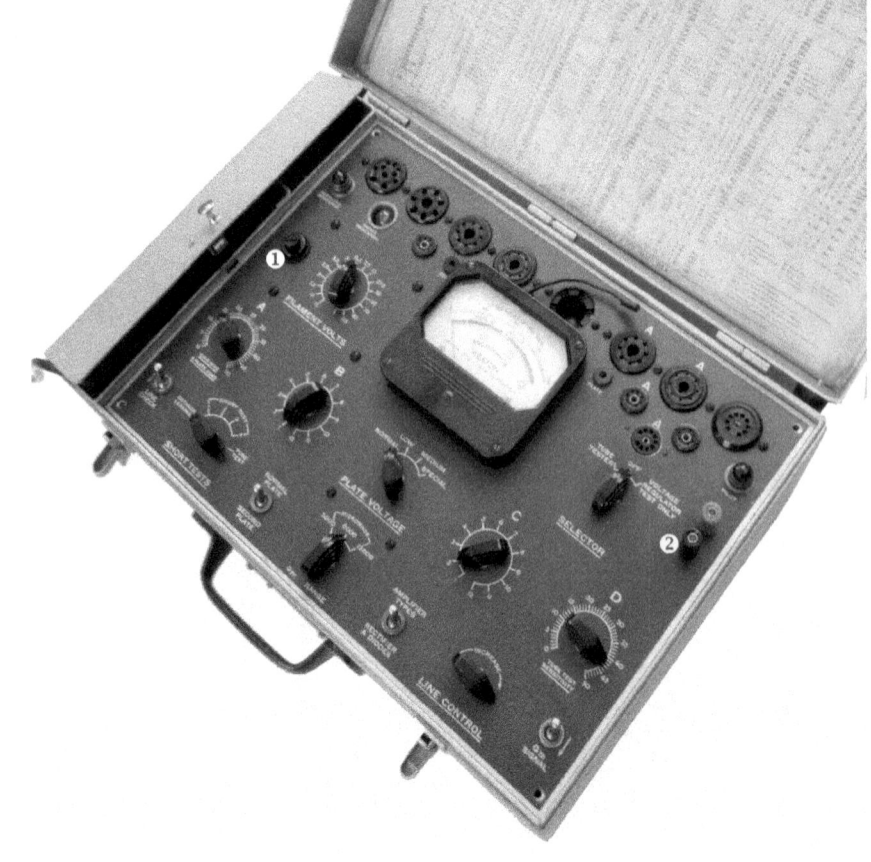

ABOVE: The operating principle behind proportional mutual conductance testers. The AC-tuning network in the metering circuit (eliminates the DC component of anode current) is not shown for clarity.
BELOW: Weston 978 with an added mains fuse (1) and anode current test posts (2)

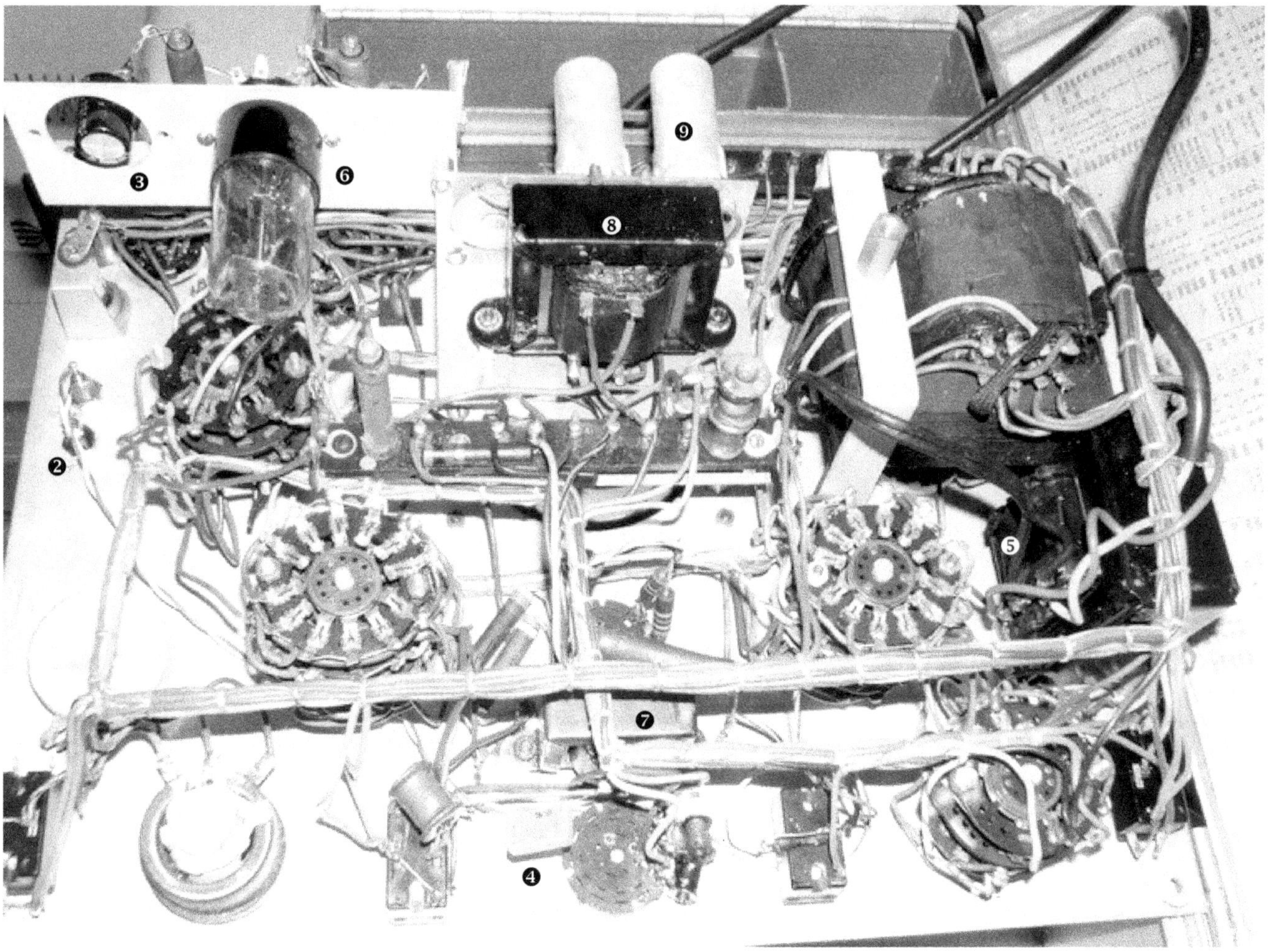

PROPORTIONAL MUTUAL CONDUCTANCE TESTERS

TThe 6X5 dual rectifier tube (6) is only used for producing a DC voltage when testing VR (voltage regulator) tubes. The major components include the anode choke in the Gm metering circuit (7), the grid signal transformer in the 5kHz oscillator circuit (8), and two 3A4 triodes (9), one used as signal voltage oscillator, the other (connected as a diode) as a rectifier for shorts testing.

Using our octal test jig, we measured the following test voltages in Weston 798: Normal $100V_{AC}$, Medium 67 V_{AC}, Low: 20 V_{AC}, and Special 150 V_{AC}. The bias voltage is adjustable from 0V at 0 on the dial to $13V_{AC}$ at the maximum, marked 50 on the dial.

Three Gm scales are available 0-3, 0-6, and 0-12mA/V, meaning this tester cannot test high Gm tubes developed in the late 1950s and 1960s, after its release!

TRIPLETT 3423

In Oct 1956, when Triplett's model 3413-B was selling at US$79.50, model 3423 was priced at US$199.95, a small fortune at that time, so no wonder there aren't that many of these testers around.

A capability to test tubes with Gm up to 36 mA/V, two grid signals (0.6 and 1.2V) and five anode voltages (10, 20, 30, 70, 100, and $250V_{AC}$) afford the user a certain degree of flexibility in choosing the best test regime for various tubes.

The bias is continuously adjustable from 0V to -40V. However, the scale of the "B" potentiometer's "skirt" (1) does not correspond to the actual DC voltage; it is very nonlinear.

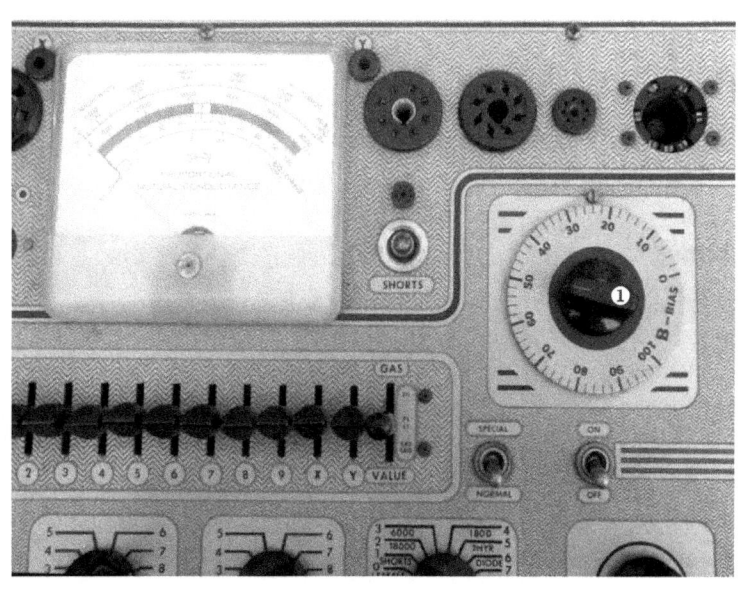

The setting of "25" on a scale 0-100 is (or should be after calibration) -2.7V$_{DC}$, "50" or half-scale is -7.25V$_{DC}$, "75" corresponds to -25V$_{DC}$, and the full scale is -40V$_{DC}$. Thus, if you need to know the actual bias voltage, the only solution is to install a pair of binding posts to measure it with a digital multimeter with high input impedance (10MΩ or higher). However, don't leave it connected during shorts and leakage tests; otherwise, the tube tester will detect this input impedance as leakage between the grid and cathode because 3423's leakage circuit is a sensitive MΩ-meter that can detect impedances below 10MΩ!

Gm ranges and test signals

Positions 2, 3, and 4 of the "G - Circuit" switch are for Gm measurements using a 1.2V$_{RMS}$ signal if the "Normal-Special" switch is in the "Normal" position. When in position 2 (18,000 micromhos range) and with the "Normal-Special" switch in the "Special" position, the grid signal is halved to 0.6V, and thus, the sensitivity of the Gm circuit is also reduced to half. The 0-18,000 scale becomes 0-36,000, so tubes with higher Gm can now be tested.

This trick can be implemented in older Gm testers. Install a 2:1 resistive voltage divider in the signal circuit plus an SPDT switch, and you will double your tester's Gm range!

The metering circuit

The anode load of the tested tube is an LC circuit tuned to the frequency of the grid signal voltage; its resonant frequency is approximately fR=1/(2pLC)) = 1/(2*3.14*0.11*20*10-9) = 3.4 kHz.

The DC component of the anode current passes only through the choke L; the 12n film capacitor prevents it from passing through the analog meter.

Diode D2 provides half-wave rectification of the AC voltage across the choke, which drives the DC meter. The 5μF elco dampens the meter movement, and diode D1 protects it from reverse voltages. D1 and D2 are one dual copper oxide rectifier unit (3-leads).

R20 is an internal adjustment for "calibrating" the meter circuit, although the factory calibration document does not mention it at all.

Triplett never released a calibration procedure for users of their testers; the factory calibration document is not practical since it calls for eight "standard" cal. tubes such as 6AL6, 6BG6, 6CD6, 6C8, 6SN7, 2D21, 1X2 and 117N7!

Testing example

Since there are only three electrode selector switches (for control grid-D, screen grid-E, and anode-F), all other electrodes are switched by the row of 11 lever switches.

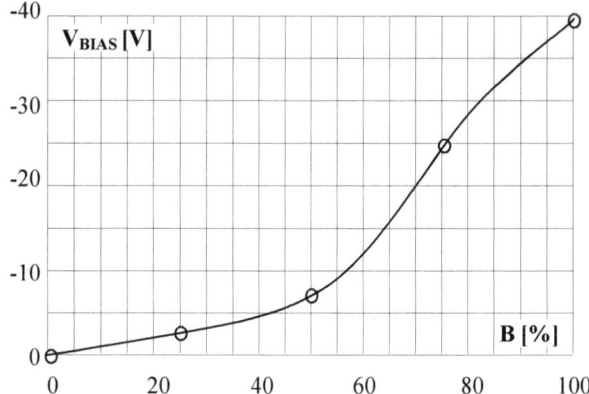

ABOVE: DC bias voltage versus the 0-100 scale of the "B" control potentiometer

ABOVE RIGHT: Halving the signal test voltage (AC grid signal) by installing a voltage divider and a changeover switch would double the maximum Gm range on any Gm tester

BELOW: The simplified Gm test circuit of Triplett 3423

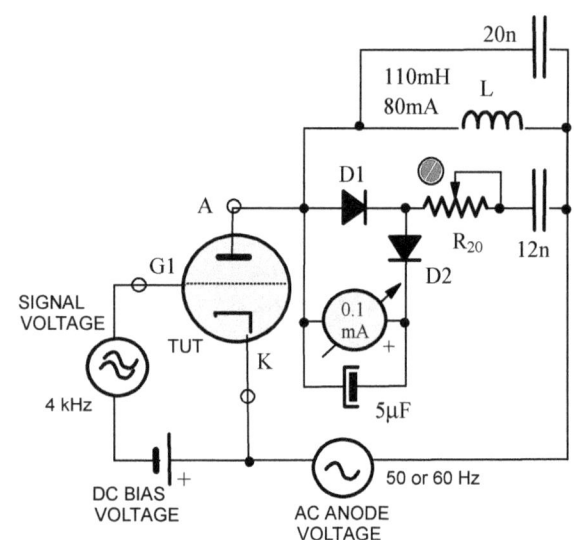

TUBE TYPE	A	B	CDEFG	UP	DOWN	READS
7581	6.3	36	55433	7	258	3250
7591 †	6.3	21	56832	2	567	6630

(Pins 4 & 8 also show short †)

The "Up" position connects the tube's pin to the heater voltage selector switch, the other side of the heater. The other end of the heater is the "Down" bus or COM point, to which cathode, suppressor grid, and control grid CG must also be switched. Pins 2, 5 & 8 are in "Down" position in 7591 example.

The CG will be disconnected from COM by its switch D, while levers for anode and screen must be left in the middle position; that is why pins 3 and 4 are not listed for either "Up" or "Down" positions!

TAYLOR 45D

Taylor Electrical Instruments Limited was incorporated in Jan 1939 and taken over in 1958 by The Automatic Coil Winder & Electrical Equipment Co. Ltd., the makers of AVO meters and tube testers. Model 45D was their last and most modern tester. Two external fuses are mounted on top, flanking the bank of 12 thumbwheel switches, one on the mains side and the other in the anode supply line. Ten tube sockets are neatly arranged in a line, and there is plenty of space for additional ones.

1) A bank of 12 thumbwheel switches
2) Heater voltage selector switch
3) Anode voltage selector switch
4) Screen voltage selector switch
5) Control grid bias voltage (RV10)
6) "Backing off" or "bucking" voltage source and "B" control (RV7)
7) "A" control (RV13)
8) PB1 - "Gas" push button switch
9) PB2 - "Meter" push button switch

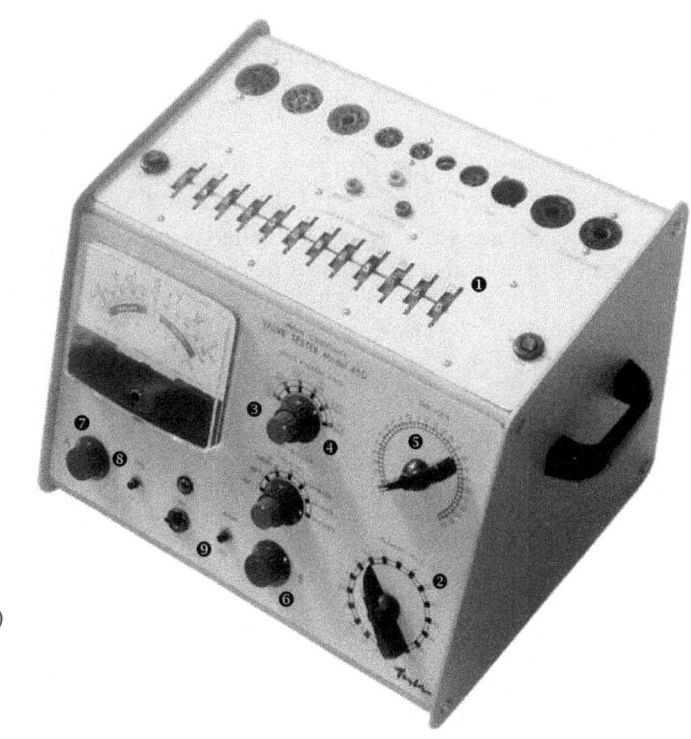

BELOW: 45D circuit diagram, © Taylor Electrical Instruments Ltd.

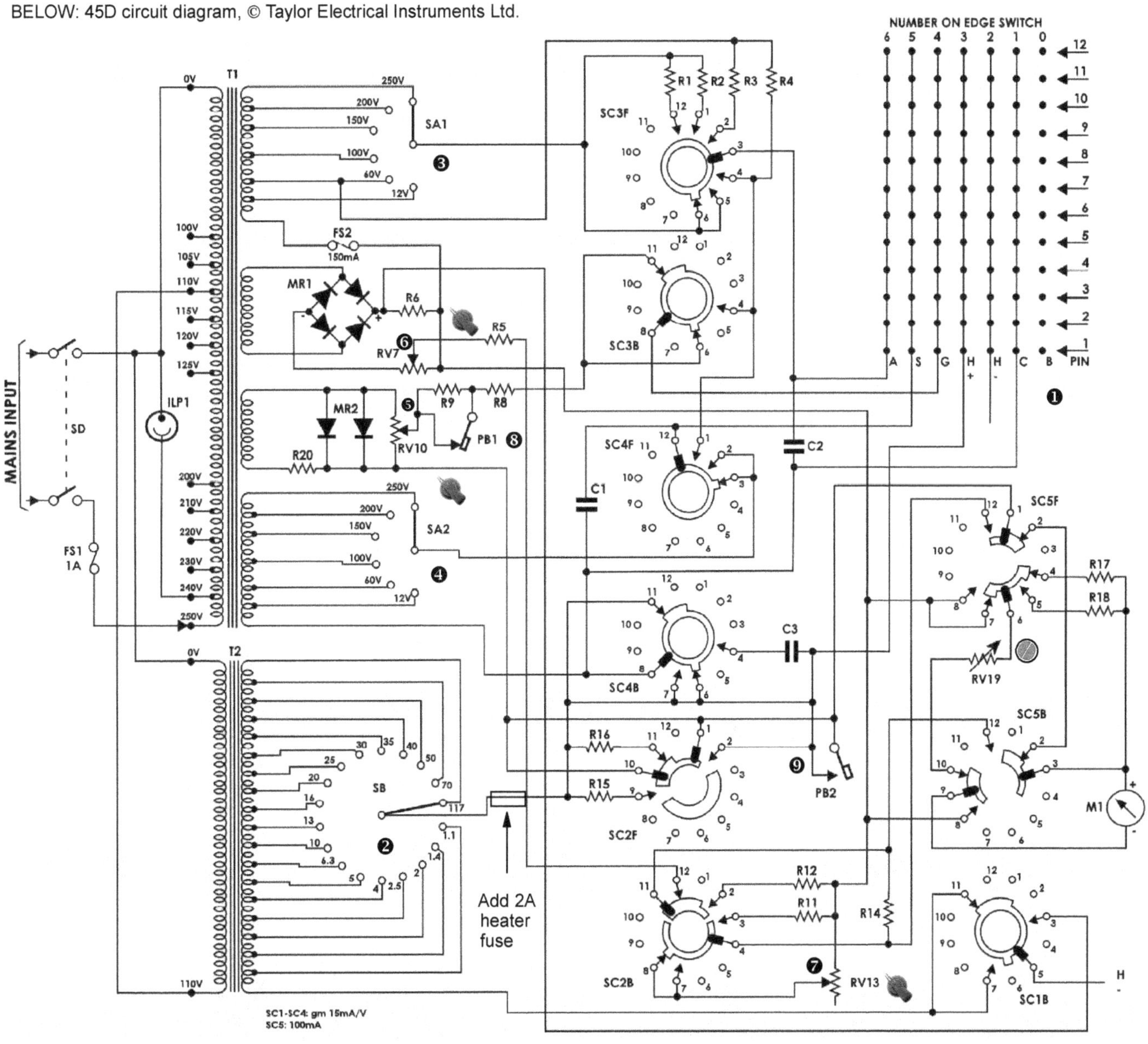

There are five selectable anode and screen voltages in Taylor 45D (12V, 60V, 100V, 150V, 200V, and 250 V_{DC}), so 25 combinations are possible. The adjustable bias voltage, two anode current ranges, 0-10mA and 0-100mA, plus a precise MΩ scale for heater-cathode leakage, make this English tester a pleasure to use. Previous versions (45B and 45C) are similar.

Notice that two separate transformers were used, so the loading of the heater supply transformer does not affect the test voltages. However, there is no "Life" test or "line adjustment."

The leakage & shorts scale is not particularly sensitive, only up to 10 MΩ, but that is enough for most practical applications. There is also a 0-100mA anode current scale and two Gm scales, 0-3 and 0-15 mA/V.

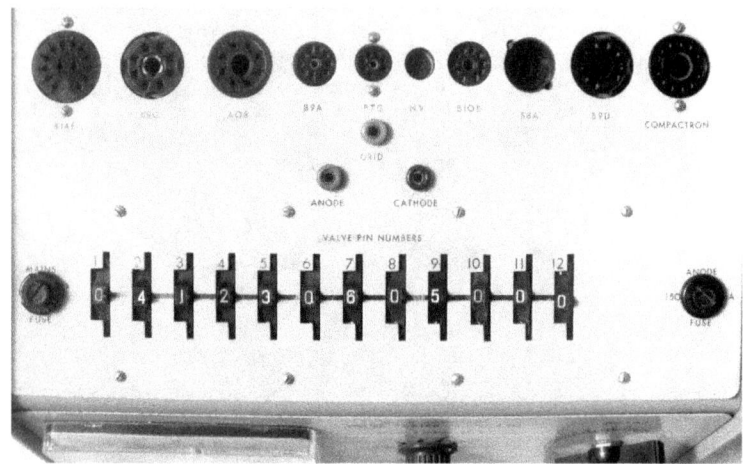

ABOVE: Using both the front and top panels of their testers was a much more efficient arrangement than that of USA-designed vintage testers that used only the top panel and thus required an enormous amount of bench space! As a result, Taylor testers were much more compact.

ABOVE: The sensitive 250µA FSD meter would be difficult to replace (300µA is standard), so its protection by two anti-parallel diodes is recommended.

Operational principle

There is an error in Taylor's documentation. PB1 is the GAS test button on the circuit diagram, and PB2 is the "Meter" test button. Here, in the explanation of its Gm test circuit, the numbers are swapped, so please ignore the PB1 & PB2 on this diagram (below).

The rectifier/diode, filament continuity, and cathode leakage tests are described in the "Operating Instructions," available online, so we will only focus on the mutual conductance test circuit shown below.

The anode and screen windings provide AC voltages which will be half-wave rectified by TUT. The control grid bias is fed negative pulses from the 30V secondary; the shorting diode MR2 cuts off positive pulses (half-waves of the sine voltage). The screen and anode are positive during the negative grid pulses, the TUT conducts, and anode current flows. When screen & anode voltages are negative, the grid voltage is zero, and TUT does not conduct.

Notice the particular way of applying voltages to various electrodes. The screen voltage is applied between screen and cathode, but the grid and anode voltages are applied with reference to the lower end of the cathode resistor (R16 or R15) and not the cathode itself!

That was done so only anode current would flow through the cathode resistor and *not* the cathode current, which is the sum of anode and screen currents. You will see why in just a moment.

BELOW: The operating principle behind Taylor's mutual conductance test © Taylor Electrical Instruments Ltd.

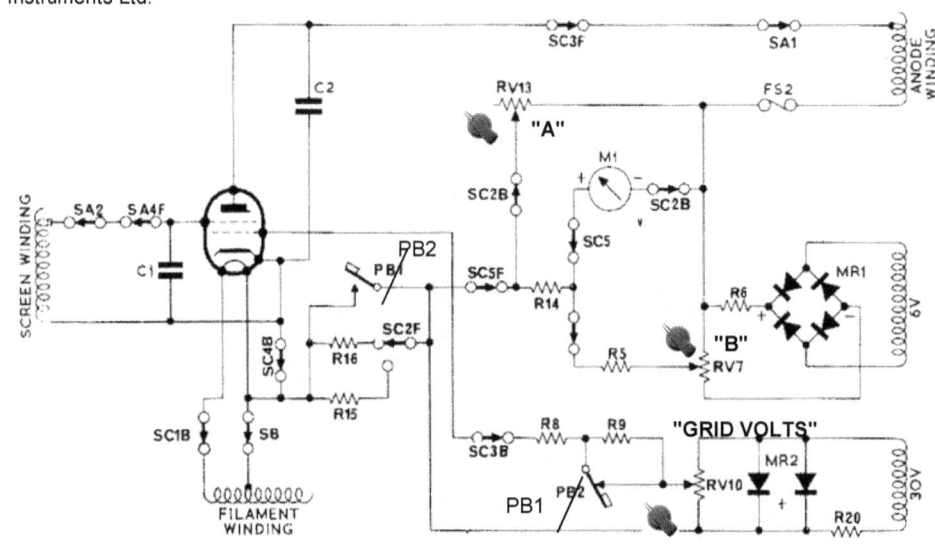

Potentiometer RV13 is control "A." When fully CCW, it shunts the anode current away from the meter, which reads zero (it shorts out the meter). Control "B" is at zero. "A" is then turned CW until the meter pointer is at the X1 mark (at 60% of FSD).

Control "B" is now adjusted until the backing voltage in provides returns the meter to zero. This way, the standing or DC anode current is "removed" or compensated for. For Gm tests, only AC or, more precisely, "an incremental" component of anode current is of interest. This "backing" voltage is a DC voltage provided by the $6V_{AC}$ winding and a bridge rectifier.

When the normally open "Meter" push button (PB2) is pressed, it shorts out the cathode resistor. It thus biases the tube less negatively, which causes an increase in anode current, indicated by the meter on the Gm scale calibrated in mA/V.

The values of the two cathode resistors, R15 and R16, are so chosen that shorting R15 results in the full-scale deflection corresponding to the FSD value of 3mA/V while shorting of R16 (which is active on the 15mA/V range) would result in the increase in anode current equivalent to FSD of 15mA/V.

Now you understand why only anode current was allowed to flow through those two cathode resistors. The screen current had to be taken "out of the picture"; otherwise, the jump in current caused by shorting out the cathode resistor would not be an indication of Gm!

This whole shebang draws mental parallels with Hickok's trick. There the steady or DC current through the tube under test is naturally and automatically "balanced out" by the bridge. However, with grid AC signal present, the TUT would have different internal resistances during positive and negative pulses, unbalances the bridge, and the meter's reading is tweaked or "calibrated" to show Gm.

There is no bridge here, so the steady (DC) anode current must be manually compensated for or eliminated from the reading first. There is no AC grid signal either - a variant of the grid-shift method is used instead.

In reality, it is not the grid voltage that is "shifted" but the cathode voltage. It does not matter if the grid's negative DC bias is reduced (made slightly more positive) so the anode current jumps a bit, or if the positive cathode voltage is made a bit lower, the effect of those bias changes is the same.

The ranges can be doubled by using the "X2" mark (30% of FSD) instead of the X1 mark for setting the anode current, so 0-6mA/V and 0-30mA/V ranges are achieved.

Conclusion

Not quite a lab-grade accuracy, but a competent, cleverly designed, well-built tester. One of the least complex and easiest to understand and repair of all Gm testers, a credit to Taylor's designers. There are no internal tubes and no internal electrolytic capacitors (only three film caps), so not much can age or go wrong.

AVO VALVE CHARACTERISTIC METER (MK III)

The Automatic Coil Winder & Electrical Equipment Co. Ltd. was formed as a private company in 1923, the same year that its first "multimeter" was offered for sale. Named "Avometer" or "AVO," it measured direct current (A), direct voltage (V), and resistance (O for ohms), replacing three separate instruments. Although AC-measuring capabilities were later added, many of its features, including the old-fashioned looks (curved meter opening), were retained right through to the Model 8, in production for an incredible period of 57 years (1951-2008)!

Over the years, the company diversified into other test gear. It acquired one of its direct competitors, Taylor Electrical Instruments Ltd., who also made a similar line of tube testers, as we have just seen.

In principle, from the operational and functionality point-of-view, there is very little difference between MkI, MkII, MkII, and MkIV. MkIII, MkIV, and CT160 (a totally different AVO tester) used a meter with FSD of only 30mA. MkI and MkII meters were 100mA FSD.

Such high sensitivity is bad news about replacement; it isn't easy to find such sensitive moving coil meters nowadays. It seems meter damage and burnout were common with AVO testers since AVO themselves suggested two anti-parallel diodes be added to the meter to protect it from overload.

AVO testers are very expensive on the used market. If a meter is damaged, has a weak permanent magnet, or is nonlinear (out-of-calibration), such an investment is completely wasted. So, extreme caution is needed when buying and transporting AVO testers!

Just as its testing method may seem strange due to its use of the "backing" principle that nulls the anode current reading on the meter (4), the bias adjustment is also done in a peculiar way. You select an approximate value (0-20-40-60-80V) using a 5-position switch (2), and then you add from 0 to 20 Volts via a potentiometer dial (3), so a de facto continuous adjustment is possible.

The heater selector switches are also strange; more mental algebra is needed. The lower figure selected on the "fine" switch (5) is added to the figure selected on the "coarse" switch (6).

While this increases the range of possible heater voltages, it can be dangerous. Just ticking the "coarse" switch one notch clockwise increases heater voltages by full 10 Volts, enough to destroy most tubes!

Two current scales, 0-25mA and 0-100mA, are fine, as is the top short and leakage scale that extends up to 25 MΩ, for very sensitive leakage tests (7).

Two Gm test modes are used. The co called "English" scale gives quick "Replace - ? - Good" results.

If the actual Gm is to be determined, that is done by backing the anode current reading back to zero, turning the "Meter Switch" to "mA/V" position, and rotating the "Set mA/V" knob until the meter indicates 1mA/V, at the center of the "Good" segment (8).

The Gm can then be read from the calibrated markings on the skirt of the "Set mA/V" knob.

The accuracy and repeatability of Gm testing depend not only on the accuracy of the analog meter but are also limited by the imprecise graduations on a plastic skirt of a potentiometer!

The "scale" is not linear; above "7" or "8" you are guessing, not measuring! For that reason, very late in the process, AVO engineers finally added a proper Gm meter on model VCM-163.

ABOVE: The control panel is morbid looking due to the old fashioned black & brown color scheme. The meter is way too small and the scale crowded.

BELOW: AVO Mk III meter faceplate

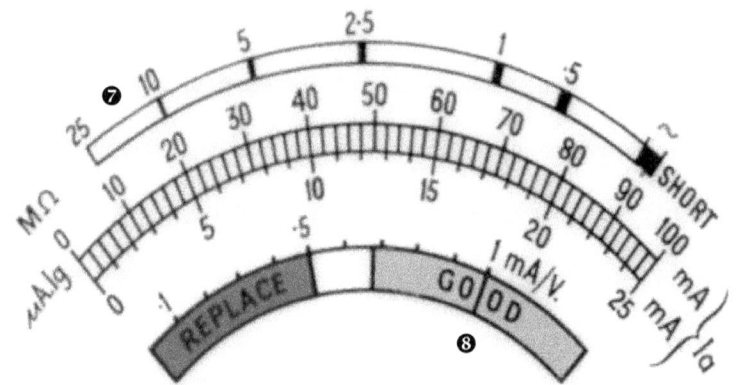

Circuit highlights

A choice of four mains voltages was very wise (1), so the "Line Adjust" in only nine steps is fine (2). The protective relay has three coils, one in the mains line and two in the anode and screen grid supply lines (3). At the current overload or short circuit, the relay trips, its latching contact opens and thus inserts lamp LP1 in series with the primary winding, the old but effective series bulb trick.

Although there are two duo-diode vacuum tubes inside the tester, they are not used to provide DC test voltages for anodes or screens. AC voltages are used for that purpose, as in all proportional Gm testers. The schematics (next page) is of the upgraded version, where solid-state diodes replaced two vacuum duo-diodes.

One diode rectifies DC bias voltage (9), the other is in the screen circuit ("screen stopper rectifier"), in series with the protective relay's coil (3).

In beam power tubes tested with AC anode and screen supplies, screen current can rise in the opposite direction to normal screen current, so these diodes prevent it from flowing backward, out of the tube back into the power supply, which would not only provide false results but could overheat the screen grid and damage the tube.

The same transformer secondary taps supply screen and anode voltages (4). The backing voltage has its own secondary winding (5), as does the bias power supply (6). The resistors around the meter switch (7) are either series (R21-R27) or shunt resistors (R28-R32), switch-selectable for I_A measurements, or D/R (Diode/Rectifier) tests.

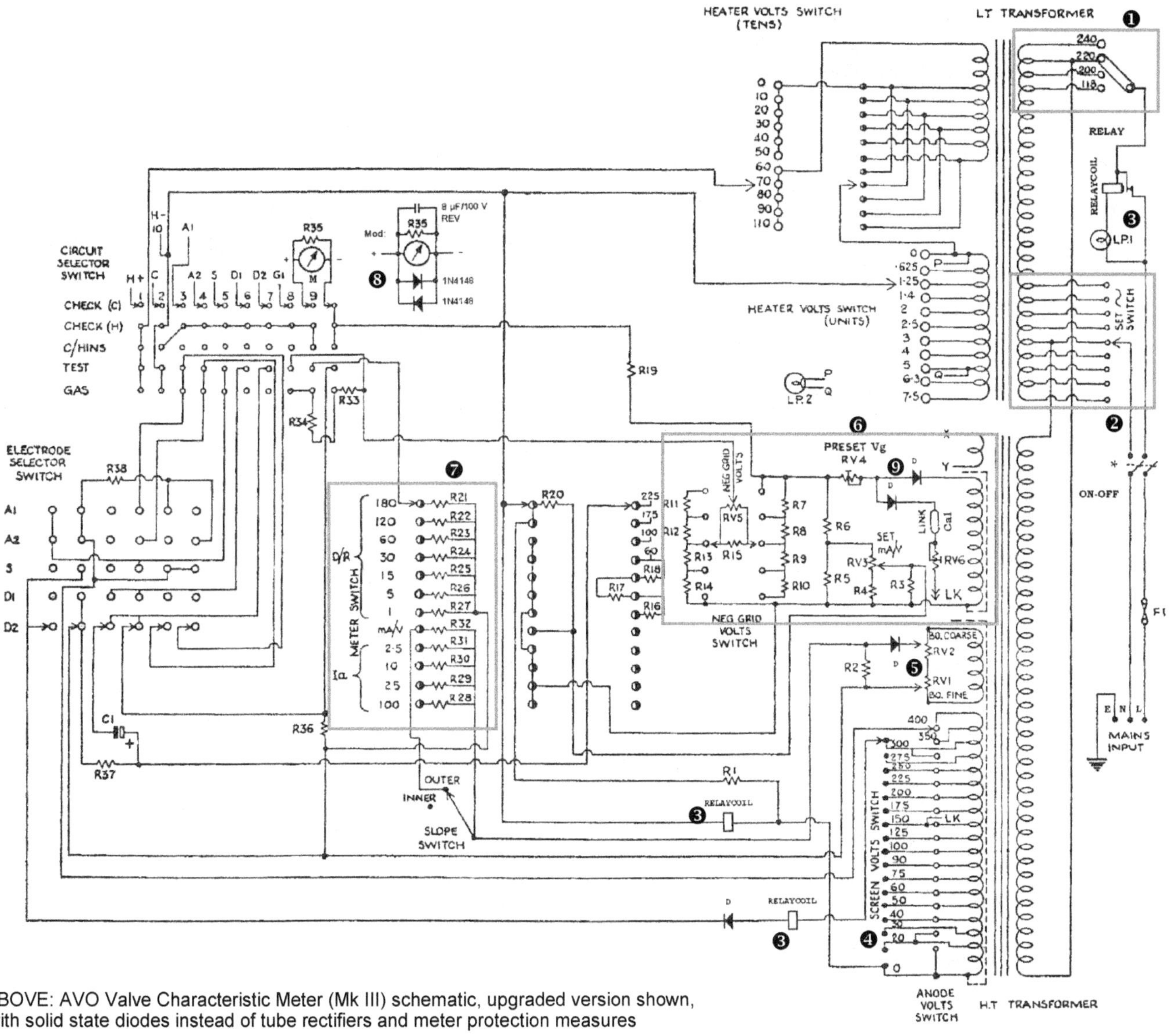

ABOVE: AVO Valve Characteristic Meter (Mk III) schematic, upgraded version shown, with solid state diodes instead of tube rectifiers and meter protection measures

Originally, the sensitive meter had no protection of any kind (8), so two anti-parallel diodes and a damping elco were added in this revision. It is very easy to forget to reset the meter range or zero the "Backing Off Control" after testing the previous tube, so extreme care is needed not to burn out the meter, a major problem with AVO testers.

The 5-position "Circuit Selector" switch selects "C" for cold and "H" for hot (heater on) electrode leakage tests. At "C/H INS," the cathode-heater insulation of the hot tube is measured. The "Test" position for Gm and the "Gas" position are obvious.

Users either love AVO tube testers or hate them with passion. Personally, I find them slow and tedious to use, complicated to calibrate, and, just as with many vintage testers that can measure Gm, way overpriced on the 2nd hand market.

The AVO Testing method

FURTHER READING

U.K. patents 480752 and 606707 discuss the design of AVO valve testers and a few issues regarding measuring mutual conductance using AC and DC voltages. Both can be downloaded from online sources.

A detailed explanation is provided on why DC bias voltages aren't used in such testers (the meter would not indicate accurate Gm figures) and why AC biasing is the only option. Integration of signal waveforms is involved in the proof of the problem, so refresh your knowledge of calculus!

Lampemètre METRIX 310CTR

Based in Annecy-le-Vieux in France, Metrix is one of the rare makers of vintage tube testers still in business today. In 1997 Metrix became part of the Chauvin Arnoux Group, specializing in industrial, laboratory, and educational instrumentation.

Principle of operation

There are no internal tubes, transistors, or other active elements, only one capacitor (in the neon short test circuit), one power transformer, a single analog meter, two toggle switches, 19 rotary switches, and no internal calibrations (trimmer pots) of any kind.

You'd think this is a description of one of those abominations, butt-of-all joke "quick checkers" designed by accountants and made in the dying days of American electronic supremacy. However, nothing can be further from the truth.

Like its predecessors, 310A, B, and C, this very capable tester was conceived by some frugal yet innovative French minds. Strictly speaking, this is not a proportional mutual conductance tester. It uses no grid signal, and its analog meter measures DC anode current, not the AC component. Yet, it's simpler, works better, and is more accurate than most proportional Gm testers.

The basic connection of electrodes is such that AC voltages of various amplitudes (switch selectable) are brought to G1, G2 and anode. Even G3 (suppressor grid) gets its own voltage, something American testers cannot do. That means tubes with a separate G3 pin such as EL34, EL156, and EL12 (all of European origin) can be tested in various ways concerning G3, instead of just having G3 strapped to the cathode as with most other testers.

The tube-under-test rectifies those voltages itself, and a pulsating sinewave anode current flows through the meter calibrated in mA_{DC}. A 4-position switch selects one of four resistors (R5-R8) and connects them as shunts across the 0.9mA/50W moving coil meter.

Primary and heater supply fuses protect the power transformer, as does the circuit breaker in the anode line ("security relay" in French); under overload, its contact disconnects the primary winding from the mains.

Ten different anode, screen, and suppressor grid AC voltages are available (0, 50, 70, 100, 150, 180, 200, 225, 250 & 300V), so numerous combinations can be achieved. The control grid ("grille") bias is adjustable across two ranges, 0-5 and $0-50V_{DC}$.

As is the case with most if not all European testers, the tester uses no roll charts (that is an American approach). Instead, tube manufacturers' data sheets are used to set the controls.

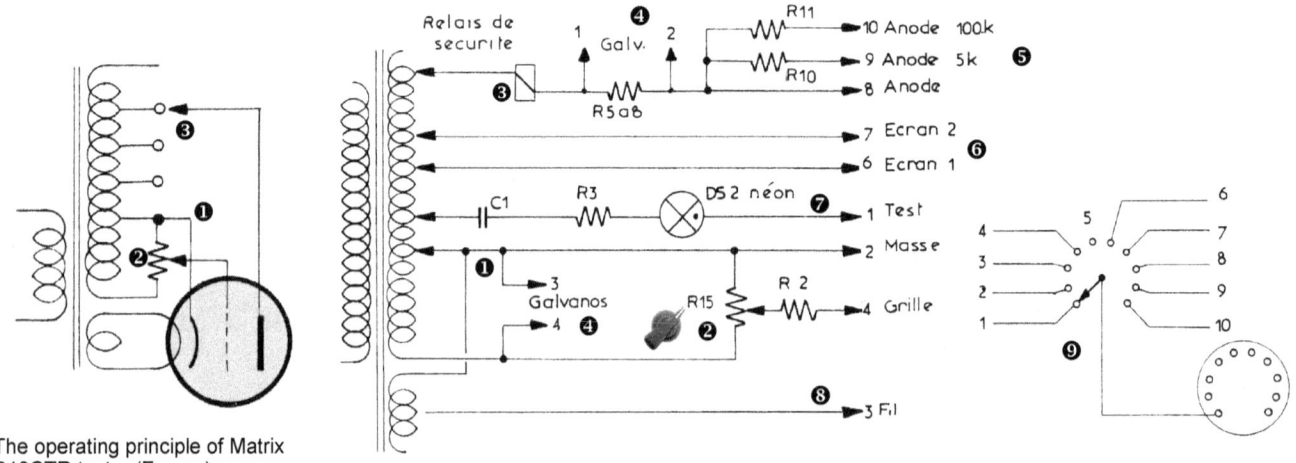

The operating principle of Matrix 310CTR tester (France)

1) GND (Masse) or referent point
2) Bias control
3) Overload protection (circuit breaker) in anode supply line
4) During idle and shorts test the meter indicates "Line voltage" (3-4), during tube merit it is connected across R5-R8 (1-2), four meter shunt resistors, only one of which is connected at a time, in the anode circuit
5) Three different anode connections ("Load" resistors): 0Ω, 5kΩ and 100kΩ
6) Separate voltage selector switches for screen (Ecran 1) and suppressor (Ecran 2) grids
7) Neon "Shorts" test circuit ("Test")
8) Heater voltage selector switch
9) How each of the element switches is wired

Measuring mutual conductance, internal resistance and amplification factor of a tube

Mutual conductance can be determined using the grid shift method. Since Gm or "S" is $Gm=\Delta I_A/\Delta V_G$, the ratio can be calculated by varying control grid bias by a small amount and noting the change anode DC current. For instance, a power tube with a bias of -12.5V passes 42mA of anode current (read on the meter). When the bias is increased to -13.5V, the current drops to 38mA. So, Gm = (42-38)/(13.5-12.5) = 4mA/1V = 4mA/V!

Note: With bias levels under 5V, instead of a 1V change in either direction, a more accurate result is obtained by a +/- 0.5V variation around the mean point. A tube's internal resistance in that test point can be calculated as $R_I=\Delta V_A/\Delta I_A$.

If the initial anode test voltage of 100V was increased to 200V, and anode current jumped from 42mA to 44mA, so $R_I=\Delta V_A/\Delta I_A$, = (200-100)/(44-42) = 100V/2mA = 50kΩ! Judging by its high internal resistance, this is obviously a power pentode. Remember, for pentodes, such a huge increase in anode voltage results in a minimal increase in anode current, providing the screen voltage stays constant.

Now we can calculate its voltage amplification factor using Barkhousen's equation $\mu= Gm*R_I = 4*50 = 200$!

Circuit diagram

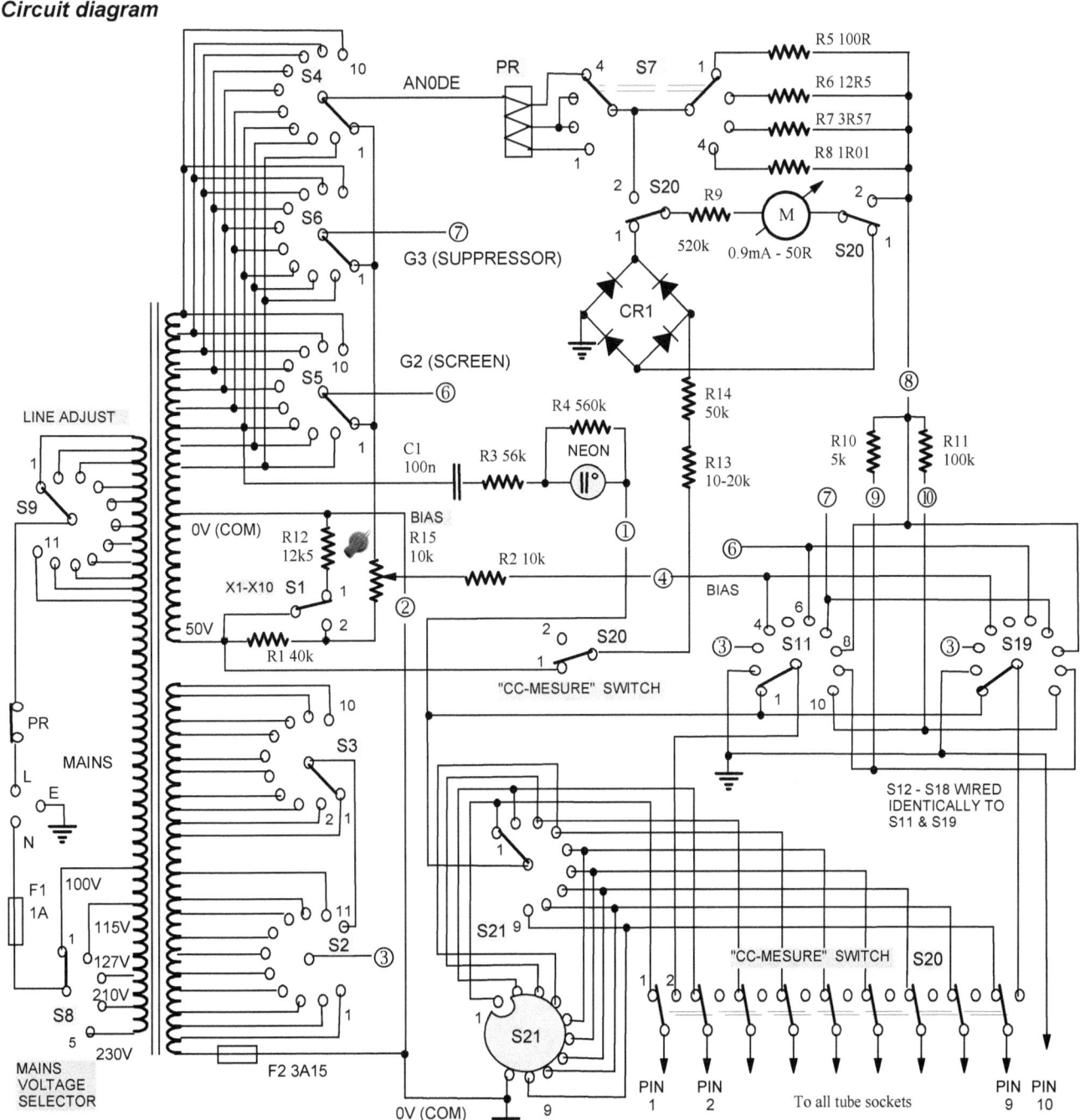

Circuit diagram of Matrix 310CTR tube tester

POS	S2 HEATER SW	S3 HEATER SW	S4-S5-S6 (ANODE-G2-G3) SW
1	1.1	13	0
2	1.25	20	50
3	1.4	25	70
4	2.0	30	100
5	2.5	35	150
6	4.0	45	180
7	5.0	55	200
8	6.3	70	225
9	7.5	90	250
10	10	117	300
11	10V+	-	-

ABOVE: AC voltages for various heater and electrode switch positions

ABOVE: Metrix 310 series are very capable tube testers

Five AC primary taps (100V-230V) and eleven "Line adjust" taps obviate the need for the hot and prone-to-failure "line adjust" rheostats. The "PR" is a contact of the protection relay whose coil is in the anode circuit. If the anode current exceeds the three preset values (that's why there are three "taps" on the overload relay), the contact will open and disconnect the transformer's primary from the mains.

S20 has only two positions but switches many circuits in or out. In position "1", one contact feeds $50V_{AC}$ to rectifier bridge CR1 and the meter, which is now part of the "line adjust" circuit, so S9 is adjusted until the meter indicator lines up with a red mark on the meter's scale.

S21 has two wafers - one shorts all but one of the tube pins, and the other connects that pin to the neon shorts circuit. Switch S20 is still in position "1" as per the diagram.

When S20 is flicked into position "2", S21 is removed from the circuit, as is the $50V_{AC}$ voltage from CR1, meter M is switched over to one of the four shunts (selected by switch S7), and the anode supply through the protection relay as mentioned.

AC bias is used; the bias pot R15 controls it through two ranges, 0-5V and 0-50V.

The element or electrode switches S11 to S19 connect each pin to one of ten points. Position "1" is for "isolated cathode," "2" is GND or grounded cathode, "3" is the filament end (the other end of the filament is at GND), "4" is "POL," which stands for "Polarisation" in French, i.e., G1 bias, in position "5" the pin remains open (not connected to anything), "6" is G2 AC supply voltage, "7" is G3 (suppressor grid) AC supply voltage, "8" connects the anode directly to the meter. In positions "9" and "10," the cathode is connected to the meter via 5k and 100k, respectively.

All in all, a relatively simple but very well thought out and trouble-free design. There is only one transformer, so voltage sags under heavy loads are an issue.

The "backing off" principle, the static Gm test and metering circuit tuned to the grid signal oscillator's frequency

You have now seen examples of how various proportional mutual conductance testers designers approached the problem of measuring mutual conductance. As the subtitle above foreshadows, there are three methods.

Metrix 310 and its class of testers such as German Neuberger and Funke models did not measure Gm but left it to the user to perform the static grid-shift test. By slightly changing the grid bias, the DC anode current would change. By dividing the difference (increase or decrease) of anode current ΔV_A by the change of grid bias ΔV_G, the operator could calculate Gm at that point on the transfer characteristic.

Taylor and AVO, of the British school of valve testing, achieved the same by requiring the user to compensate or null the DC anode current by using the backing off voltage that pushed an equal current in the opposite direction through the meter, and then applying a small AC voltage to the grid and calibrating either the meter's scale or a separate control pot in Gm (since both ΔV_A and ΔV_G are known). This could also be considered a static Gm measurement method, but one that eliminated mental calculation by the user.

Finally, Simpson 330 (to be covered next) and Triplett 3423 solved the problem of eliminating the DC component of anode current from the metering circuit by using a higher frequency grid signal and making the tester's meter respond only to the rectified AC component of the anode current, which is proportional to Gm. That AC component must be rectified since moving coil-meters respond only to DC currents.

SIMPSON 330

Despite its advanced age (dating back to the 1940s), this granddad of testers is a quality instrument, one of the best mutual conductance testers in its day, and still a very competent tester 80 years later. It competed directly with the earlier Hickok models such as 532 and 533. The top panel (molded Bakelite) was almost completely black, including the meter's faceplate. All buttons (except the white reset button), knobs and thumbwheels, and sockets were black Bakelite.

Circuit analysis

The power transformer has seven secondary windings, from left they are heater winding with 17 taps and S-position (for cold cathode tubes), heater winding for 6J5 triode in signal oscillator (2.5kHz), anode supply voltage for 6J5 oscillator, bias voltage winding, anode/screen supply winding ($90V_{AC}$ and $180V_{AC}$), anode supply for 6SN7 meter amplifier, and heater supply for the 6SN7 duo-triode.

Whenever a dedicated oscillator is used to provide a grid signal, you can immediately conclude that you are dealing with a mutual conductance tester. There are no rectifiers or electrolytic capacitors, AC voltages are fed to the grid, screen, and anode, so it is a proportional Gm tester.

BELOW:
1) Signal amplitude calibration (trimmer)
2) Signal frequency calibration (capacitor)
3) Bias level calibration
4) "Bias" external control
5) Output circuit calibration - 0% reading (trimmer)
6) Output circuit calibration - 120% reading (trimmer)
7) "Range" - external control
8) 2.5kHz grid signal frequency calibration (tuning), resistor + capacitor
9) Line adjustment ("Adjust meter to zero")

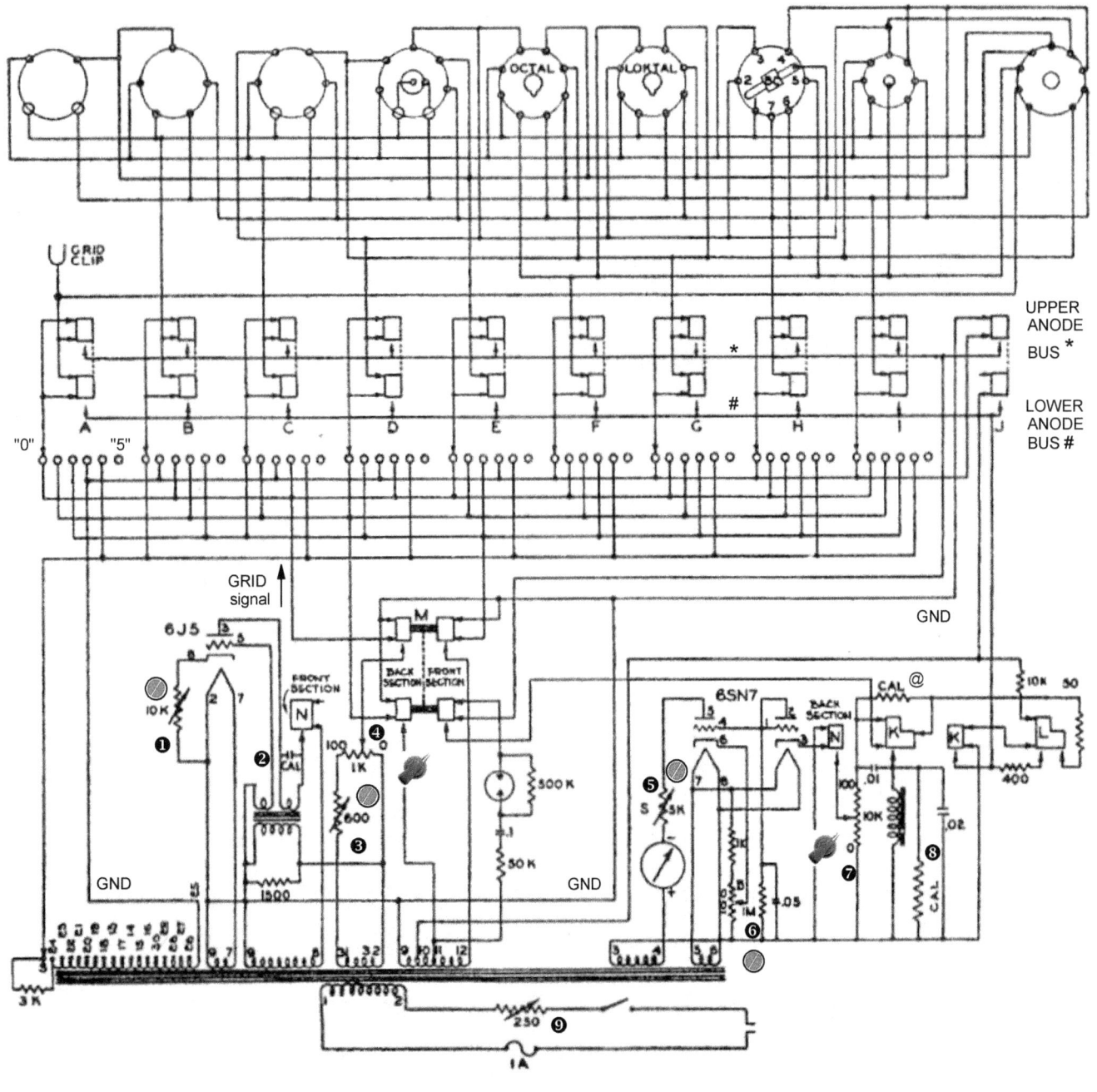

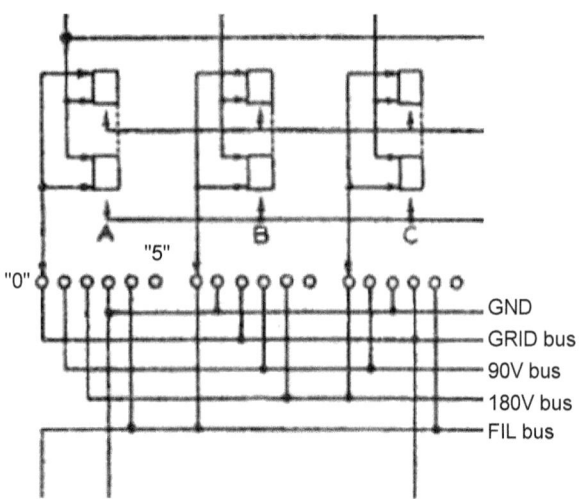
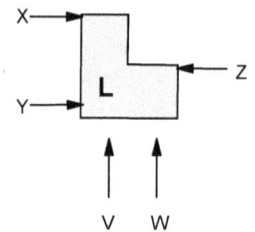
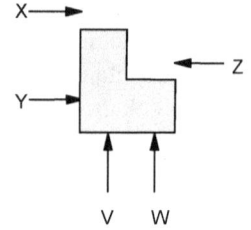

Pushbutton in its normal state: Y connected to X and Z, V and W open

Pushbutton pressed: Y connected to V and W, X and Z open

How to read the diagram

There are five lines or "busses" to which various contacts of thumbwheel switches (TWS) A-I are connected. Each TWS has six positions, marked "0" through to "5". The five busses are (top-to-bottom): GND or common, grid bus (signal + bias voltage together), $90V_{AC}$ anode/screen supply, and $180V_{AC}$ anode/screen supply. The filament buss (the other side of the filament is connected to GND).

MODEL 330
Mutual Conductance Tube Tester

The Simpson Model 330 tests tubes in terms of PERCENTAGE of rated DYNAMIC MUTUAL CONDUCTANCE, a direct indication of tube performance with reference to the manufacturer's STANDARD MICROMHO rating. The colored zones on the dial coincide with the percentage scale to indicate good, fair, weak or definitely bad tubes. Tubes are tested at audio frequency (2500 cycles) with voltages applied automatically over the entire operating range, reproducing more completely than ever before the actual conditions, under which a tube normally functions. A compact assembly of ten push button switches and nine rotary switches of six positions each provide infinite combinations for tube circuit selection.

When you have finished a tube test, ONE BUTTON* returns all switches to the normal position.

Size: 16" x 12½" x 6¾" *With Simpson "No Backlash" Roll Chart*

Dealer's Net Price, complete with Operator's Manual.............**$132.50**

Also, we need to understand the symbol for the push-button switches, also marked with letters A-J. Each TWS is connected to its dedicated PB switch, A to A, B-B, etc. Push-button "J" is different; it does not have its TWS.

Each "square" or "L-shape" represents a moving contact (blade) of a push-button switch. It seems that all 3 or 4 terminals (fixed contacts) of the switch (arrows) are connected to it, but that is not the case. There is a small gap ("g") between the bottom fixed contact and the blade. Using PB switch "L" as an example (left), when not pressed (normal state), L's blade connects fixed contacts X, Y, and Z together, V and W are open (not connected to anything).

Once the PB "L" is pressed, X and Z are disconnected from Y, and Y is now connected to V and W (illustration on the left).

During the "Quality" test, the function of buttons A-J is to connect the right pins to the anode (plate) circuit (anode bus marked *). Buttons K and L select the right load for each tube, a combination of three resistors - 50Ω, 400Ω, and 10kΩ, anode choke, and the "RANGE" control pot. Button M controls the anode, screen, and grid voltage supply, while button "N" is the test button since it is pressed last and switches the meter circuit from "line test" to "tube quality test." When PB "N" is released, buttons K, L, and M automatically return to the open position, a safety feature but one that quickly becomes annoying, since every time you want to test a tube, you need to press "KLM" again and again!

Example: Testing 6L6

Apart from the KLM and N pushbuttons, only two other settings are needed for testing a 6L6 tube. "Selectors: D1" means that screen (pin 4 or "D") will be connected to the 180V bus.

The anode still needs to be connected, which is done by pressing the button "C." Its upper contact connects pin 3 (anode) to the upper bus, and its lower contact to the lower anode bus, which goes to the load circuit.

SETTINGS: 6L6
- Bias: 32 Range: 21
- Selectors: D1
- Quality: CKLM-N

SETTINGS: 12AU7
- Bias: 18 Range: 25
- Selectors: C1, G5
- Q1: DKM-N
- Q2: CKM-N

SETTINGS: 12AT7
- Bias: 0 Range: 30
- Selectors: C1, G5
- Q1: DKLM-N
- Q2: CKLM-N

SETTINGS: 12AX7
- Bias: 0 Range: 64
- Selectors: C1, G5
- Q1: DKLM-N
- Q2: CKLM-N

Example: Testing 12AU7

Except for "Bias" and "Range" figures and different loads used, the test settings for the Noval family of tubes are identical. For 12AU7 only load button "K" is pressed, meaning none of the three load resistors is in the circuit.

For 12AT7 and 12AX7, both K & L load buttons must be pressed, connecting the 400Ω and 50Ω resistors in series and then in parallel with the "Range" control pot (via the upper CAL resistor, marked @ on the diagram).

For the Q1 test (1st triode), pressing D selects pin 6 as the anode, and for Q2 test (2nd triode), pressing C selects pin 1 as the anode. C1 connects 90V test voltage to the lower anode bus, and G5 disconnects pin 9 (heater center tap) and leaves it free (not connected to anything) because 12.6V heating is used between pins 4 and 5, which are already connected as they should be, pin 4 through B0 and pin 5 through I0 to GND (the other side of the heater winding). The two grids, pins 2 & 7, and the two cathodes, pins 3 & 8, are also already connected the right way, pin 2 through A0 and pin 7 through E0 to grid bus, pin 3 through F0 and pin 8 through H0 to GND.

330's manual (available online) gives all the internal connections of pushbuttons and controls in a table, so one doesn't need to trace the connections on its diagram.

Conclusion

Shorts and leakage tests are done through a primitive neon circuit with a low 300kΩ sensitivity, and there is no provision for gas/grid leakage testing, a serious shortcoming of this tester.

Early serial numbers didn't even have a Noval (mini 9-pin socket) so double-check before purchase. Although such a socket can always be installed, the top Bakelite panel is brittle, and caution is required not to crack it during drilling. Punching a hole for the socket is out of the question. The safest way is to remove a socket you don't need and install a required socket in its hole.

Finally, notice that the analog meter's needle rests on the right end of the scale, the opposite of most other moving coil meters. So, if this tester's meter is damaged or burned out, forget about finding a replacement!

HICKOK-TYPE TESTERS

- THE HICKOK BRIDGE CIRCUIT AND HOW IT MEASURES TRANSCONDUCTANCE
- HICKOK TESTERS
- B&K 500
- B&K 550
- B&K 650
- B&K 675
- B&K 700 & 707
- B&K 747
- MERCURY 1000, 1200 & 2000
- PRECISE 111
- PRECISE 116
- DYNAMATIC DM456
- SIMPLE FIXES & UPGRADES FOR HICKOK-TYPE TESTERS

THE HICKOK BRIDGE CIRCUIT AND HOW IT MEASURES TRANSCONDUCTANCE

Job R. Barnhart's US patent titled simply "Tube Tester" was approved on April 30, 1935, under the number 1,999,858. Over the years, it became known as "The Hickok circuit" or the Hickok bridge.

Two identical transformer secondary windings, S1 and S2, with between 100 and 200 V_{AC}, are in two branches of a bridge. One type 83 mercury vapor rectifier (on earlier models) or two solid-state diodes on later models are in the other two branches of the bridge.

A tube-under-test is just like a fixed resistor with only DC bias on the control grid and no AC grid signal. Every half-cycle of the mains voltage, the direction of current through the meter changes, its needle pointer is subjected to equal and opposite forces, and there is no deflection. The bridge is balanced.

With the grid AC signal present, the grid voltage becomes more positive during one half-cycle and more negative during the other half. With a more positive grid voltage, the tube conducts, and its internal resistance drops significantly, so the current through the meter rises. When the signal voltage makes the grid more negative, the tube conducts less current, its internal resistance rises, and the current through the meter drops.

The opposing currents through the meter are not equal anymore. The bridge is unbalanced. Due to its inertia, the meter cannot follow fast current changes; its needle will show a deflection proportional to the average increase in current. It can be shown (see the mathematical proof in Barnhart's patent) that the deflection of the indicator is proportional to the mutual conductance of the tube-under-test (TUT).

A more detailed schematic below shows another rectifier tube, usually 5Y3GT, providing screen and bias voltages for TUT. Another transformer winding, E5, is a source of grid signal, which is selectable, either 1 V_{AC} or 5 V_{AC}. All windings (E1-E5) are secondaries of the same mains transformer.

Every half-cycle, the current flowing through the meter changes direction, depending on which anode is positive, A1 or A2. When A1 is positive, electrons flow from Z (cathode of 83 tube), through E1, R1, R4, cathode K of TUT, to its anode A, through the closed S3, and back to Z.

However, a portion of this current gets diverted from point "X" through the shunt resistor R3, meter M, R2 and lower part of R4 back to cathode K.

With A2 positive, electrons flow from Z through E2, R2, R4, cathode K of TUT, to its anode A, through the closed S3, and back to Z. Again, a portion of this current gets diverted from point "Y" through the shunt resistor R3, meter M, R1 and upper part of R4 back to cathode K.

Notice that the current through meter M is now flowing in the opposite direction. This flip-flopping happens 60 times a second in the USA and 50 times a second in Europe and Australia (whatever the mains frequency is).

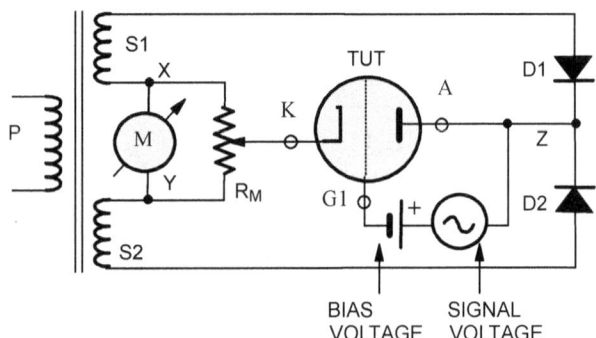

ABOVE: The Hickok bridge in principle
BELOW: The Hickok bridge in more detail, emphasizing the pulsating nature of anode and screen voltage supplies.

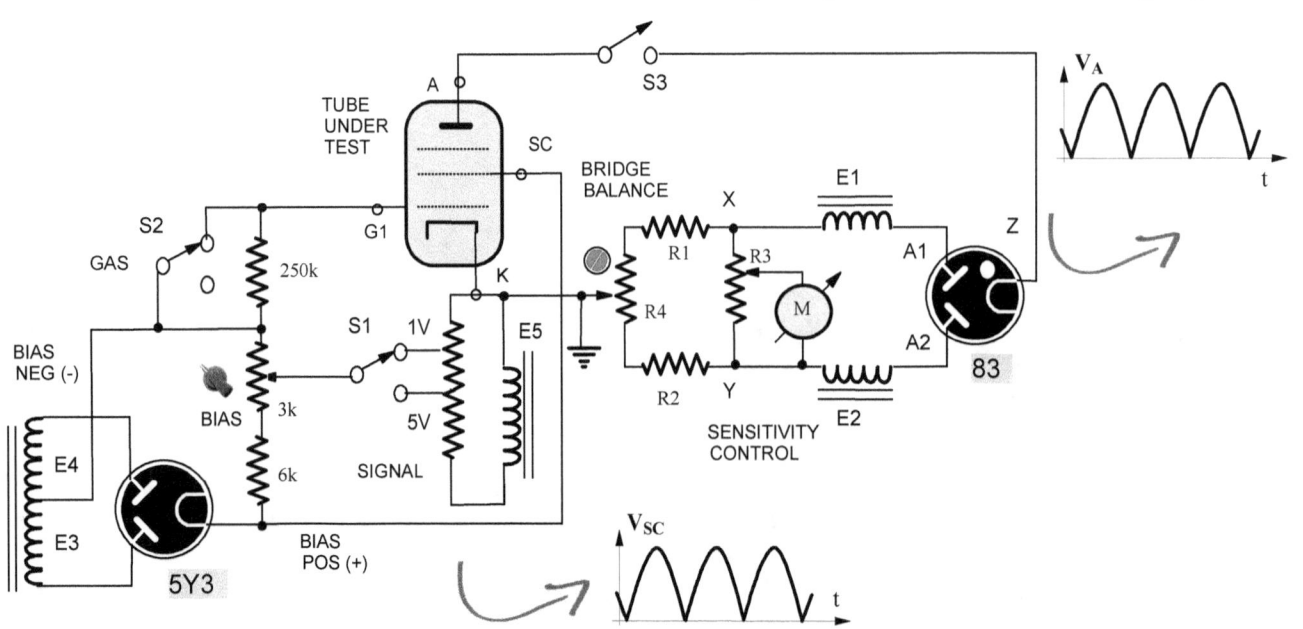

ns testers.

Resistor R4 is an internal trimmer pot that balances the bridge without a signal. Potentiometer R3 is an external "Shunt" or "Sensitivity" control used on Hickok-type testers by Mercury, B&K and Precise, and some Hickok testers.

Gas test

When S2 ("Gas test") is pressed, its contact opens and switches a 250k resistor into the grid circuit. If any grid current flows due to gas or grid emission, it will create a DC voltage drop on that resistor, and the grid will become more positive, causing a meter deflection.

Problems with Hickok's approach to Gm testing

There are a few problems with the Hickok method of Gm testing. First, the two transformer windings have to be balanced. A significant error will be introduced into the measurement if they are not.

Second, if there is only one pair of secondary windings (E3-E4), as it is on most Hickok tester models, all tubes are tested using the same anode voltage. Therefore, the user has no control over the test conditions, and any tester that does not give its users such a control is a mere "quick" tube checker and not a laboratory-type instrument.

Third, nothing is regulated or stabilized here. As the mains voltage varies, so will all secondary voltages, E3, E4, signal voltage E5, the bias voltage, the screen voltage, and the anode voltage. Again, setting "line adjust" to one value will do nothing to prevent or compensate for these fluctuations.

Fourth, because the signal is scaled down mains AC voltage, any leakage between the heater and cathode or any other electrode of TUT will introduce hum. Such hum voltage will be added or subtracted (depending on its phase relationship with the signal voltage), thus increasing or decreasing the mutual conductance figure indicated on the meter, again seriously reducing the accuracy of the test.

This introduction from Triplett 3423 manual outlines the four main reasons Triplett Gm testers are better than their Hickok competitors. They are not just marketing fluff but genuine issues Hickok testers suffer from.

RIGHT: Hickok bashing from Triplett 3423 manual

1. This refers to the Hickok bridge circuit, which is a clever trick, far removed from the standard way of testing tubes for Gm.

2. The Hickoks work on mains frequency only (50 or 60 Hz) and can pick up stray electromagnetic fields, which impact the results.

3. Most Hickoks use only one plate voltage (usually 150-170V), much lower than the maximum test voltage on Triplett 3444 (250V_{DC}).

4. Older Hickoks use an incredibly high grid signal of 5V! 0.6V in Triplett 3423 is a much better choice for preamp tubes. Triplett 3444, a better tester than 3423, uses four grid signal levels, 33mV, 100mV, 0.333V and 1.0 V, way superior to highly revered Hickok 539C, which only goes down to 250mV, and only on higher Gm ranges!

5. This point isn't mentioned by Triplett, but, in most Hickok testers the user needs to change lots of switch settings to test dual tubes, while on Triplett all that is needed is a simple flick of the test lever switch from T1 into T2 position. That saves significant time and prolongs the life of switch contacts and increases the reliability and longevity of the instrument.

> Mutual conductance has been a common rating applied to tubes for many years. Unfortunately, the equipment required to give actual mutual conductance readings at specified operating conditions has been so complicated and bulky that testing has been confined to manufacturers and laboratories. Portable testing equipment generally has been confined to elementary tests or to hybrid circuits which obviously do not duplicate conditions under which the tube was originally rated.
>
> To provide a portable tube tester which more nearly duplicates the basic requirements of mutual conductance, the Model 3423 has been developed. The circuit consists of a straight forward system of applying an audio signal (4KC) to the grid and detecting the effects of this signal in the output by a
> ❶ tuned indicator. Hybrid or trick circuits are thus eliminated and the deviation from laboratory practice essentially reduces to the proportion of ideal conditions imposed on the tube. While each tube may not operate at identical conditions to those of the original rating, it is possible to produce conditions which yield proportional readings.
> ❷ The use of a 4KC signal on the grid and tuning the output indicating circuit eliminate the possibility of false readings from pick-up of the 60-cycle line frequency. Actually, it has been found that with this method 60-cycle voltages can be applied to the plate and screen without disturbing the proportional readings.
> ❸ Five different plate and screen potentials are used which permit closer approximation to original rating conditions than is possible in hybrid circuits where only one voltage is available.
>
> The detecting circuit is isolated from the normal plate current since it is responsive only to the 4KC component. Tubes of widely varying plate current requirements therefore can be effectively tested.
> ❹ Two grid signals one of .6, the other 1.2 volts are employed. Overloading of the grid is avoided while retaining an extremely wide range of Gm readings.

The bias pot adjustment is not linear. For instance, at 20% of the dial, the bias voltage is -3V, while on 100% it is -40V. The only way to be sure about the bias voltage during testing is to monitor it with an external DC voltmeter.

Another inconsistency and a source of confusion is that Gm figures specified by Hickok on smaller (basic) testers' charts were expected average readings for "bogey" tubes, while those on models such as 539 series and 750/752 were the lowest acceptable limits or reject points.

If you study the more detailed depiction of the Hickok bridge a few pages back, you may notice how the cathode of the tube-under-test is connected with regard to the screen grid and the bias voltage. Since the cathode is practically connected to the wiper of the "Bias" pot, as grid bias increases, the already low (compared to real-life situations in amplifiers) screen-to-cathode voltage will be reduced by the same amount!

Triplett's fluff

After all this criticism of Hickok's way of testing tubes, just in case you thought I was some kind of die-hard Triplett fanatic, let me show you how even reputable companies can not only "stretch the truth" but print outright lies. Triplett 3444 manual specifies "Twenty-three AC voltages, closely regulated" (our underline).

The heater AC voltages in Triplett 3444 (or any other tester for that matter) are not regulated in any way, let alone "closely" regulated! They will vary as the mains voltage varies and will drop depending on the heater current and affect the Gm indication.

Just because you cannot see it does not mean that Gm or anode current do not fluctuate during testing, so there is an impact on Gm test results!

If you install the anode current measuring facility (as described in this book) and use a precise digital multimeter to measure it, you will see digits fluttering all over the place. The same is happening to the Gm indication; it's just that the inert and slow responding analog meter does not show it.

HICKOK TESTERS

Early post-WWII units and basic models

A few Hickok tube testers were made before WWII; however, they are now too old and were not very good even when new. Even model 533, released in 1948 and followed in 1951 by 533A, used only one signal level, one anode/screen voltage, and three Gm ranges (0-3,000, 0-6000, and 0-15,000 micromhos), so it cannot test modern tubes with Gm above 15mA/V.

Apart from a similar test "engine" to model 533, models 534 and 534A have a multi-tester to measure voltage, current, resistance, and capacitance. Such low-spec multimeter test features are now obsolete, so these models are not recommended due to their increased complexity and even messier internal wiring.

There is also a neon shorts test and noise test terminals so the tube's noise and microphony (if any) could be heard on headphones.

RIGHT: : The top panel of our Hickok 533A (benchtop version) said "533" (1), but the meter's scale was marked "533A", and, indeed, the internal components were as per 533A diagram and parts list!

We installed recessed insulated test sockets so anode current could be measured (2).

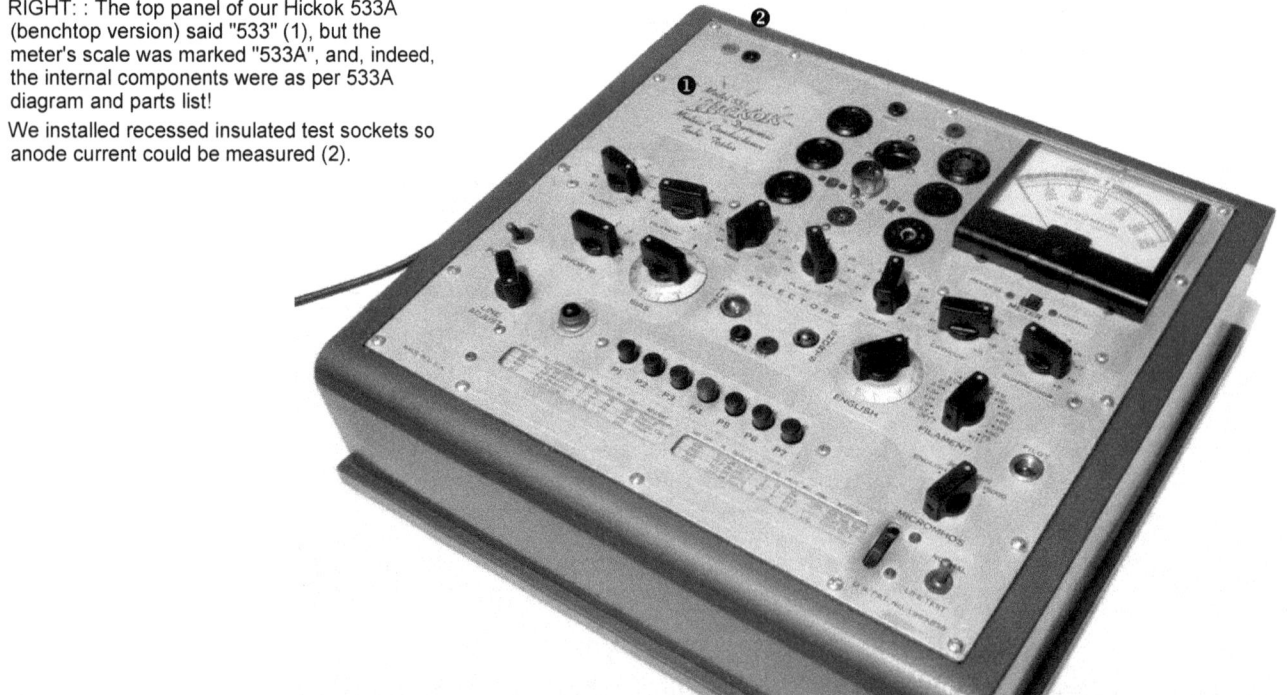

The 600, 800, and 8000 series by Hickok are the most common Hickok testers available for sale. They include models 600, 600A, 605, 605A, 800, 800A&K, 6000 and 6000 A&B. The accuracy of these models is +/- 15%, so they are only quick checkers, far from serious testers.

Models 6000 and 6000A tube testers are updated versions of 600 and 600A, with transistor testing capability and a different way of socket mounting.

All sockets are mounted on a common base or sub-chassis, which in turn plugs into one base socket between the "Bias" and "Shunt" controls (photo below). Notice how the socket base overlaps the two controls and makes their operation awkward (1), one of those "What were they thinking?" moments!

Model 6000 has 4,5,6,7 pin, 7 & 9 pin miniature, octal and Loctal sockets, so it's better for checking older tubes, while 6000A has 7 & 9 pin miniature, octal, Compactron, and nuvistor sockets, thus geared towards more modern tubes. Model 6005 has an integral analog multimeter (voltage, current, resistance, and capacitance).

Like most other tube tester makers, Hickok testers use a grid signal that is way too large, $5V_{RMS}$ on models 533, 600 & 605, and $2.5V_{RMS}$ on 533A, 600A, 605, and 800, so test settings and results are not compatible.

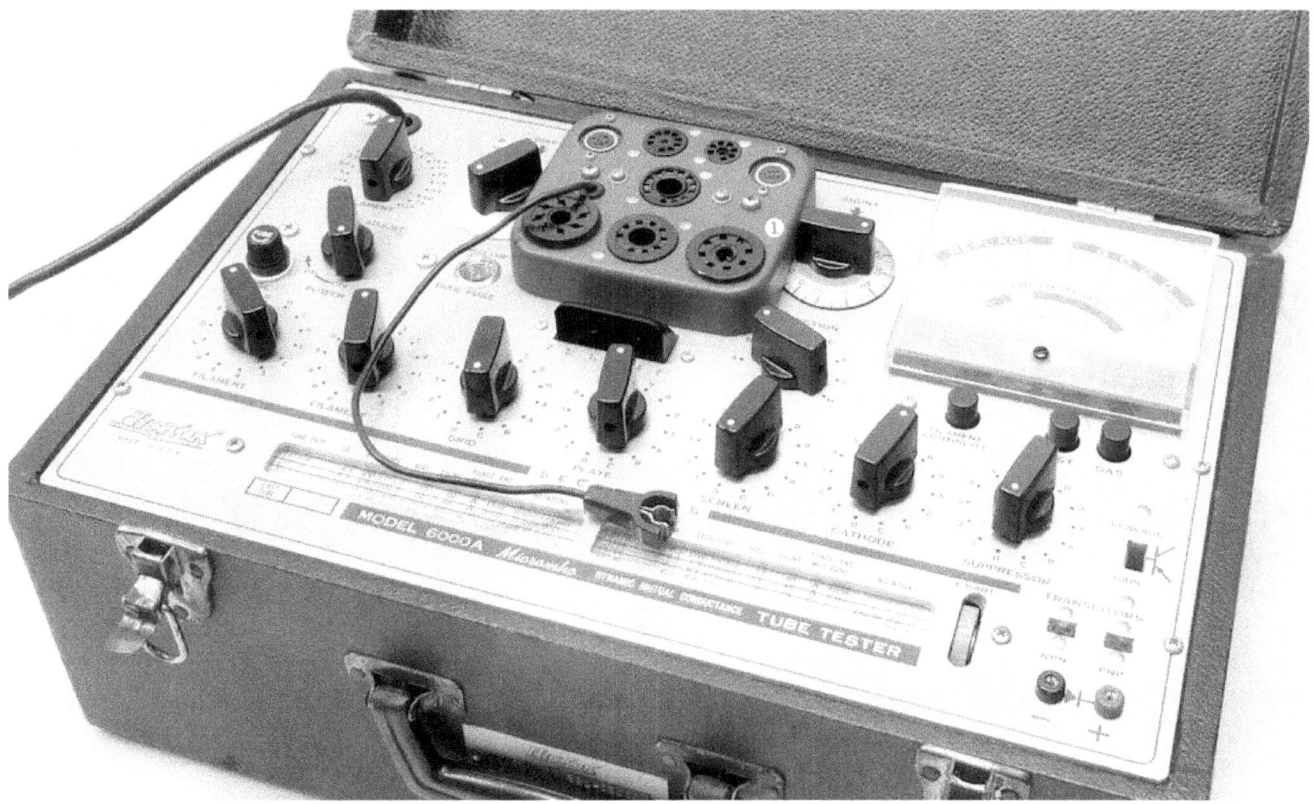

750, 752 & 752A

While not as good as 539B & 539C, Hickok 750, 752 and 752A had five Gm ranges, up to 30,000 micromhos, two anode (75 & $150V_{DC}$) and screen voltages (65 & $130V_{DC}$) and four signal levels (from 0.25V to 2.5 V_{AC}).

To test duo-triodes, most Hickok testers required selector switches to be reset. Thankfully, on 752 & 752A, that was not necessary any longer; both sections of duo-triodes could be tested with only one setting. While model 750 uses the primitive "yes-no" neon shorts test circuit, some later 750 units and 752 & 752A had a high voltage megohmmeter for such tests and could display the leakage figure on the main meter's MΩ scale.

There is only one power transformer and no AC meter for line voltage monitoring as on 539B&C.

539A, 539B & 539C

These models are the most sought-after Hickoks and fetch US$1,500-2,500 on eBay, an incredible price for an instrument that wasn't great even when new, and now, 50+ years later, with all those drifted and out-of-tolerance components, is even less accurate. The prices increase in alphabetical order, with 539C the most expensive.

The main attractions are seven test signal levels and Gm ranges, up to 60,000 micromhos, a separate meter for monitoring the line voltage while line adjustments are performed, and a bias voltage meter.

However, the lowest test signal is 0.25V, still too high for some low-bias tubes such as 12AX7 and 6DJ8 (ECC88), and that signal is only used on higher Gm ranges, not for those particular tube types.

As for the tiny AC meter for monitoring the line voltage, that was meant to instill confidence in users that the results were stable, repeatable, and consistent. They are not! The mains voltage will fluctuate during testing, so the whole thing is pointless. Only fully regulated plate, screen, grid, and heater voltages will ensure reliable and repeatable results. However, very few commercial vintage tube testers included all of these.

Hickok 799 ("The Mustang")

Although The Mustang is not a typical Hickok tester, we'll start with its test "engine" since it is simpler and its diagram is easier to follow and comprehend than the more standard models that use #83 and 5Y3 rectifier tubes.

Model 799 is not typical for three reasons. First, it uses solid-state rectifiers. Second, it has no switching panel, meaning its tube sockets are prewired in a fixed configuration, just like B&K's Dyna-Quik testers.

Finally, there are no separate anode and screen grid voltages; all tetrodes and pentodes are tested as triodes.

The Hickok bridge is at (1), CR3 and CR6 are the two silicon diodes that replaced the type 83 rectifier tube.

Notice a very low single test voltage of 130V$_{RMS}$!

Rectifier diodes CR1 & CR2 replaced the 5Y3 rectifier tube in the bias power supply (2).

The secondary winding (3) provides a signal voltage of 3.1V$_{RMS}$, which then gets scaled down through R2 ("Sig. Cal."), and the resistive voltage divider R11-R12 to 2.5V.

Again, a very high signal cannot properly test low bias tubes such as 12AX7 and 6DJ8.

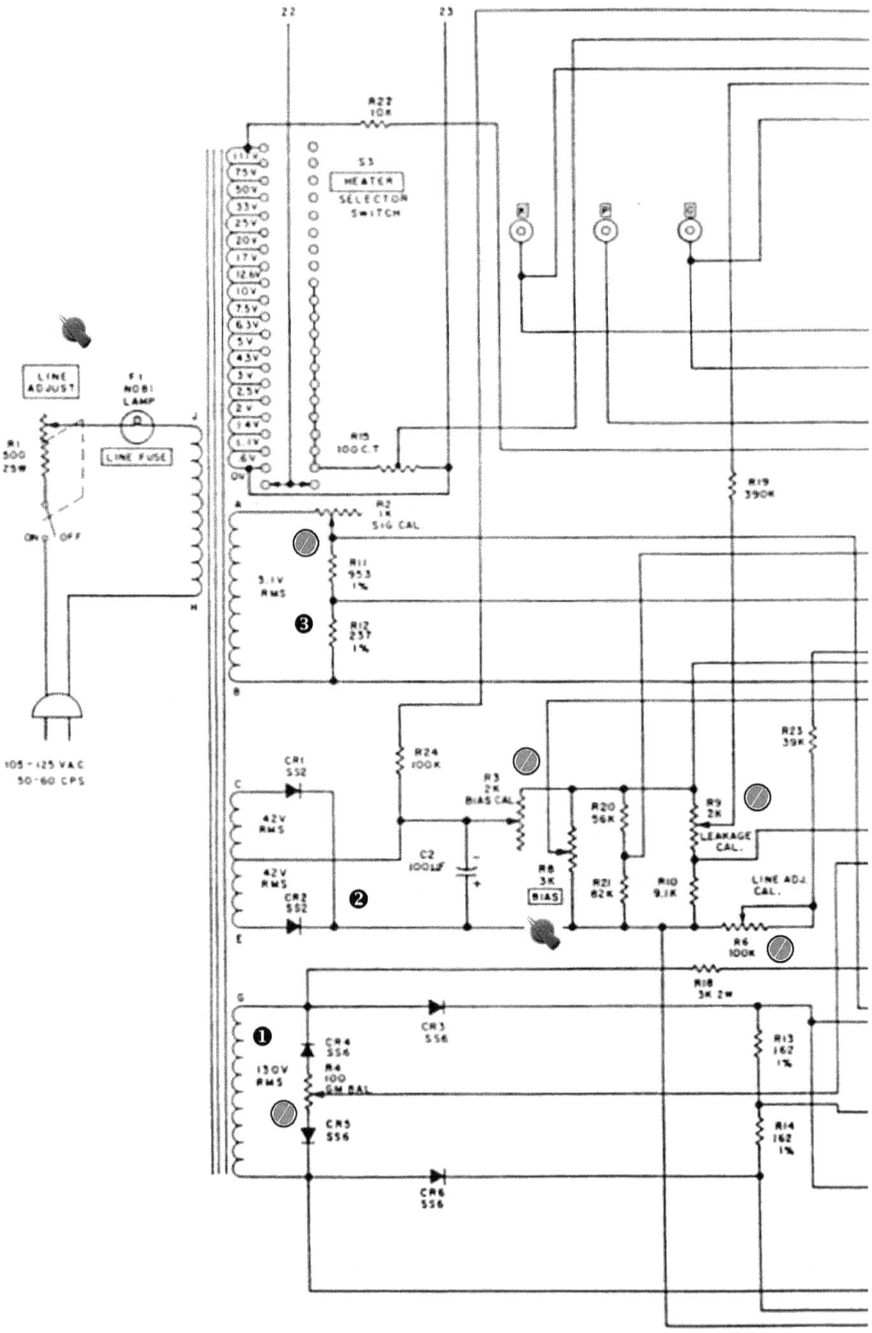

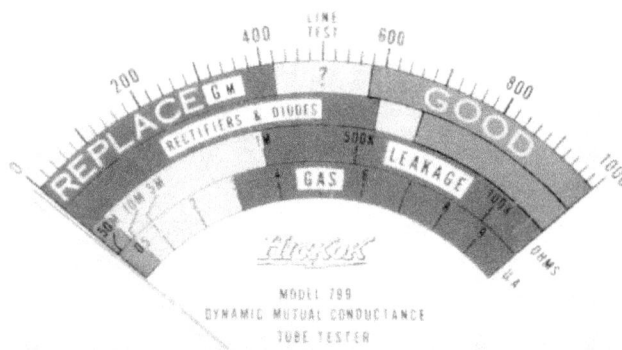

ABOVE: The power supply transformer, Hickok bridge and the associated circuitry of Hickok 799 tester © The Hickok Electrical Instrument Co.

LEFT: The meter scale is certainly colorful in red, yellow and green, plus there are separate "Gas" and "Leakage" scales and "rectifiers & Diodes" indicating segments (could not be called a "scale" since there are no graduations or values marked).

HICKOK-TYPE TESTERS

Hickok 580A

By 1964, when model 580 was released, Hickok had been in the tube testing business for almost 40 years. 580A followed six years later, in 1968. You'd imagine their engineers had learned a thing or two in all that time, so let's see what they came up with. It is certainly Hickok's most modern tester, but is it the best one?

On the plus side, three separate power transformers were a good start, one supplying heater voltages, one for the anode power supply (the Hickok bridge itself), and one for all other voltages, namely screen HV supply, bias supply, and signal voltage. That way, a high load on one transformer would not cause voltage droop on others.

A much smaller signal voltage of $0.28V_{RMS}$ was used with 12 switch-selectable screen & anode voltages, ranging from $6.3V$ to $300V_{DC}$.

Just as in Triplett 3444, a 0-5 V and 0-50 V bias adjustment was now possible and, as in 752 & 752A models, duo-triodes could finally be tested with only one setting of the selector switches.

Alas, even this final attempt by Hickok in the dying days of tube testers was inferior to Triplett's 3444, not to mention the even better 3444A! We've never had the pleasure of scrutinizing model 580 on our test bench, but according to many tube tester experts, including Alan Douglas and the late Chris Haedt, 580A suffered from a long list of issues.

While the earlier Hickok's testers at least worked reasonably well, apparently, model 580A never did. The zero point and grid signal weren't stable. High-frequency oscillations were common and, together with heater hum, affected Gm readings. Some testers could not be properly calibrated to the required signal and voltage values.

BELOW: 580A partial circuit diagram © The Hickok Electrical Instrument Co.

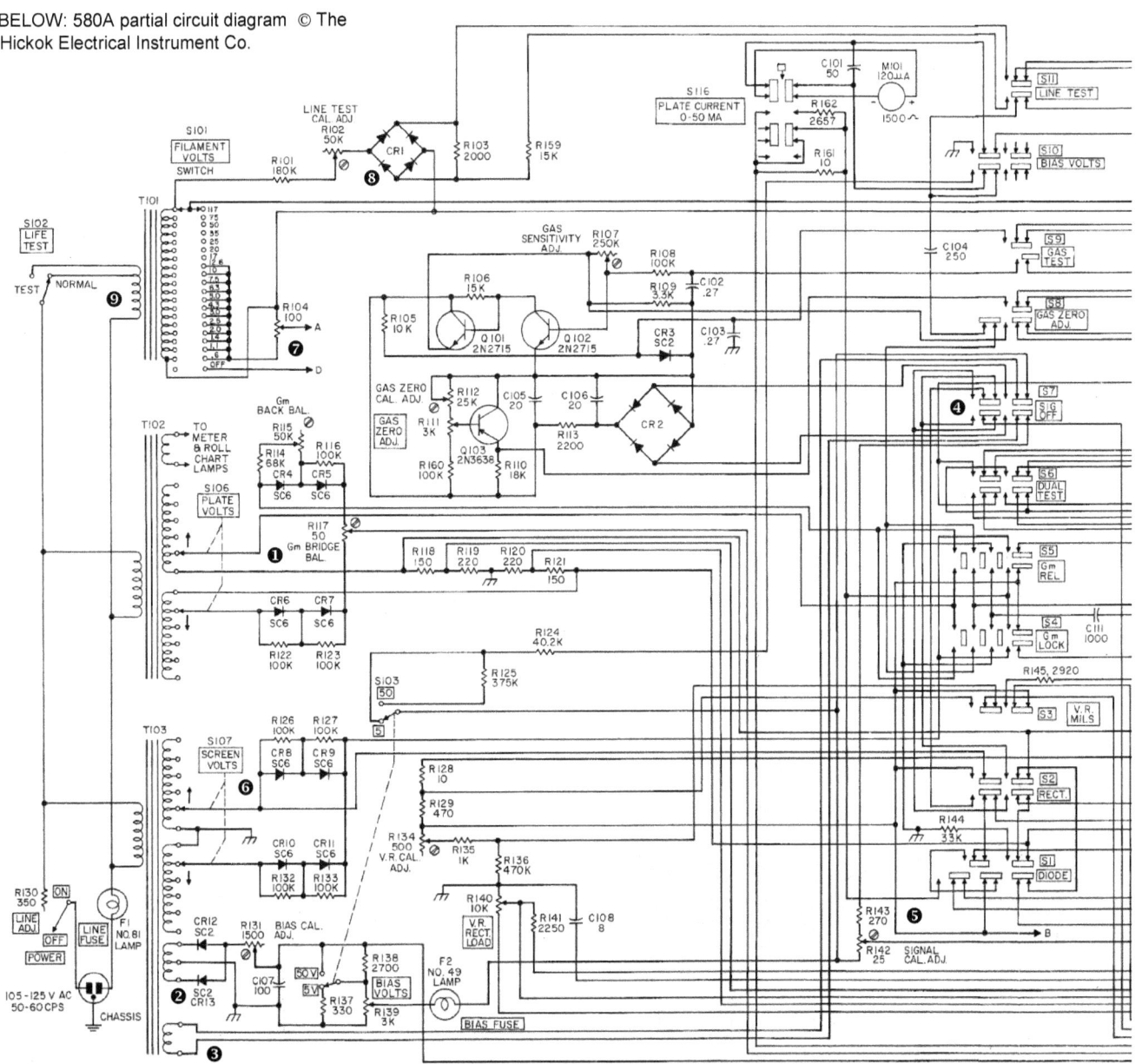

The Hickok bridge balanced at one anode voltage would be unbalanced at another. This wasn't an issue when Hickok testers used only one pair of windings to provide high voltage; the bridge could be manually balanced. Now, all pairs of taps had to be balanced (have the same number of winding turns between them), which obviously wasn't the case. It seems either Hickok's transformer suppliers had never heard of bifilar windings (that would ensure perfect balance for all taps), or Hickok did not want to pay for the additional cost of such transformers.

As in "The Mustang," we see heaps of silicon in the main bridge (anode supply) at (1), the screen grid (6), and bias power supplies (2). The secondary (not marked in any way) at the very bottom is the signal source.

If you follow its connections through the "Sig. Off" switch (4) and "Signal cal. adj." at (5), it is a fairly simple circuit."Filament center"(7) is akin to the "Hum balance" adjustment on some vintage tube amps, where a user could find a point of minimum hum by adjusting the rheostat and, instead of grounding one of its ends, grounding one point between its extremes thus reducing hum. Here it was done not for aural reasons but because in Hickok testers, hum seriously affects test results.

The "Line test cal. adj." adjusts the AC voltage at the input of the CR1 diode bridge that feeds the meter (8). The "Life test" changes the number of primary taps the mains voltage is feeding (9) and affects only the heater transformer T101. In the "Test" position, the same mains voltage (105-125V) is applied across more primary windings compared to the "Normal" position, thus reducing all heater voltages by a certain ratio or percentage.

The gas/grid emission test circuit is unusual. Most testers switch a high resistance of a few MΩ into the grid circuit of the tube-under-test and measure the voltage any grid current would develop across it, either by a triode or FET. Both are voltage amplifying devices, so they detect grid current indirectly by a voltage drop it creates on such a high-value resistor. Here Hickok used a bipolar transistor's base to detect grid current directly, amplify it, and display it on the main meter.

Normal transconductance test voltages are applied to the electrodes (cathode, screen, and anode) of the tube under test (not shown on the sketch), as is the grid bias supply.

"Gm lock" and "Gas zero adj." switches must be pushed to apply those voltages and keep the meter in the bridge circuit.

If there is any grid current present as a result of gas, leakage, or grid emission, it will flow straight into the base of the NPN transistor, a current amplifier whose collector and emitter currents will increase proportionally (depending on transistor's current gain or b). Its internal resistance will change, the bridge will become unbalanced, causing current to flow through the meter.

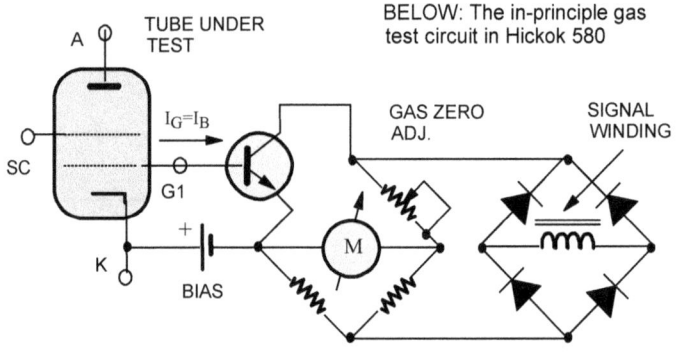

This current is proportional to the degree of imbalance, which reflects the amplitude of the grid/base current and the seriousness of the problem (gas, leakage, or grid emission).

B&K 500

Dating back to 1955, this was the first in B&K's line of "Dyna-Quik" Gm testers. There is only one Gm position of the test switch (SW-3 on the circuit diagram), meaning duo-triodes are tested in parallel. In that case, their mutual conductance figures add up, so if each triode has Gm of 1,200 micromhos (1.2mA/V), the tester should indicate 2,400 micromhos (2.4mA/V). There are two scales, 0-6,000 and 0-18,000 micromhos.

The first version used a 6AX5 rectifier tube; the later ones opted for 83 rectifier, as in most Hickok testers. Selenium rectifiers were used for bias and screen supplies. A 6AT6 duo-diode+triode tube was an amplifier for gas/grid emission tests, but its two internal diodes were unused.

Older versions had no "Bridge balance" adjustment (R-19).There were only 30 prewired sockets on model 500, so the 510 panel was added with additional 16 sockets, but even that was not nearly enough, so some units were shipped with a switchable universal panel, model 610. Models 650, 700, and 707 followed within the next decade, five similar models in ten years.

The first limitation (parallel tube tests) can be overcome by installing an additional "Tube 1-Tube 2" switch (1), marked SW-5 (diagram on the next page). Anodes must be separated at the two sockets that test duo diodes (rectifiers) and the two sockets for duo triodes. Additional wiring runs are installed for the separated anode pins.There is no provision for measuring anode current during tests, so a 10Ω resistor R_X can be added in the "plate" line with two test points (TP) brought out to binding posts (2).

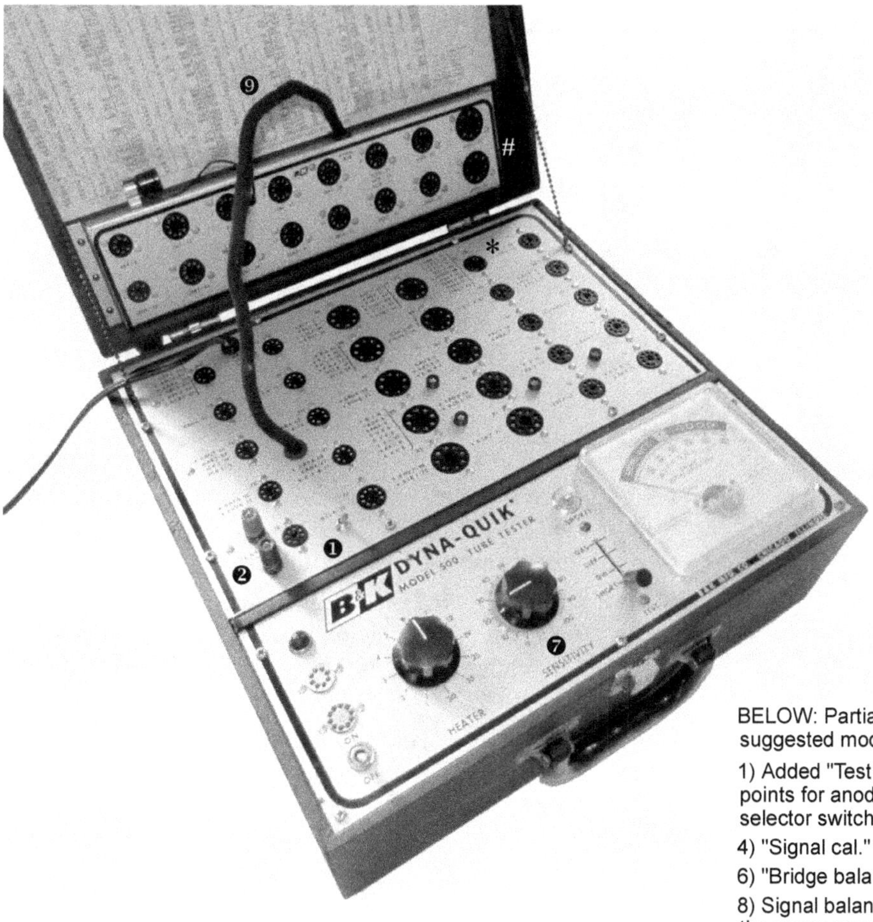

HICKOK-TYPE TESTERS

Instead of three fixed bias voltages (terminals 5, 6 &7), a switch can be installed (SW-4) that will enable each tube to be tested at any of the three bias levels (-2.5, -7.5, and -13V_{DC}).

The annoying thick & ugly umbilical cord (9) could be repositioned to exit the 510 panel at # and enter the main chassis at *, thus shortened and moved out of the way. This kind of thoughtless design is common in the tube tester field.

Notice that diodes are tested with around 200V_{AC}, but high voltage rectifiers such as 5U4 with only around 10.5 V_{AC} - an incredible design mistake!

BELOW: Partial circuit diagram of B&K 500 with a few suggested modifications © B&K

1) Added "Test 1-Test 2" switch 2) Added resistor and test points for anode current measurement 3) Proposed bias selector switch (not implemented on the tester pictured)
4) "Signal cal." potentiometer 5) "Bias cal." potentiometer
6) "Bridge balance" potentiometer 7) "Sensitivity" control
8) Signal balancing bridge 9) The umbilical cord connecting the upper panel (model 510) to the main prewired panel.

$10.5V_{AC}$ Actual measured values

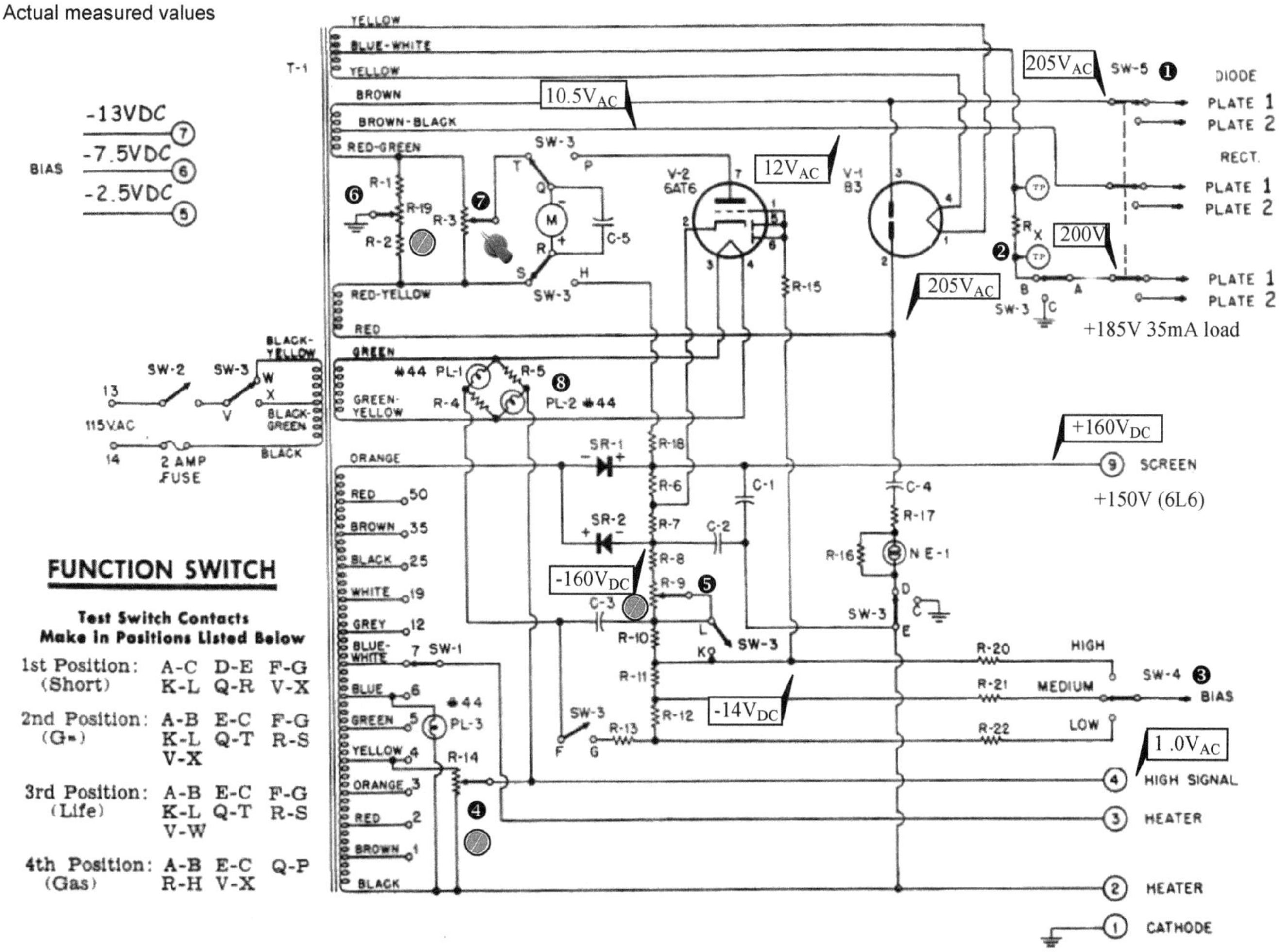

FUNCTION SWITCH

Test Switch Contacts Make in Positions Listed Below

1st Position: A-C D-E F-G
(Short)   K-L Q-R V-X

2nd Position: A-B E-C F-G
(G=)   K-L Q-T R-S
               V-X

3rd Position: A-B E-C F-G
(Life)   K-L Q-T R-S
               V-W

4th Position: A-B E-C Q-P
(Gas)   R-H V-X

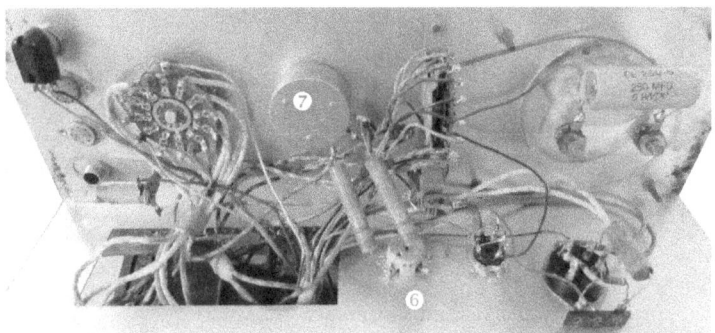

ABOVE: 1) Type 83 rectifier tube   2) 6AT6 tube   3) Selenium rectifiers (SR-1 and SR-2)   4) "Signal cal." potentiometer (R-14) 5) "Bias cal." potentiometer (R-9) 6) "Bridge balance" potentiometer (R-19) 7) External Sensitivity control (R-3) 8) Two electrolytic capacitors (in bias and screen filtering circuits) 9) Terminal strip

Rewiring the sockets for independent testing of duo-triodes

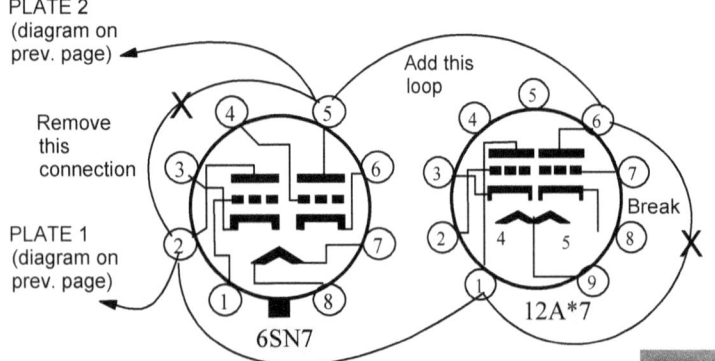

Calibration

SIGNAL: Test Switch in Gm position. Connect a multimeter (AC Volts) to terminals 5 ("Low bias") and 1 ("Cathode") on the internal terminal strip. Turn R-14 to get a 1.0 V_{RMS} reading.

BIAS: Change the multimeter's range to "DC Volts". Turn R-9 to get a 2.5 V_{DC} reading. This sets the "Low bias" line. The medium and high bias line's voltage levels will depend on the precision of resistors in the voltage divider circuit and cannot be adjusted except by replacing those resistors (R8, R10, R11 & R12).

SHORT SENSITIVITY: Cannot be adjusted, fixed by the 470kΩ resistor. GAS SENSITIVITY: There is no provision for such adjustment on B&K 500 tester.

METER BALANCE: Connect a 5k/7W wire-wound resistor between terminals 10 (plate) and 1 (cathode or GND). Turn "Sensitivity" control to 100%.

Move the "Test" Switch into the Gm position. If there is any meter deflection, adjust R-19 for zero meter reading.

ABOVE: The anode current measurement posts and Gm1-Gm2 switch added. The location was chosen because the terminal strip inside the upper panel (not strip #9 on the photo above) is just to the left of the binding posts. Wire runs are thus minimized.

LEFT: How to rewire B&K 500 tester for independent testing of duo-triodes

BELOW: The octal socket 8A (1), and the Noval socket, marked 9A (2). The bare wire jumpers across each socket are clearly visible (p2 to p5 on 8A and p1 to p6 on 9A).

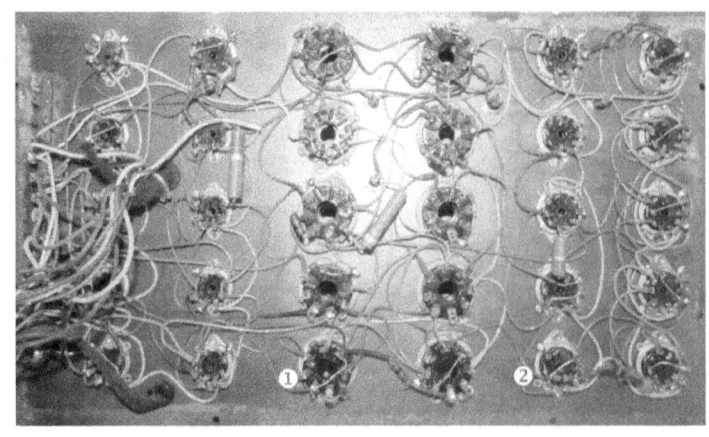

After installing a 3PDT switch SW-5, wire it as per the diagram on the previous page. Two anodes (octal pins 2&5 and pins 1&6 on the Noval socket) must be separated, and another loop/line run to contact P2 of the new switch. The existing line, disconnected from contact A of SW-3, needs to be connected to contact P1 of the new switch.

Switch SW-3 remains the test switch, but it will test either triode #1 or triode #2, depending on the position of the switch SW-5.

B&K 550

Despite its large dimensions, B&K 550's wiring is messy; most components are crammed into the bottom right corner, making it difficult to work with. The detachable power cord entry point (1) was too close to the test switch (5), and there was no On/Off switch at all!

A new power cable with a 3-pin plug was installed (2). The on/off switch was installed where the 2-pin power inlet socket used to be (1).

B&K testers don't have any older sockets, such as UX4 (4-pin) socket for directly heated triodes such as 45, 2A3, and 300B. A UX4-to-octal adapter solved that issue, but when testing those tubes, the meter went backward, so a "Meter reverse" switch was installed (3).

As indicated, the current measuring test points (4) are across a series 10Ω resistor.

B&K 500, 550, 650 and all other Dyna-Quik testers use preset bias voltages. There are 3 bias levels in model 500, "low" (-2.5V), "mid" (-7.5V) and "hi" (-13V). Model 550 uses 4 bias levels, -0.2V, -2.5V, -7.5V and -19.5V.

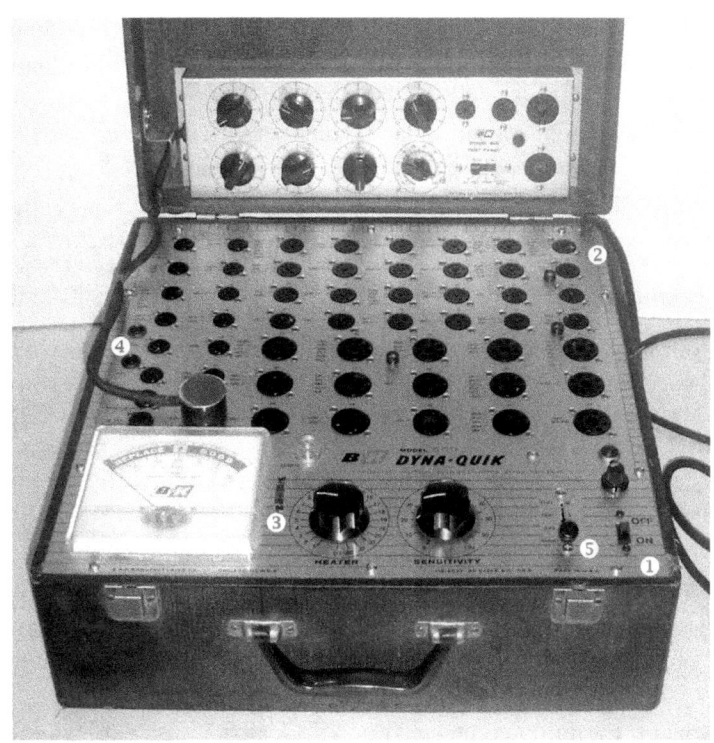

ABOVE: B&K 550 after a few essential upgrades
BELOW: Partial circuit diagram of B&K 550 © B&K

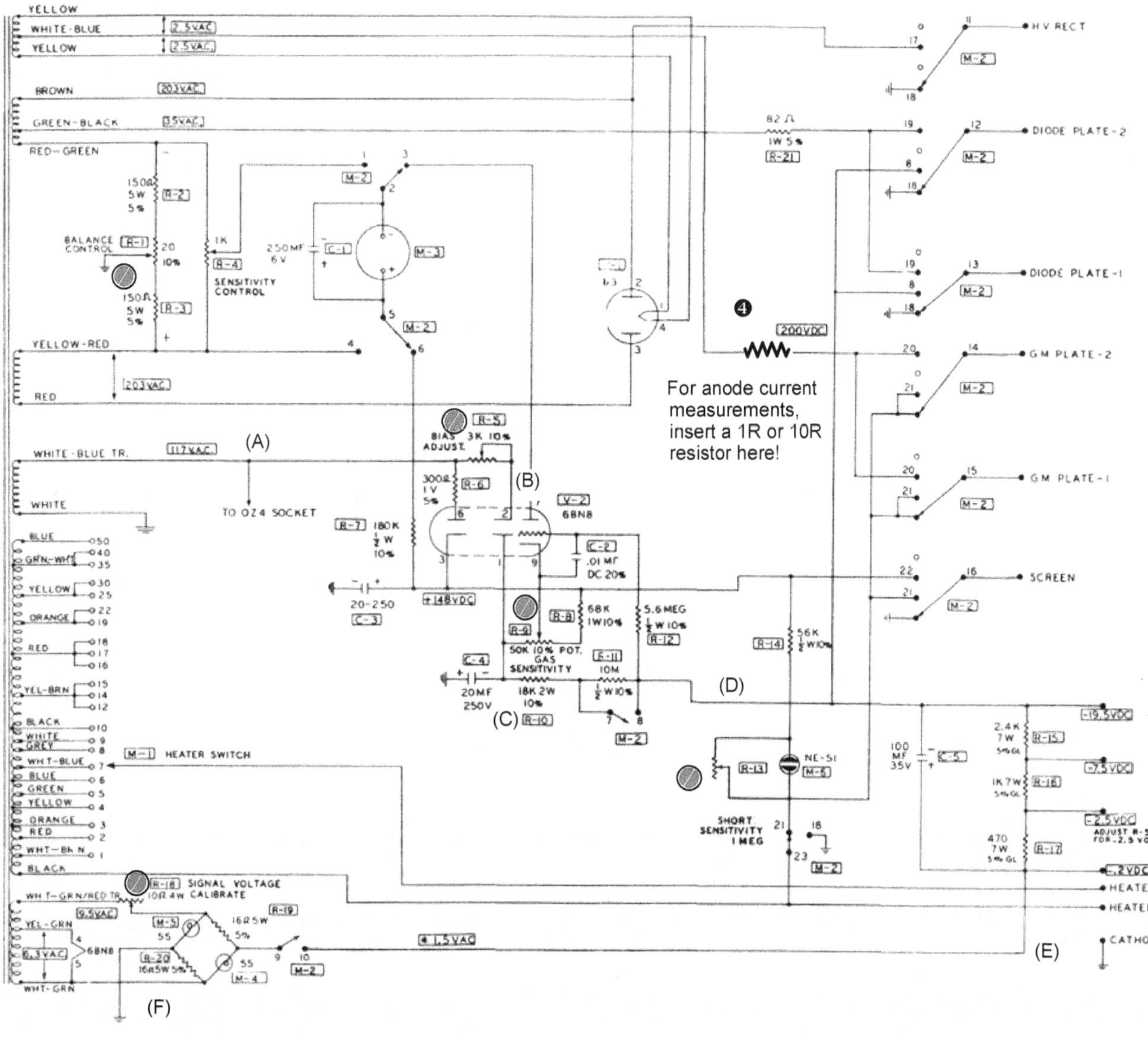

While the test signal in model 500 was obtained by subtracting the voltage from the 5V heater tap from the 6.3V output of the bridge (that is where 1.3V came from), the signal voltage here comes from a 9.5V secondary (that also has a 6.3V tap for heating the 6BN8 tube).

Bias adjustment was now made with a series trimmer R-5 on the AC side of the rectifier diode (pins 1 and 2 of 6BN8), and a "Gas sensitivity calibration" pot was added (R-8).

Model 500 tested diodes with $205V_{AC}$ and HV rectifiers with $12V_{AC}$ (10.5 V_{AC} measured), which made no sense. That design mistake was corrected on 550. Diodes are now tested with $35V_{AC}$ and HV rectifiers with $203V_{AC}$.

Making bias adjustable

Four bias voltages may be fine for some tube types, but, as we have seen in Chapter 2, may not be optimal for others, those that are tested at the tail of the transfer curve, close to the cutoff, where Gm would vary significantly with a very small change in bias voltage. Should you wish to choose the testing point, and since the plate voltage is fixed in Hickok-type testers, the only way to vary the plate current at which you want your test done is by adjusting the bias. The procedure is very similar for all testers with fixed bias voltages, not just B&K or Hickok-type testers. We will use Model 550 as an example.

This modification will not change your tester's calibration, but since tubes will be tested at different operating points, it will change the displayed Gm values according to the position of the bias potentiometers.

After this modification, there will be only two instead of four different bias lines, but they will be fully adjustable. The sockets with -0.2 and -2.5 V will be connected together, and their bias will be adjustable from 0 to -4 volts, beyond the current -2.5 V.

Sockets currently connected to -7.5 or -19.5 Volts will be joined together, and their adjustable range will be from -4 V to -19.5 Volts. By using different values for the two bias potentiometers, you can change those ranges to suit your preferences.

Since there is no grid current flowing (or very low levels, in mA), the power dissipation on the pots is low, so there is no need to use rheostats; ordinary potentiometers will be fine. You could install a #49 globe in each bias line (wiper of the pot) to act as a bias fuse.

When you mount the potentiometers, make sure you graduate the scale so you can achieve repeatable measurements. Alternatively, to set the desired bias voltages precisely every time, you could install a small DC voltmeter (0-20V), but most testers would not have that much free real estate on their control panels. Again, the easiest way is to install test points (binding posts or terminals) and use an external multimeter set on "V_{DC}."

Let's follow the biasing circuit on the schematics to fully understand what is involved. The $117V_{AC}$ (A) is half-wave rectified by one of 6BN8's diodes (B). If it weren't for R-5, there would be $-148V_{DC}$ in point (C), but now the voltage in that point is much lower.

After passing the 18k series resistor (R-10) and test switch M-2, which shorts the 10M series resistor during Gm tests, we end up with -19.5V in point (D). That is the "HIGH" bias. The triple voltage divider R-15, R-16, and R-17 then produces the MID (-7.5V), the low (-2.5), and "zero" bias (-0.2V).

The current through the voltage divider is I=(19.3V-0.2V)/(2.4+1+0.47)kΩ = 4.987mA, say 5mA. The voltage drop on R-10 is V_{R10}=I*18k = 5*18 = 90V. Since 90+19.5 = 109.5V, the voltage in point (3) is -109.5V.

The voltage in point (E) is not zero because the 5mA current has to close its loop to the ground (F), and to do that, it must pass the signal bridge, the "signal voltage cal." trimmer, and 9.5V secondary winding in parallel to the bridge. Since 0.2V/5mA=40Ω, that is the equivalent resistance of that part of the bias circuit.

Going back to calculations, we need to keep the "HIGH" bias point at -19.5V or thereabouts (19-20V), so the total resistance of two pots must remain the same - 2.4+1+0.47=3.9kΩ. If we use a 1k pot for the lower bias points, the upper one needs to have a resistance of 2.9kΩ, so a 3kΩ pot would be perfect. A 2.5kΩ pot would also work, in series with a 390R fixed resistor.

Alternatively, use a 5kΩ pot in parallel with a 15k fixed resistor, which will make it behave as a 2k9 potentiometer!

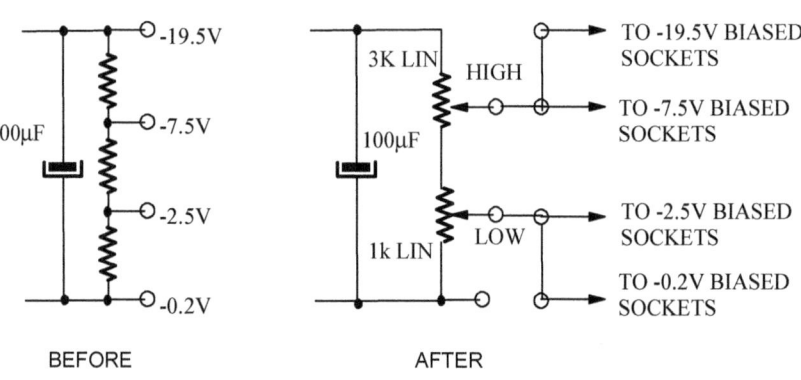

Calibrating B&K 550

B&K 550 calibration instructions are quite clear, although they refer to socket #1 for testing 6AU6 pentode and similar tubes. We didn't have a mini 7-pin plug, so we converted all instructions to socket #43, an octal socket for testing 6L6, a much more practical solution.

A. SHORT SENSITIVITY: Connect 1 MΩ resistor between pins 7 & 8 on the octal socket. This simulates heater-cathode leakage. Adjust R13 (mounted inside the tester, across the neon shorts light) until the neon just lights up. To change the tester's shorts sensitivity, change the 1M resistor to a different value!

B. GAS SENSITIVITY: Connect 10 MΩ resistor between pins 5 (grid) & 8 (cathode). Adjust R9 (flat screwdriver through the center of socket #52) for a meter reading of 2000 on the black Gm scale. Our resistor was 12 MΩ, so we adjusted for the reading of 1200!

C. SIGNAL: Connect an AC voltmeter between pins 5 & 8, with test switch in Gm Test 1 position. Adjust R18 (flat screwdriver through socket #46) for 1.5 V_{RMS}.

D. DC BIAS: Connect a DC voltmeter between pins 5 & 8. Adjust R5 (under the panel, inside the tester) to get a reading of -7.5 V_{DC}.

E. BRIDGE BALANCE: To check the static balance of the bridge (without the grid signal), connect a resistor of between 2 and 6 kΩ (3-5 Watts rating) between pins 3 (anode) and 8 (cathode) on the tester's octal socket. Set Sensitivity control to 100 (maximum) and "Test" switch to "Gm Test 1". There should be no meter movement. If there is, adjust R1 (located inside the tester) for zero reading on the tester's meter.

F. GM TEST: To check if the tester's Gm circuit is working properly, connect a silicon diode of 200V or higher PIV rating in series with a 100R resistor between pins 3 (anode) and 8 (cathode). That simulates the rectifying action and internal resistance of a vacuum tube. Flick the "Test" switch to "Gm Test 1". Depending on the setting of the *"Sensitivity"* control, there should be a smaller or bigger meter deflection.

Calibration module for testers based on the Hickok bridge

DIY PROJECT

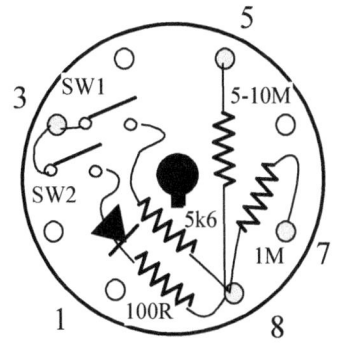

Instead of mucking around resistors and diodes, building a calibration module is a much more elegant solution. Connect the four resistors, two switches, and a silicon diode (as per diagram) inside an empty octal tube base (available for sale on eBay).

SW1 would be turned on in step E) above, the BRIDGE BALANCE TEST, and would then be switched off, after which SW2 would be turned on, for test step F), the GM TEST.

The two resistors between pins 5&8 and 7&8 can remain connected during test steps E) and F)

ABOVE: Hickok calibration module inside an octal socket, 6L6 pinout

---

B&K 650

B&K 650 is a very large tester. When opened up, the two halves of the case take lots of space on a test bench.

A very basic junction transistor test circuit and a 3-position Gm "Test" switch were included, so model 650 could test triple tubes.

Otherwise, for some reason, 650 was closer to model 500 than model 550 since it used the same internal tubes as model 500 and selenium rectifiers as well.

RIGHT: Model 650 has a 3-position Gm "Test" switch, so it can test triple tubes

Automatic line voltage compensation

One feature of all B&K's gm testers is "automatic line voltage compensation." Model 650's user manual states, "A voltage sensitive bridge monitors the line voltage at all times, and automatically adjusts the sensitivity of the gm bridge to compensate for these line voltage fluctuations."

The service notes for model 650 are more precise: "The voltage sensitive bridge has an inverse current characteristic. When the line voltage drops the resistance of the bridge circuit decreases causing an increase in signal voltage."

Likewise, an increase in line voltage that results in the higher heater, screen, and anode voltages, would reduce the amplitude of the signal voltage, and, B&K claims, such a reduction in grid signal "offsets the increase of all other tube voltages and causes the instrument meter to read approximately the same as it did if the tube were tested at 115 volts." The emphasis is on *approximately*!

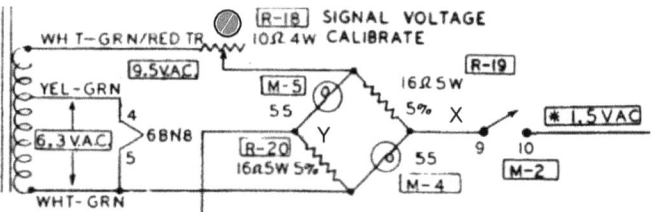

ABOVE: B&K's "voltage sensitive signal bridge", model 550 shown

RIGHT: Small incandescent globes such as #55, used by B&K, are still available, but get some spares, just in case they go out of production. The same applies to #81, #83, and #49 miniature bulbs used as fuses and indicators in Hickok testers.

B&K 675

From a functional point of view, one could argue that since all B&K testers up to that point used prewired sockets, nothing would be gained by releasing a card-based model.

The "Sensitivity" and "Heater" controls still had to be set manually, and tests weren't any faster on this card machine compared to its predecessors.

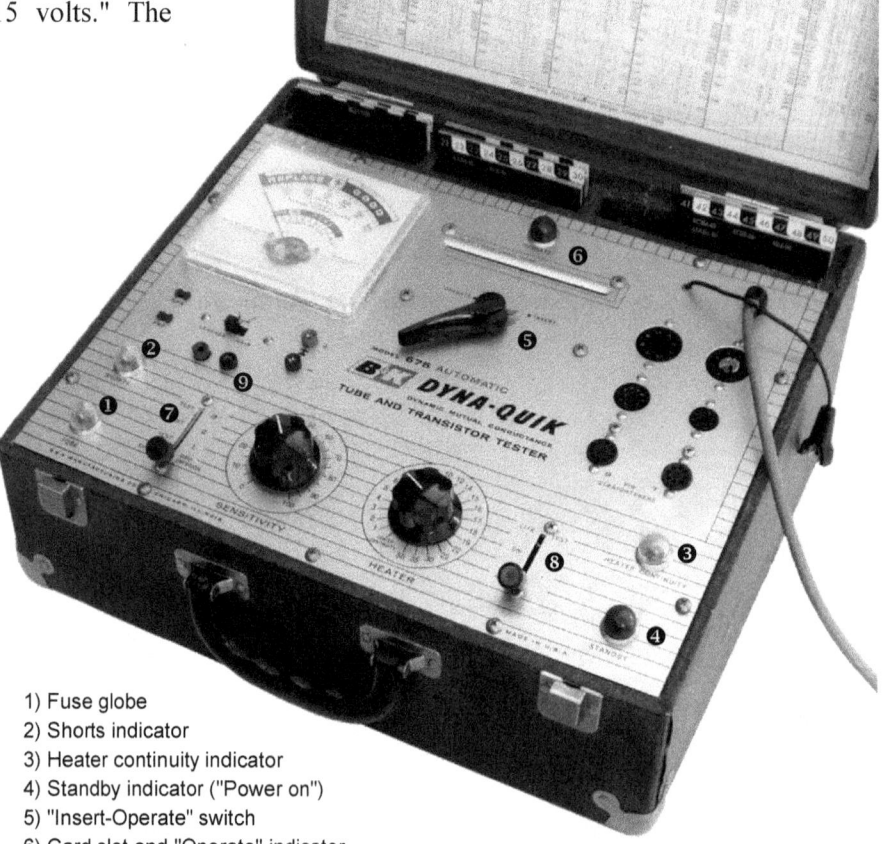

1) Fuse globe
2) Shorts indicator
3) Heater continuity indicator
4) Standby indicator ("Power on")
5) "Insert-Operate" switch
6) Card slot and "Operate" indicator
7) "Shorts/Grid emission-Test 1- Test 2 - Test 3" lever switch
8) "Off - On - Life" test lever switch
9) Two added test points for measuring anode current

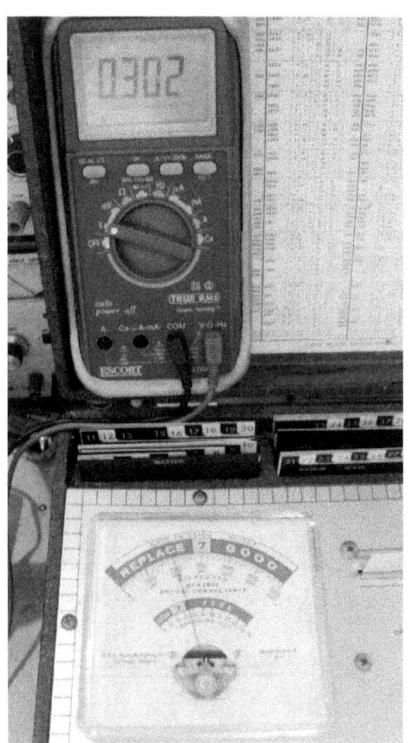

Internally, 675 is a well-laid-out tester, just like model 500. Compared with the messy internals of models 550, 650, and 700/707, the wiring here is neat and easy to follow. It makes one speculate that different design teams worked on various B&K testers since lessons learned on one model were not implemented on subsequent models; only the mistakes were perpetuated.

LEFT: Testing a 6L6 beam power tube on B&K 675. The mutual conductance is just over 6,000 micromhos (lower scale 0-18,000) and the anode current is 30.2 mA.

B&K 675 tube tester - Calibration

The only two calibration pots accessible from the outside are the 10-ohm "Signal level" pot (accessible with a flat screwdriver through the Octal socket) and the 50 kohm "Gas sensitivity" pot under the Loctal socket.

The 1.5 Mohm "Shorts sensitivity" pot is located adjacent to the main "Test" switch, soldered onto the holder of the "Shorts" neon light. The "Bridge balance" pot is located on the underside of the mainboard and cannot be seen. Only its slot is accessible with a flat screwdriver through a hole in the board. It's located between the two #55 globes for the signal circuit above the two 16 ohms signal bridge resistors.

The "Bias" pot is to the right of the "Bridge balance" pot, next to the selenium rectifier, and is fully visible (metal body).

Before calibration, replace all electrolytic capacitors. The dual selenium rectifier (1) should also be replaced by modern silicon diodes. Selenium rectifiers age, and their internal voltage drop increases with time.

Finally, test the 6AT6 tube (2) and replace it if necessary. The tube's mutual conductance can even be tested on the 675 tester itself since the 6AT6's triode is only operational in the "gas or grid emission" test, and its two diodes aren't used at all.

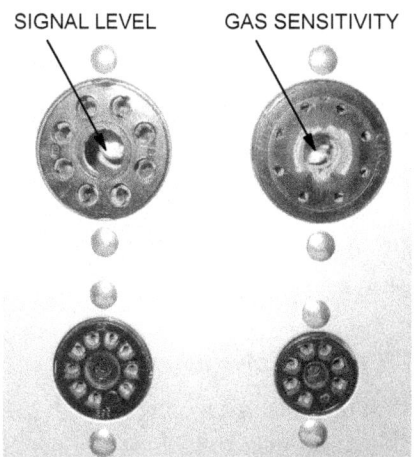
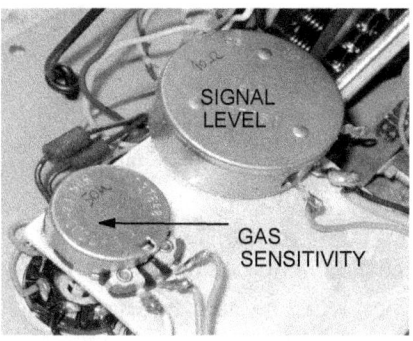
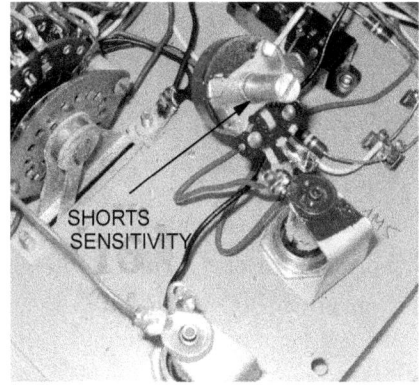

STEP 1: Choose a calibration card

Any card that tests an amplification tube can be used for calibration. We have chosen Card #14, with settings for common octal power tubes such as 6V6, 6L6, EL34, 6550, KT88, and many others.

Unscrew the top plate or panel from the case and place it in a vertical position, securely supported, so you have access and a good view of both sides of the panel. Turn the tester on. Be careful not to touch any exposed terminals with mains (line) voltage.

Place "Test" switch in "Short/Grid emission" position. Insert card #14.

STEP 2: Shorts sensitivity adjustment

Connect a 1 Mohm resistor between pins 5 & 8 of the octal socket. This simulates a grid-cathode short circuit or leakage.

Turn the "Insert-operate" switch into the "Operate" position. Adjust the "Shorts" potentiometer until the "Shorts" neon light just lights up.

If you want to make the shorts test more sensitive, use a higher value resistor for this adjustment. If you want to make it less sensitive, use a lower value resistor, 820 Kohm, 590 Kohm, etc. Remove the resistor.

STEP 3: Bridge zeroing ("Balance control adjustment")

B&K, in their other testers' calibration instructions (models 550, 650, 700, 707), calls this "Balance Control Adjustment." Connect a 6 kohm (or similar value, such as 4k7 or 5k2, the value is not critical) resistor between pins 3 & 8 on the octal socket. This simulates a plate-cathode resistance of a tube.

Since this resistance is the same in both half-cycles, the Hickok bridge should show zero reading. When testing a tube, which has a higher internal resistance when not conducting during one half-cycle, and a lower resistance when conducting, during the other half-cycle, the meter will show the difference, which is proportional to Gm (transconductance).

1. Place "Test" switch in the "Test 1" position.
2. Adjust the "Bridge balance" pot to get a zero reading if the meter moves away from zero.
3. If the meter stays at zero, there is no need to adjust anything. Return the "Test" switch into the "Short/Grid emission" position.

STEP 4: DC bias adjustment

With the "Insert-Operate" switch still in the "Operate" position, connect a DC Voltmeter or multimeter on a "Volts DC range" between pins 5 & 8 of the octal socket (control grid and cathode).

1. Place "Test" switch in the "Test 1" position.
2. Adjust the "Bias" potentiometer until the meter reads 7.5 Volts. This assumes that the meter's red lead is on pin 8 and the black lead on pin 5. If the leads are reversed, the meter should read -7.5V (the grid voltage is negative with reference to the cathode).
3. Return the "Test" switch into the "Short/Grid emission" position.

You have now finished with the adjustment of internal calibration controls and can close the tester (put the top panel back onto the timber carry case). Turn the tester off before you do that and unplug it from the mains, so you don't accidentally touch the live wiring inside while flipping the top panel and placing it back onto the carry case.

The preceding adjustments need to be done rarely - after components are changed (service) or after years have passed, and some components might have drifted significantly from their original values. That is why these trimer potentiometers are under the panel. The next two adjustments are performed much more often, and all adjustments are made from the outside, without the need to open the tester up!

STEP 5: Grid signal level adjustment

1. Turn the tester on.
2. Place "Test" switch in "Short/Grid emission" position. Insert card #14.
3. Turn the "Insert-Operate" switch into the "Operate" position.
4. Connect an AC Voltmeter or multimeter on a "Volts AC range" between pins 5 & 8 of the octal socket (control grid and cathode). The polarity of the leads does not matter.
5. Place "Test" switch in the "Test 1" position.
6. Adjust the "Signal" potentiometer until the meter reads 1.5 Volts. This is an RMS (Root-Mean-Square) or effective value of the grid signal voltage.

If you are using a VTVM (Vacuum Tube Voltmeter) that is calibrated in peak values, the reading should be 1.41 times higher or 2.115 V_P. If your meter is calibrated in peak-to-peak values, the value should be double the peak value, or 4.23 V_{PP}.

The "Signal level" trimmer pot is accessible with a small flat screwdriver through the keyhole of the octal socket.

Return the "Test" switch into the "Short/Grid emission" position.

STEP 6: Gas or grid emission sensitivity adjustment

1. If not already there, place the "Test" switch into the "Short/Grid emission" position.
2. Connect a 20 Mohm resistor between pins 5 & 8 of the octal socket. If you don't have or cannot get a 20 Mohm resistor, connect two or more resistors in series to get the desired value.
3. Adjust the GAS trimmer pot, so the meter reads right on the line between good/bad (green and red) on the "Gas" scale. The "Gas sensitivity" trimmer pot is accessible with a flat screwdriver through the keyhole of the Loctal socket.

To make this test more sensitive, use a lower value resistor to get the same meter reading, for instance, 15 Mohm, 10 Mohm, 4.7 Mohm, etc.

Alternatively, simply connect a digital multimeter or VTVM between pins 5 & 8 of the octal socket. Most meters have an input impedance of 11 Mohms, which should produce a meter deflection well into the "Grid emission - Reject" sector, at the half-scale mark, as illustrated on the next page.

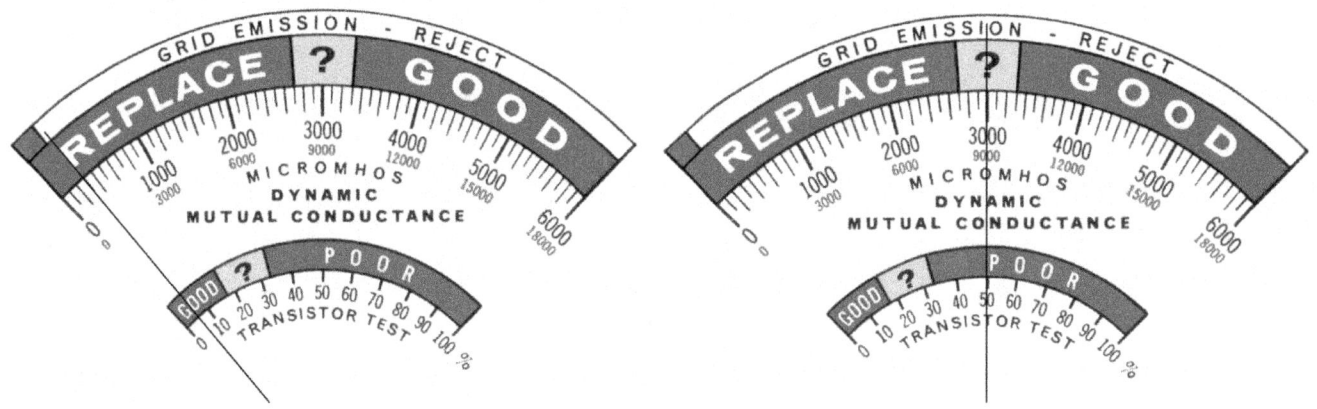

ABOVE LEFT: 20 Mohm indication should be right between the good (left) and bad sectors on the upper scale
ABOVE RIGHT: 11 Mohm indication should be around the middle of the scale

B&K 675 tube tester - making cards for tubes that are not listed in the tester's book

Making test cards for this tester is not difficult, but to know which holes to punch, you must understand how the tester's switching matrix works.

Horizontal strips 1 to 9 (top to bottom) correspond to tube pins 1 to 9. Strip 10 is connected to the top cap J-7 via a 47-ohm resistor.

Vertical strips represent the functions or electrodes; two are heater connections (H), two are diode testing connections (DIODE 1 and DIODE 2), two are Gm (transconductance) strips (GM-1 and GM-2).

Screen grid, control grid (G-1), and cathode each have their own vertical strip. The G1 or control grid strip can also be connected to five different bias voltages (0V, -2.5V, -7.5V, -15V and -25V) via pins 1, 11, 21, 31 and 41.

Each of the 100 possible contacts (crossings of horizontal and vertical busbars) is numbered.

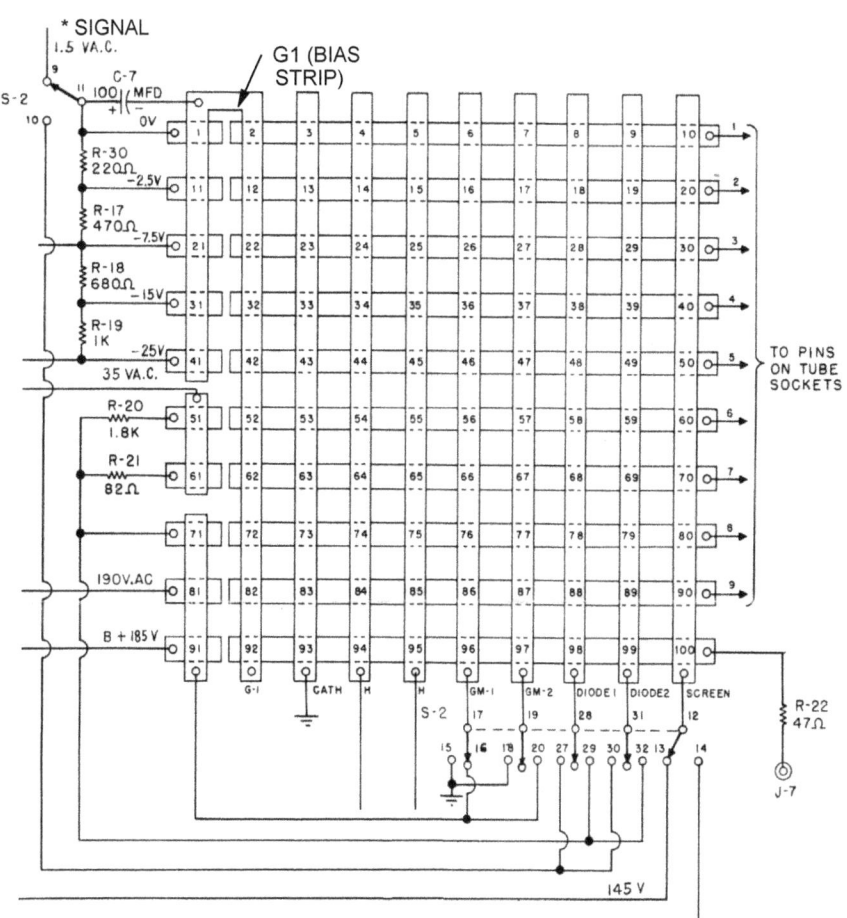

Listed example #1: Octal power tubes (6K6, 6V6, 6L6, EL34) Card #14

The three examples that follow should be sufficient for you to understand the logic behind the punched numbers.

These tubes share the same pinout but differ in power rating, internal resistance, amplification factor, and mutual conductance. The logic here is that despite these differences, even using the same test voltages (signal, plate, screen, and bias), correct mutual conductance readings are obtained using different sensitivity settings for each tube type.

Contact C21 selects -7.5V grid bias. C42 brings that voltage onto Pin 5, which is the control grid.

The signal is always connected to the biasing strip (see the * SIGNAL mark on the diagram), so we don't have to worry about the signal.

C73 connects P8 to the cathode (ground).

C14 connects P2 to one side of the Heater.

C65 connects P7 to the other side of the Heater.

C26 connects anode (P3) to GM-1 vertical strip to test transconductance. For that test, we also need C91, which brings +185 V_{DC} to the anode.

C40 connects the screen grid (P4) to the SCREEN vertical strip connected to 145V_{DC}.

Listed example #2: Noval duo triodes (12AU7, 12AT7, 6BQ7) Card #2

Contact 11 selects -2.5V_{DC} grid bias.

C12 brings that voltage onto Pin 2 (control grid) for triode #1, and C62 connects it to P7, which is the control grid of triode #2.

C73 connects P1 (cathode) to the cathode vertical strip, which is always grounded (reference level).

C23 and C73 connect P3 and P8 to the cathode (ground).

C34 and C45 connect P4 and P5 to Heater strips.

C6 connects triode #1 anode (P1) to GM-1 vertical strip to test transconductance, and C57 connects triode #2 anode (P6) to GM-2 vertical strip.

We also need C91 to bring +185 V_{DC} to the anodes when either "Test 1" or "Test 2" lever switches are activated.

Listed example #3: Rectifiers (5V4, 5U4, 5Y3, 6X5, 6AX5) Card #3

First, notice that both directly- and indirectly-heated rectifiers are tested using the same card. That is possible since the cathode for indirectly heated tubes such as 5V4 is connected to pin 8.

Also, dual rectifiers with anodes at pins 4 and 6 can be tested, as well as those with pins 3 and 5, such as 6X5 and 6AX5, since those pins aren't used on the other rectifier types, what a fortunate situation.

C14 and C75 connect the heater pins P2 and P8 for 5V4 and 5U4 tubes. Contact 65 connects one side of the heater to pin 7 as well, for 6X5 and 6AX5 tubes. Pin 7 is not used on 5V4 and 5U4 tubes, so that is OK.

C26 and C36 connect one anode (P3 and P4) to Gm-1 (or "Test 1" lever).

C47 and C57 connect the other anode (P5 and P6) to Gm-2 (or "Test 2" lever).

All that is missing now is the rectifier plate test voltage, for which we need AC voltage. C91 would connect the plates to 190 V_{AC}, but for some reason, B&K chose to test these tubes with 35 V_{AC}, contact C61 via contact C71. Not the best choice; the 35 V_{AC} test voltage is way too low for testing high voltage rectifiers!

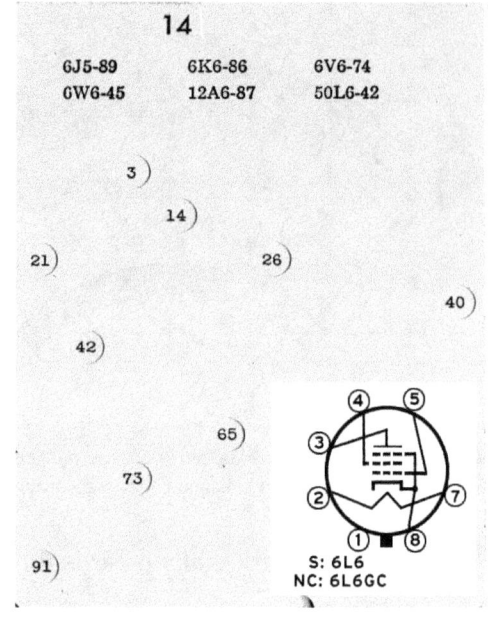

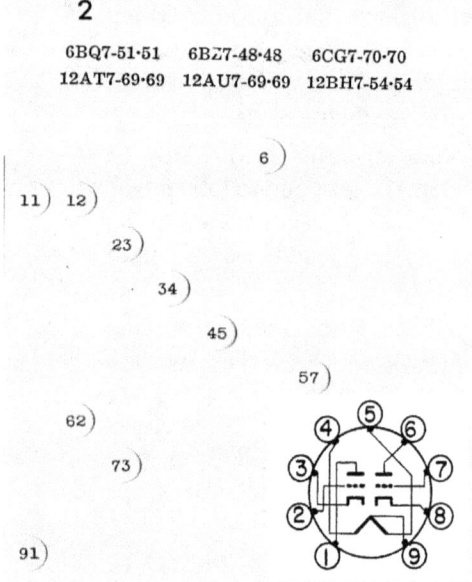

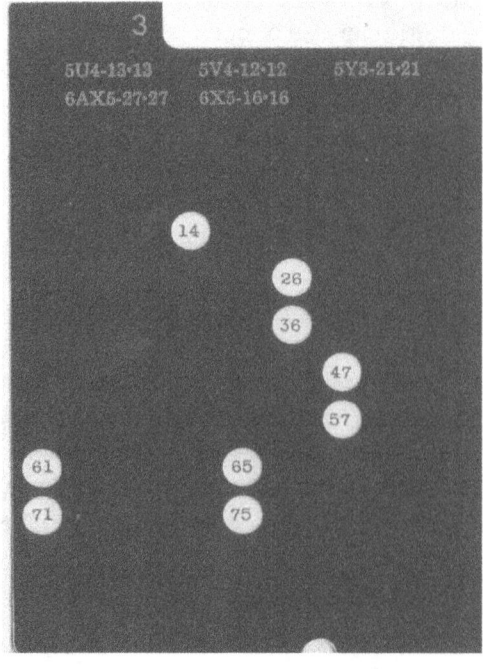

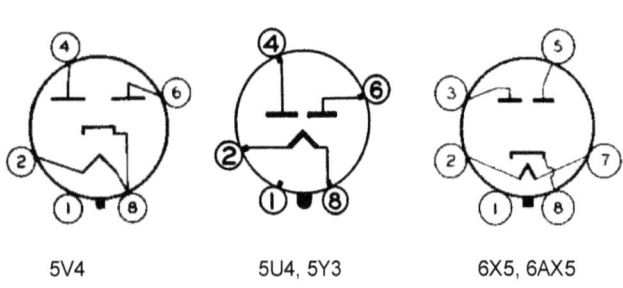

5V4   5U4, 5Y3   6X5, 6AX5

New example #1: 6BQ5 (EL84), 7189 Card #61

In our card stack we only had cards #1-57, so although some Model 675 owners may have Card #61, here is how to determine the settings for this power tube (6BQ5 a.k.a. EL84).

Contact 11 selects -2.5V grid bias. This bias voltage has to go to Pin 2, the control grid, so we need to activate contact C12.

The cathode is pin 3, so contact C23 will connect pin 3 to the cathode vertical strip.

Heater pins are P4 & P5, so we need C34 & C45 OR C44 & C35; the polarity of the heater pins does not matter (AC heating).

The anode is P7, so C66 will connect it to Gm-1 vertical strip to test transconductance. We also need C91 to get +185 V_{DC} to the anode (plate).

The screen grid is P9, so that C90 will connect it to the SCREEN vertical strip.

The listed data for 6BQ5 in B&K Model 675 tube chart specifies that the Sensitivity setting of 46, using the RED scale, would have an average tube read 11,300 micromhos or 11.3 mA/V.

New example #2: 6CA4 (EZ81) dual rectifier

Heater pins are P4 & P5, so we need C34 & C45 or C44 & C35; the polarity of the heater pins does not matter.

We need to connect the cathode (pin 3) to one side of the heater; contacts 24 and 25 will do that, so we choose only one of them!

C6 will connect one anode (P1) to Gm-1 (or "Test 1" lever).

C67 will connect the other anode (P7) to Gm-2 (or "Test 2" lever).

To use 35 V_{AC} test voltage, we need C61 and C71, but to test using a higher voltage (190 V_{AC}), we need C81 and C91.

How to punch holes

The holes on model 675 cards are between 7 and 8 mm in diameter, most likely (being a US size) 5/16" (7.9375 mm). The original cards seem to be 250-300 gms (grams per square meter) card stock.

The first problem is that 5/16" (approx. 8mm) pliers-style hole punches are comparatively rare; most punch 6mm or smaller holes.

The second issue with most small pliers-style punches sold today is their extremely small reach; they can only punch holes in the first two columns on each side of the card and in the last two rows. All other holes would have to be drilled out or punched manually using a hammer and a "hollow hole" steel punch.

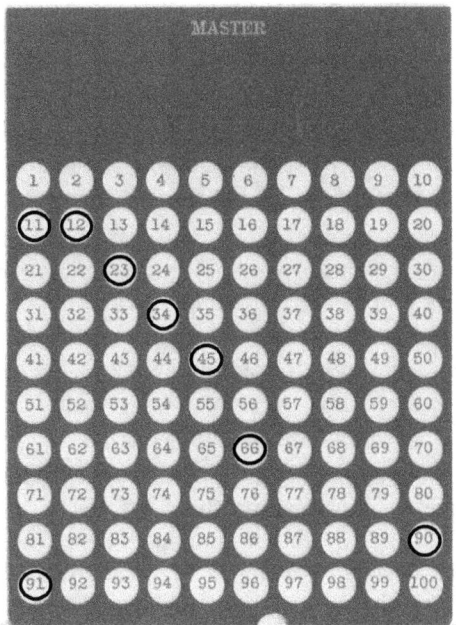

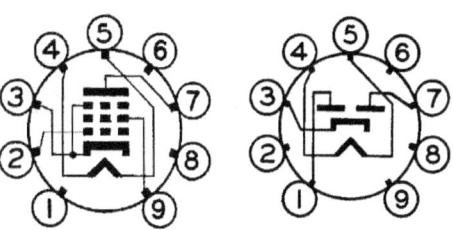

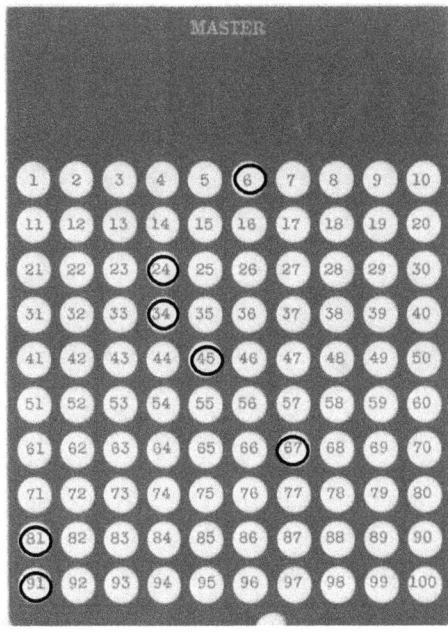

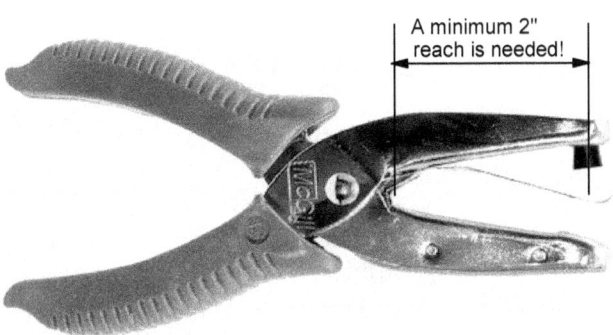

McGill does make a punch with a 2" reach, which is just enough to reach the middle of the B&K card, pictured on the right.

If you are making more than one card, it takes about 20 minutes to make (cut and punch) 10 cards, or only 2 minutes or so per card. This assumes that you have already determined the test settings, so you know the numbers to be punched out.

B&K 700 & 707

After a quick succession of three prewired testers (500, 550, and 650), a card-only model (675), and a weird 685 hybrid (some tubes tested with cards, others on a prewired panel), you'd think B&K designers had learned a thing or two and finally got it right with models 700 and 707, but alas, in many ways these tube checkers were a backward step. Why? First and foremost, the meter was not "calibrated" in true Gm but presumably in percentages. The scale was marked 0-120, but the "%" sign was missing.

There were now fewer prewired sockets, although that restriction was somewhat compensated for by including the bottom switchable panel. Just when you thought, great, I don't need the 610 universal panel, think again. To test for Gm, you'd need at least one switch for a heater pin (assuming the other side of the heater is prewired to the COM or GND terminal) and one switch each for G1, G2, anode, and cathode, a minimum of 5 selectors in total. The bottom panel has only four rotary switches. On the tester's circuit diagram, two switches are marked "Heater #1 switch" and "Heater #2 switch" (Switches "A" and "B" on the control panel), followed by the "Lockout switch" (marked "C") and "Selector switch" marked "D.

How can tube's emission be tested by selecting only one electrode? The only way is by bundling all active electrodes together and picking the control grid with the "Selector switch." So, to make matters worse, the emission test performed on the bottom panel is not the full cathode or anode conductance, but some small cathode-grid current, just as with all grid "analyzers, such as B&K 600, 606, and 625. So, since they test all tubes for Gm, models 550 and 650 with 610 panel are more capable tube checkers than models 700 and 707.

1) Heater voltage selector switch
2) "Sensitivity" control (meter shunt)
3) Five test push buttons
4) Lower switchable test panel
5) One option for installing anode current test points is at the bottom right corner of the lower test panel
6) Large meter but lousy graduations in % only, no true Gm scales

Notice that I used "checkers" instead of "testers," meaning these Gm testers are suitable for quick "yes-no" checks, not for tube matching or serious tube analysis. There are no minor graduations, no absolute Gm values, only one test and signal voltage, and no adjustable bias.

Strictly speaking, any tester without adjustable bias and anode/screen voltage (if not continuous, then at least in switchable steps) is a mere tube checker.

Internally, models 700 and 707 are almost identical to model 550, with type 83 mercury vapor rectifier and 6BN8's triode section used as a gas/grid leakage amplifier. The only difference is the addition of a $60V_{AC}$ tap on the power transformer's secondary winding, which is fed to the lower emission testing panel.

Apart from the black case and black & silver overall color scheme of model 700 versus a more modern light blue & silver looks and rectangular meter of model 707, the two are functionally identical.

RIGHT: Testing a 12AU7 duo-triode on B&K 707. The anode current is 21.9mA and the mutual conductance is 98% of "who knows what" !

B&K 747

The last, and for many the best B&K tester, model 747, was released in 1971, 16 years after the first model 500. It retailed for US$249.95, in the same ballpark as Sencore's MU-150.

On the one hand, it was a different beast: fully solid-state, light and sleek, in a molded plastic case, and with a bottom switched panel that finally tested all tubes for Gm.

However, it was still a very basic tester, with one grid signal ($1.5V_{AC}$, too high for many preamp tubes), a single screen & anode voltage level (although there was also a low Gm anode voltage of $90V_{DC}$), and one fixed DC bias voltage for each tube, chosen by its designer. There are four bias levels, -0.5V, -2.5V, -7.5V, and -19.5V. Again, users had no choice in which point a tube would be tested.

ABOVE & LEFT: Black meter, black test buttons, black sockets ... the morbid looks of B&K 747 are not to everyone's taste.

One either likes the "all black" color scheme (although, strictly speaking, black is not a color, it is the absence of any color!) or hates it.

I've found the tester's lack of contrast makes it difficult to "read" the controls and identify sockets. Luckily, the switching bank is silver in color and easier on the eye.

Like models 700 and 707, the meter lacks minor graduations and displays only some kind of 0-120 figure, presumably the percentage of an average tube's Gm.

The first 747 series had an internal fuse but obviously proved troublesome, so B&K replaced it with an external resettable overload circuit breaker (6), located between the "Heater" voltage selector switch and "Sensitivity" control pot, straight under the meter.

Three calibration adjustment trimmer pots are on the PCB, "Bias adjust" (1), "Shorts adjust" (2), and "Grid emission adjust" (3). The other two, "Signal voltage adjust" and "Bridge balance," are mounted on the angled bracket attached to the heater selector switch and "Sensitivity" pot.

We added an internal 10Ω resistor (4) and ran two wires from its ends to a pair of binding posts (5), so anode current could be measured by an external DC voltmeter during Gm tests.

If planning to buy one, beware - many units offered for sale suffer from broken hinges at the back and cracked or broken plastic case.

Also, the PCB tracks are thin and brittle, so care is required while replacing components. Don't heat the joints for too long, or the copper tracks will lift off the substrate, and the PCB will be ruined.

A better way is to cut the old components out and leave their "legs" as long as possible and solder their replacements onto the protruding leads instead of the PCB!

Calibration instructions are outlined in the "Instruction manual', the procedure is the same as for tube-based B&K testers. As mentioned, shorts sensitivity, grid emission sensitivity, bridge balance, signal level, and bias voltage are five parameters for which calibration controls are available.

An overview of B&K mutual conductance and grid circuit testers

500 ❶
- ✗ Tests duo-triodes, duo-diodes and rectifiers as one - matching not possible
- ✓ Scale in true Gm units (micromhos)
- ✗ Very limited choice of prewired sockets - most newer tubes cannot be tested
- ✗ Serious design errors, such as 10V test voltage for HV rectifiers

550 ❷
- ✓ Tests duo-triodes separately
- ✓ Scale in true Gm units (micromhos)
- ✗ Misleading "Heater" switch - 23 positions but only 16 voltages!
- ✗ No ON/OFF switch
- ✗ Power cable in an awkward spot
- ✗ No "Life Test"
- ✗ Limited choice of prewired sockets - many newer tubes cannot be tested

650 ❸
- ✓ Power switch, power cable and "Life test" issues from model 550 fixed
- ✓ Scale in true Gm units (micromhos)
- ✓ Transistor testing
- ✗ Very large - double panel
- ✗ No 4-pin & older sockets

747 ❻
- ✓ Solid state
- ✓ Small footprint
- ✓ Universal switching panel + prewired section
- ● Morbid looks (all black) and plasticky feel
- ✗ Meter only 0-120 (no true Gm reading) and no minor graduations
- ✗ Fixed bias voltages
- ✗ No 4-pin & older sockets

700/707 ❺
- ✓ Universal switching panel + prewired section
- ✗ The "universal" switching panel only tests for emission (grid circuit only)
- ✗ Meter only 0-120 (no true Gm reading)
- ✗ No transistor testing

675/685 - *Card models* ❹
- ✓ 685 - common tubes don't require a card
- ● 675 - all tubes need a card
- ✗ Heater voltage & "Sensitivity" levels still have to be set manually
- ● No advantage over non-card models
- ✓ Scale in true Gm units (micromhos)

LEGEND:
- ✓ A clear advantage or useful feature
- ✗ A clear drawback or deficiency
- ● A feature that has some good and some bad aspects - opinions differ

1-6: historical (temporal) progression from the older to the newest model

600/606/666/675/607/667
- ● Grid circuit testers
- ● Similar in design and testing philosophy to Sencore TC142 and TC162
- ✓ Supersensitive grid leakage/gas test
- ✗ No 4-pin & older sockets

COMMON TO ALL or MOST MODELS:
- ● Hickok circuit
- ✓ Fast - no switching for common tubes
- ✗ Fixed bias levels
- ✗ Large grid signal amplitude
- ✗ No 4-pin (cannot test 2A3 or 300B tubes), Magnoval or Nuvistor sockets
- ✗ No noise/microphony test
- ✗ Primitive neon shorts circuit of fixed sensitivity

B&K 610 test panel

Unless you test only common tubes such as the 12A*7 series, 6SN7/6SL7 and 6L6, you should not buy a B&K 500, 550, or even 650 tester without the auxiliary Model 610 test panel. It greatly expands the testing capabilities of B&K testers. Since it plugs into an octal socket, it can also be used with other tube tester models and brands.

Each of the seven 12-position rotary switches (A-G) switches one of nine pins and top cap to one of the electrodes (H1, A, G2, G1, H2, K1 & K2). Switch J selects the type of tests required (1-Gm, 2-rectifier, 3-diode, 4-HV rectifier).

HICKOK-TYPE TESTERS

B&K model 610 adapter - inside wiring

You can learn a lot (many good ideas and some bad ones, too!) by studying commercial products. The 610 test panel illustrates the feedthrough principle well. See how all pins on the A-B-C-D-E-F-G rotary switches are fed through.

If you put switch C in position 6 (as illustrated for testing the pentode section of the 6BM8 tube), all the pins #6 before switch C (meaning switches A and B) are still connected, carrying the signal back to the tube under test. Notice ferrite beads in each main feed to test sockets to prevent oscillation.

Hickok testers also used the same type of rotary switches and feedthrough principle. A serious disadvantage of this approach is that voltages and signals must pass through many contacts (in some cases through all the switches!) before reaching the tube-under-test. Only one bad contact would render the whole setup inoperative, which not only reduces reliability but makes signal tracing and troubleshooting in those testers a true nightmare.

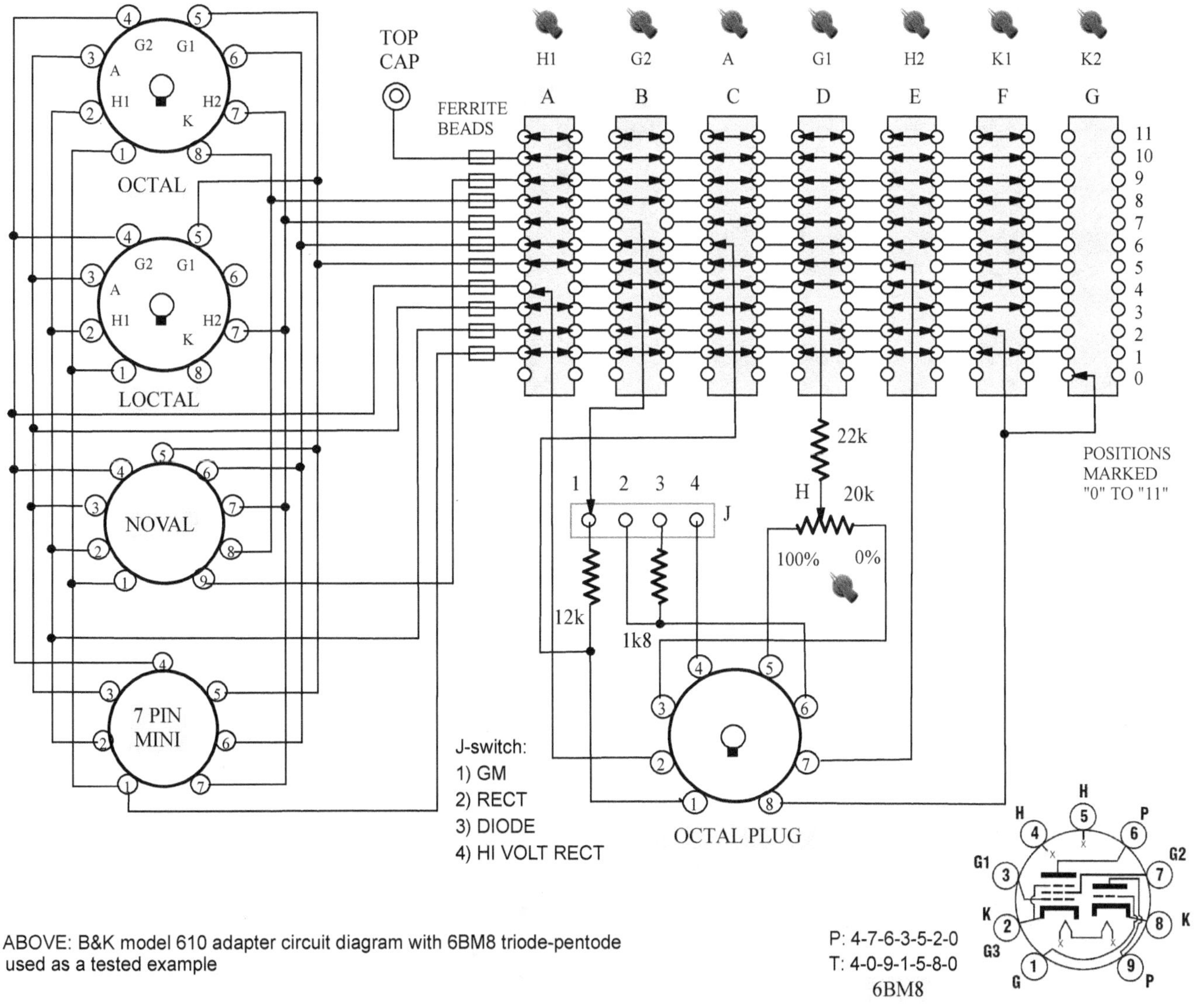

ABOVE: B&K model 610 adapter circuit diagram with 6BM8 triode-pentode used as a tested example

P: 4-7-6-3-5-2-0
T: 4-0-9-1-5-8-0
6BM8

MERCURY 1000, 1200 & 2000

Mineola, New York-based Mercury Electronics Corporation, hit the jackpot when Hickok patent expired, and they jumped on the bandwagon together with B&K, Precise, and a few other test gear makers. Despite their shortcomings, I still prefer their Gm testers to B&K's and even Hickok's! Although equally capable, these testers are much smaller, simpler, and less intimidating than the competing B&K, Precise, and Hickok testers.

Apart from the transistor testing capability of model 2000, models 1000 and 2000 are functionally identical. Although it looks different from those two, model 1200 is also functionally identical, as we will ascertain soon.

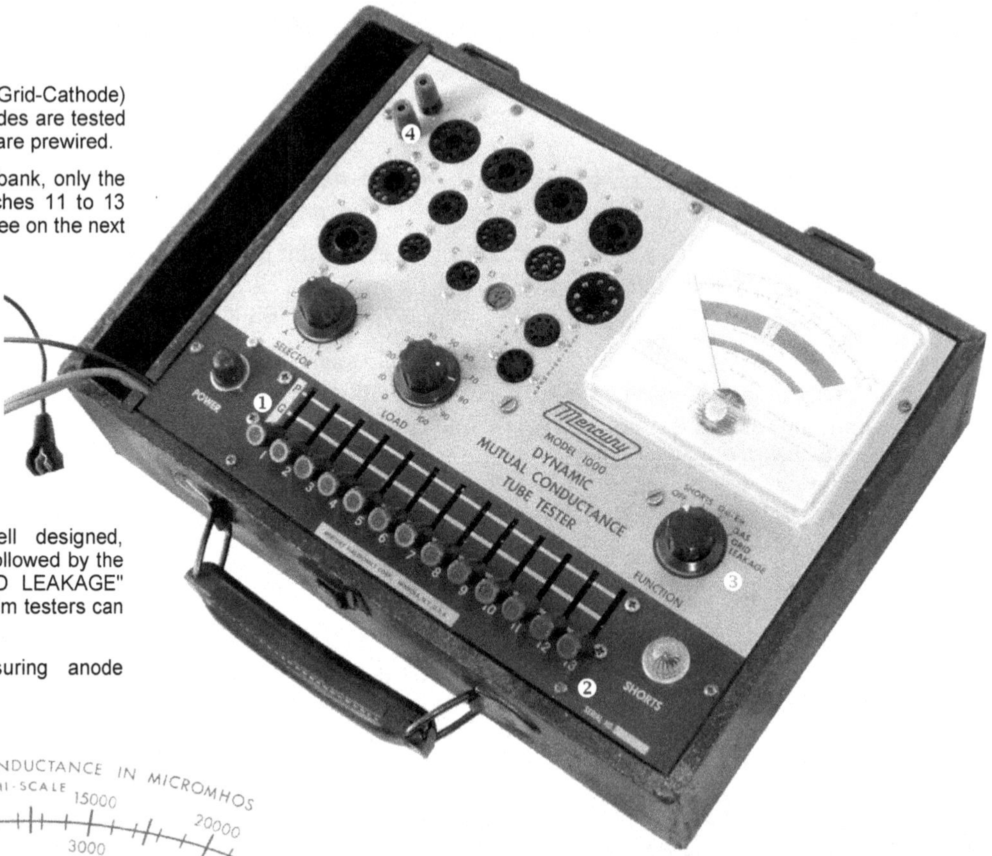

1. Only three test positions (Plate-Grid-Cathode) indicate that tetrodes and pentodes are tested as triodes, and that heater pins are prewired.

2. Despite 13 switches in the test bank, only the first 10 are for tube pins; switches 11 to 13 perform other tasks, as we will see on the next page during circuit analysis.

3. The "Function" button is well designed, "Shorts" are checked for first, followed by the "Gm-Em" test and "GAS-GRID LEAKAGE" test. This means that Mercury Gm testers can test for both Gm and emission!

4. Added test points for measuring anode current

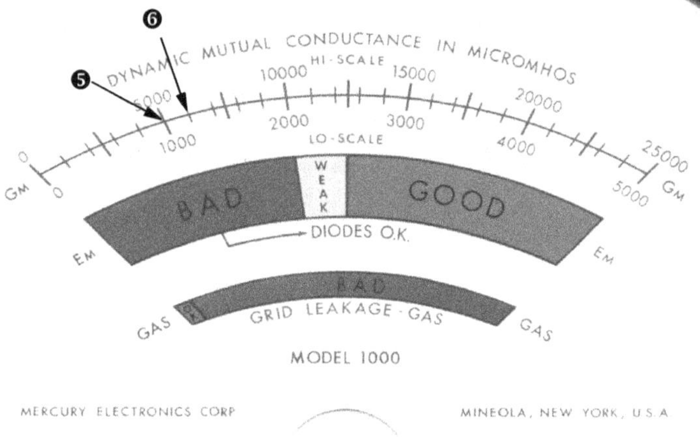

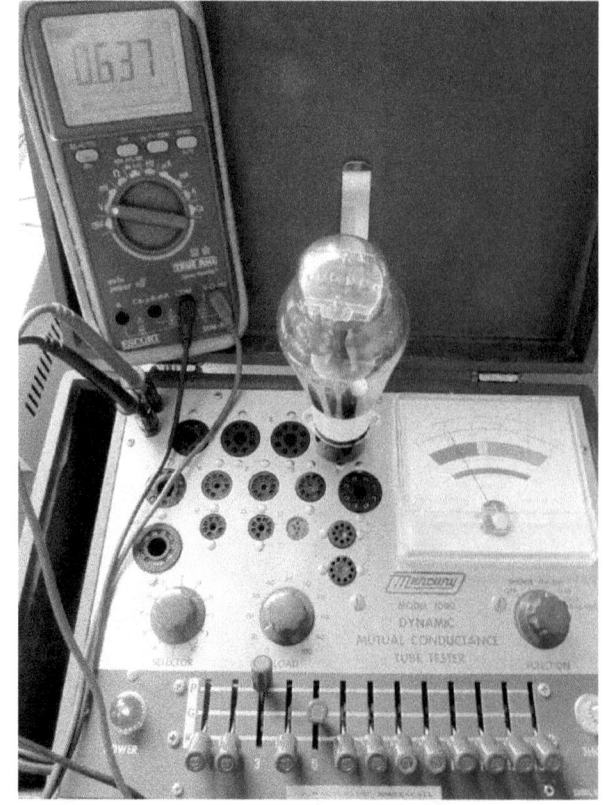

There are only two Gm scales. While the bottom one (0-5,000 micromhos or 0-5mA/V) is useable, the resolution of the top scale (0-25,000) is very low. There are no graduations between thousands of micromhos, such as 5,000 (5) and 6,000 (6), so some guesswork is needed. If such results are used for tube matching, the results are questionable.

RIGHT: Mercury Gm testers don't have a 4-pin socket, and due to its compact size and crowded innards, there is no room to add one. Here we use UX4-to-octal adapter to test a 300B tube.

Notice two binding posts added to enable a simultaneous anode current measurement. The resistor added was 10 ohms, so a DC voltage of 0.637V means the anode current is 63.7 mA. Such a high current is due to a low bias voltage on the octal socket (for testing tubes such as 6L6, 6V6, and EL34), but it is quite fortunate because it is close to currents 300B tubes would pass in SET amplifiers.

HICKOK-TYPE TESTERS

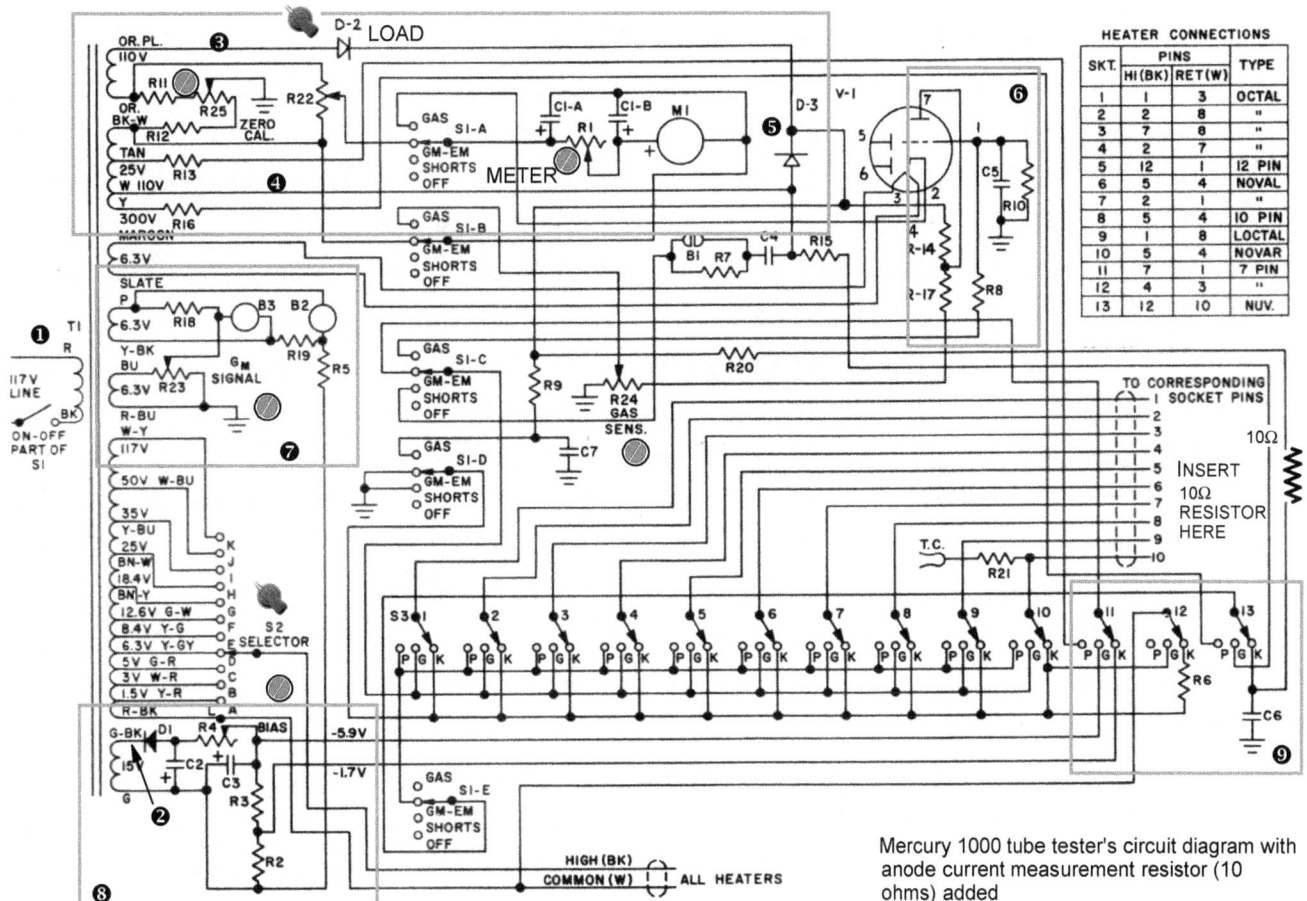

Mercury 1000 tube tester's circuit diagram with anode current measurement resistor (10 ohms) added

The primary of the mains transformer is not fused, so to avoid future transformer replacements, install a fuse there as a matter of priority. You could also add a bias fuse between the 15V bias winding and D1 rectifier (2) and anode fuses in two secondary windings in the Hickok bridge, in points (3) and (4).

The heart of the Hickok bridge are diodes D-2 & D-3 with their cathodes strapped together (5) and the gas/grid emission triode amplifier (6), as in B&K testers. 6AT6 triode-duo-diode was used. The signal circuit is also along the B&K lines of thinking, a bridge with two #44 bulbs, fed by two 6.3V secondary windings (7).

Diode D1 rectifies the $15V_{AC}$ voltage, and trimmer R4 calibrates the DC bias voltages to the required $-5.9V_{DC}$ and $-1.7V_{DC}$. Switches 11 to 13 don't control individual tube pins but connect grid, plate, and heater/cathode circuits in the prescribed manner (9). Lever switch #11 brings one of the two grid bias voltages to the "G" bus, $-1.7V_{DC}$ in position "K" or $-5.9V_{DC}$ in position "G." In position "P, $25V_{AC}$ is fed to "G" busbar, that is used only in Em (emission) tests.

Lever switch #12 connects the "common" leg of the heater supply to GND in position "G" and via R6 in position K. Position P makes no connection at all. Lever switch #13 is only active when the "Function" switch is in the "Gm-Em" position. It connects the "P" busbar to $25V_{AC}$ in position "P," to the cathode of diode D3 via R15, the point marked as (5) on the diagram, in position "G" and to the anode of D2 and D3 (via R20) in position "K."

Emission test setting for 6L6 specifies grid lever switch #11 to be placed in position "P." Since pins 3 (control grid), 4 (screen), and 5 (anode) are connected to the "G" busbar, this switch then brings $25V_{AC}$ to that busbar and test the 6L6 tube as a diode.

Its pin 8 (cathode) is in the "K" position by default since all lever switches are in the "K" position if not given a specific G or P position in the test data book.

RIGHT: Perplexingly, only emission (Em) test data is given for 6L6. However, Gm settings are specified for an identical tube, 5881, and for EL34! Such glaring omissions and inconsistencies (even dangerous errors) are rife in tube tester manuals!

TUBE	Selector	Load	Socket	SHORTS Glow O.K.	FUNCTION Gm – Em	NOTES Rated Gm Micromhos	GAS
6L6	D	35	4		G-3/4/5 P-11		G-5
5881	D	95	4		G-5/11 P-3/4	4000**	G-5
EL34	D	12	4		G-5/11 P-3/4	6000	G-5

Notice also that 5881 is tested on the lower, more precise Gm scale (0-5,000 micromhos) so its "Load" setting is very high (95% or 95 on the scale 0-100), while EL34, whose Gm is expected to exceed the 5,000 limit for that scale (6,000 micromhos nominal) is tested on the upper scale, 0-25,000 micromhos. That is why its "Load" setting is very low (12% or 12 on the scale 0-100)!

Mercury 1000 tube tester calibration

Preliminary setup

1. Set all levers to "K".
2. Turn the "Load" control to zero.

STEP 1: Hickok bridge - Zero calibration

1. Set the "Function" switch to Gm-Em.
2. Rotate the "Load" control fully clockwise (CW) to 100. The meter should stay on zero.
3. If the meter moves away from zero, adjust the "Zero" trimpot to get the zero reading on the meter.
4. Rotate "Load" control back to zero.

STEP 2: Meter sensitivity calibration

1. Set the "Function" switch to Gm-Em
2. Place a silicon diode with its cathode on Pin 8 and its anode on Pin 5 of socket #4, for testing 6L6, 6V6, etc. The diode should be rated at 500 mA and 100 Volts minimum.
3. Set lever 5 to G and lever 11 to P.
4. Rotate the "Load" control to 50 and adjust "Meter sensitivity" trimmer to get the meter to read full scale.
5. Return the "Load" control to zero and levers 5 & 11 to K.
6. Remove diode.

STEP 3: Gas or Grid leakage calibration

1. Set the "Function" switch to "Gas - Grid leakage"
2. Connect a 100 MΩ resistor between pins 5 & 8 on socket 4.
3. Set lever 5 to G and adjust the "Gas" trimpot so that the meter reads right on the line between good/bad on the "Gas" scale.
4. Return lever 5 to K.
5. Move lever 3 to G. The meter should stay in the green zone on the lower scale marked "Gas". If the meter moves to the "Bad" area, replace the internal 6AT6 tube.
6. Return lever 5 to K and remove the resistor.

STEP 4: Signal calibration

1. Set the "Function" switch to Gm-Em
2. Connect AC voltmeter between pins 5 & 8 on socket 4
3. Set lever 5 to G
4. Adjust the "Signal" trimpot (100Ω) to get a reading of 1.00 V_{AC}.

STEP 5: Bias calibration

1. Set "Function" switch to "Gm-Em"
2. Connect a DC voltmeter with its red or positive lead at the bottom of R2 and its black or negative lead on top of R3 (the wiper of the "Bias cal." trimpot).
3. Adjust the "Bias cal." trimpot until the voltmeter reads +5.9 V_{DC}.
4. Move the negative voltmeter lead to the junction between R2 and R3. The voltmeter should read around 1.7 V_{DC}.
5. Should the reading be below 1.6V or above 1.8V replace resistor R2 and/or resistor R3 in the voltage divider circuit until the two readings are accurate (1.7 V_{DC} and 5.9 V_{DC}).

VERSION 1:

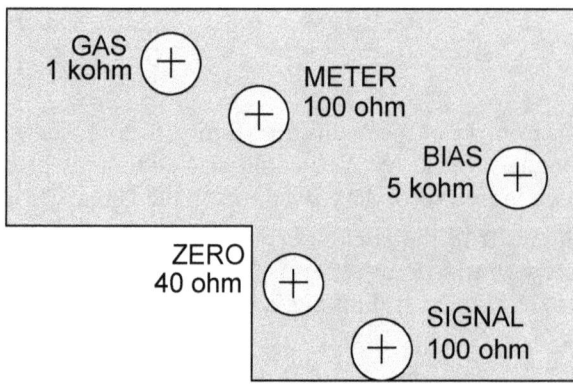

VERSION 2:

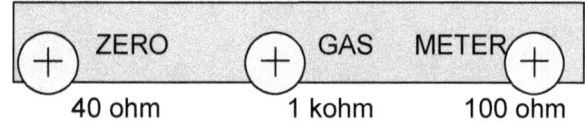

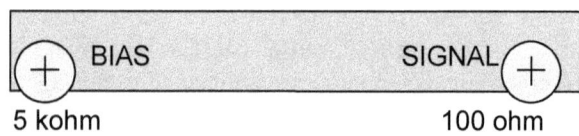

ABOVE: The location of calibration potentiometers on the board for two versions of Mercury 1000 tester

MERCURY 1200

Although functionally identical to models 1000 and 2000, Mercury 1200 features one of the largest analog meters we have ever seen. It is hard to comprehend just how large it is from photographs. However, instead of simple and generally trouble-free lever switches, model 1200 uses a bank of 26 push-buttons (1) of the same type used in many Jackson testers.

These aren't as user-friendly as the lever switches. Lever switches are independent - you can troubleshoot and replace them individually. With a push-button bank of switches, it is difficult to see which ones are depressed and which aren't, and the (mechanical) reset button has to be used often.

We've had many Eico and Jackson testers with faulty switch assemblies of the same kind. It isn't easy to troubleshoot and fix once the switching mechanism develops a mechanical or interlocking fault. Personally, no matter how much I like Mercury 1200, I'd never buy or use a tester with such switches.

The left compartment was used to store the mains cable and the CRT test loom, but once we replaced the mains cable and disconnected the CRT cabling, we covered the compartment with an aluminum fascia and installed additional tube sockets (2), RimLock, Magnoval, small 4-pin (for 2A3, 300B and similar DHT lovelies), and German Steel Y8A (for testing EL12, EL156, F2a11, and similar European power tubes).

The internals of Mercury testers are among the messiest and most amateurish-looking of all vintage testers.

While B&K and Hickok were large and relatively successful organizations (both still exist), Mercury was a micro business that most likely outsourced the manufacture of many of their testers' major components.

The cases of Mercury's tube testes are made of craft board, closer to paper than to particleboard; that is why they are so lightweight. The covering is also very thin and easy to scratch/mark/damage.

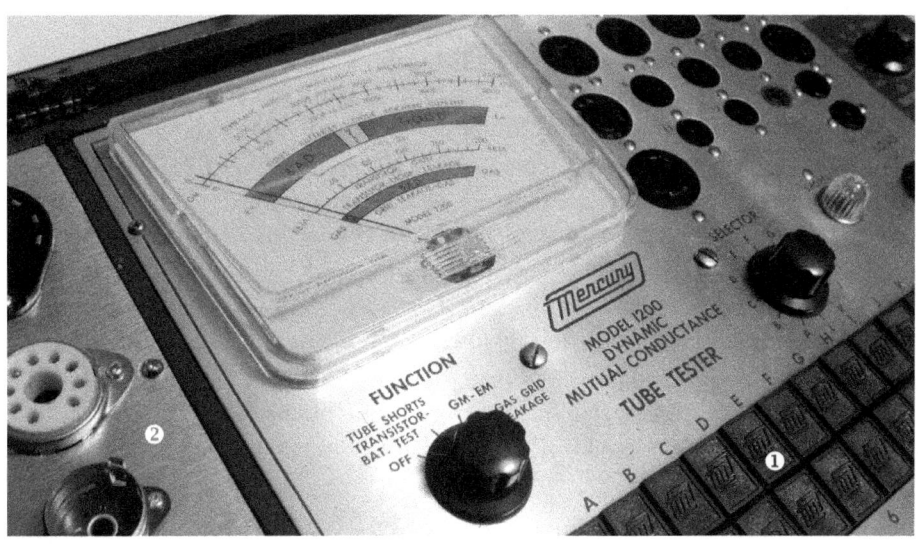

ABOVE: Mercury 1200 is functionally identical to model 2000, and, apart from its transistor test section, to Model 1000 as well

BELOW: The messy wiring of Model 1200. Luckily, the schematics is relatively simple

Notice the universal prototyping board used to support most of the componentry (3); resistors and capacitors were positioned in a willy-nilly, slapdash manner.

The power transformer (4) is tiny, and voltages sag significantly under load when larger power tubes such as EL34 and KT88 are tested.

The transistor and battery test circuit (5) is simple but effective. It displays current gain (beta) on a 0-200 scale and leakage on a qualitative (no numerical markings or graduations) scale.

For some reason, model 1200 used a 6BJ8 triode-duo-diode instead of 6AT6 in models 1000 and 2000.

Mercury 1200 tube tester calibration

The calibration of Model 1200 is identical to that of model 1000, except all references to "levers" should be replaced by "buttons."

Step 2-3 should read "Press buttons E and K," and steps 2-5 should read "Press both RELEASE buttons."

Steps 3-3 to 3-6 should read:

3. Push button E and adjust the "Gas" trim pot (1,000 ohms) so that the meter reads right on the line between good/bad on the GAS scale.

4. Press the upper RELEASE button.

5. Push button C. The meter should stay in the green zone on the lower scale marked "Gas'. If the meter moves to the "Bad" area, replace the internal 6BJ8 tube.

6. Press the upper RELEASE button.

Step 4-3 should read: "Push button E"

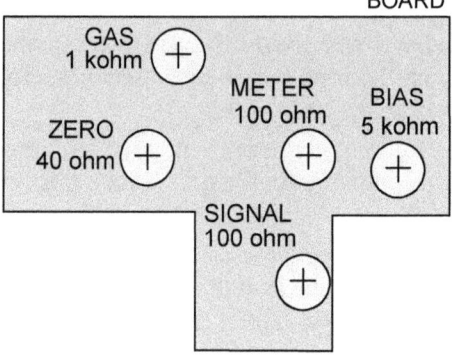

ABOVE: The location of Mercury 1200 calibration potentiometers on the board

PRECISE 116

Precise Development Corp. was a small business based in Oceanside, New York, not to be confused with the Precision brand. In 1955, as soon as the 1935 Hickok patent expired, they released two mutual conductance and emission testers, both based on a Hickok bridge circuit and sold both in kit form and as factory-wired units.

111 was their fully-featured transconductance tester (next page). Its smaller brother (not in a physical sense, but with regard to fewer controls and lower price), model 116 (full name "Precise 116 Gm and Em Ultra Fast Tube Tester"), seems to be an attempt to compete directly with B&K's fast Gm testers, marketed under their "Dyna-Quik" label.

Just to confuse users, while Precise 111 has a "Shunt" control, model 116 also has the same control, but for some strange reason, this time called "Function"(1), while the switch that should be called "Function" is called "Selector" here! Ah, those pot-smoking 1960s engineers!

The most impressive and desirable feature of Model 116 is the five banks (2), each with three test sockets: 7-pin miniature, 9-pin miniature (Noval), and octal. Each bank was prewired for the most common tubes that use that socket. The 7-pin sockets can test voltage amplifying pentodes such as 6AU6, 6AK5, 6AG5, 6BC5, and 6BC6.

The 9-pin sockets can test common duo-triodes such as 12AU7, 12AT7, 12AX7, 12BH7, 12AV7, and 12AY7. Finally, the octal bank is for duo-triodes such as 6SN7, 6SL7, 12SL7, and 12SN7.

Up to five identical tubes can be plugged in, saving the user lots of time that would otherwise be spent on unplugging a tube, plugging the next one, and waiting for it to warm up and stabilize.

Now, once they are all plugged in and preheated, testing them all together involves a few flicks of the "Tube bank" A-B-C-D-E selector switch (3), great for tube resellers and anyone with hundreds of identical tubes to be tested.

There is no provision for balancing the rectified peaks of the dual 6AX5 rectifier in the Hickok bridge, so that should be the first addition I would make.

There are only three internal calibration adjustments, bias level, gas sensitivity, and signal amplitude, so this tester isn't difficult to calibrate.

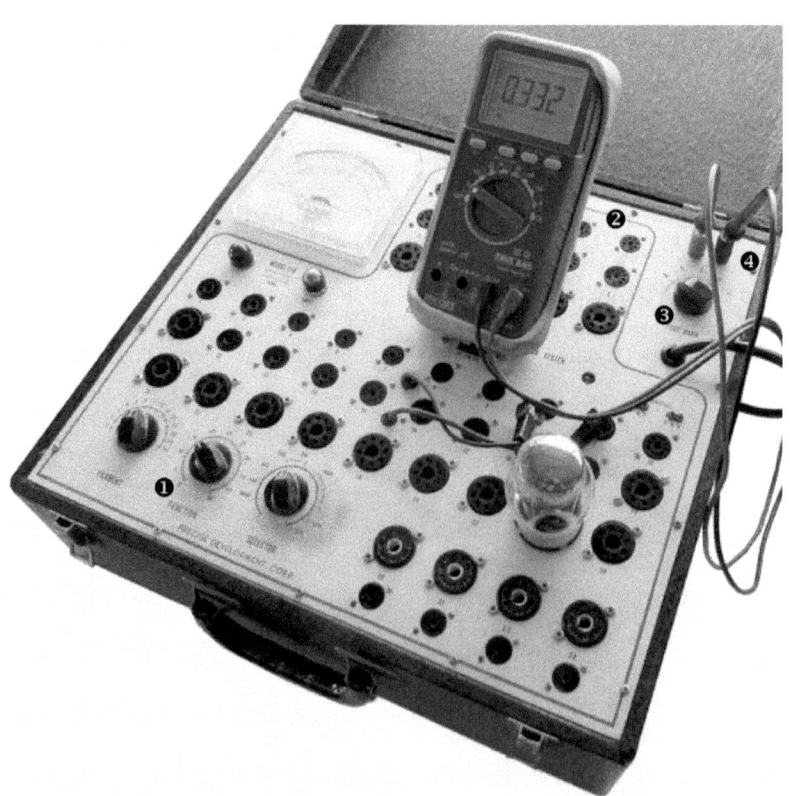

ABOVE: Testing a well-used Philips 6550 tube. The anode current measuring binding posts were added (4), the anode current is 33.2mA$_{DC}$.

RIGHT: The Hickok bridge and associated circuitry of Precise 116 Tester

1) Incandescent bulb used as a fuse
2) "Off - Life - On" switch
3) "Function" control ("Shunt" or "Sensitivity")
4) "Gas sensitivity calibration"
5) Bias High-Med-Lo terminals
6) "Signal cal." control
7) F1(common) and F2 (active) heater terminals
8) "Bias calibration"
9) Hickok bridge using 6AX5 duo-diode tube

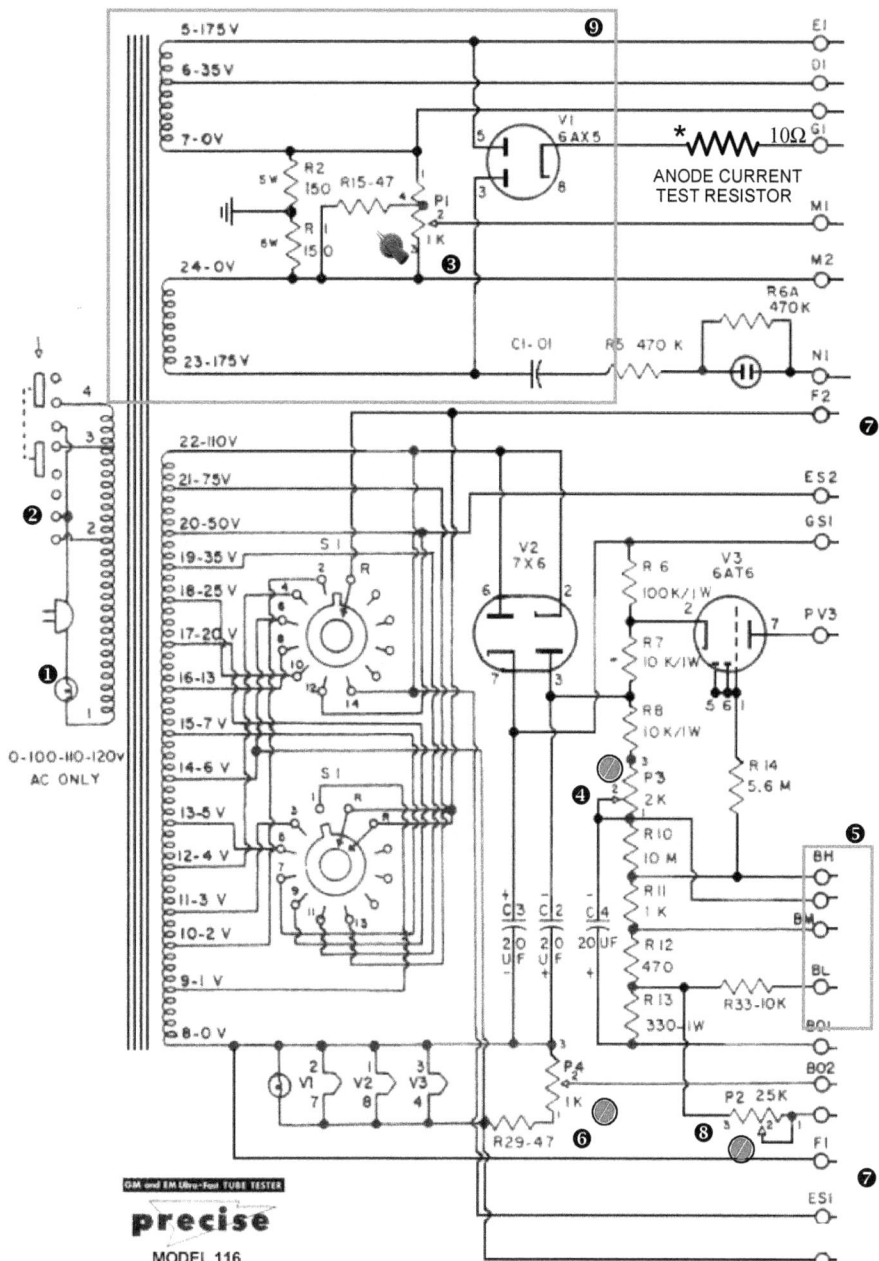

There are no aging selenium rectifiers, but there are three 20μF elcos that need replacing.

To measure anode current via a voltage drop on a precise resistor, insert a 1Ω or 10Ω resistor as indicated by * on the diagram (right).

Two vacuum diodes inside the 7X6 tube act as half-wave rectifiers, one in the DC bias supply, the other in the anode voltage supply for the gas amplifier, the 6AT6 triode. These can be replaced by standard 1N4007 silicon diodes.

Again, only one bridge voltage (175V_{AC}) means only one anode/screen test voltage, one grid signal voltage, and three DC bias voltages.

PRECISE 111

Functionally, Precise 111 is one of the better Hickok bridge-type testers. From a construction and reliability perspective, though, its build quality was in the B&K ballpark, better than Mercury's atrocious and haphazard wiring, but below Triplett and Simpson standards.

In 1955 the kit (111K) sold for US$69.95, and the factory wired unit (111W) sold for US$139.85. At that time, Hickok 750 was priced at US$229.- and Jackson 648P (a dynamic emission tester) was US$114.95, so 111 was a true bargain!

Later units had two more switches and two additional sockets (Novar and Compactron) where the "Precise" logo is on the photo below. They sold for US$115.95 as a kit and US$199.95 prewired.

The later model was also released as Realistic 113 and sold through Radio Shack outlets, with an added transistor testing section.

ABOVE: Realistic 113 was made by Precise, and apart from the additional transistor testing capability, was identical to the later model 116 version. It had a translucent plastic meter cover instead of black Bakelite of the earlier version, and two additional sockets and selector switches in the middle of the upper panel.

Internally, the wiring isn't that messy despite the tester's complexity. Wires of various colors were used, so tracing the connections isn't that difficult.

Separate transformers for the heater and all other signals and voltages are always a good sign, something most Hickok testers didn't have. Incredibly there are no less than 13 internal trimmer potentiometers; I dutifully marked them on the photo of the internals (next page), so you can locate them on the circuit diagram (two pages forward).

The two rectifiers (5U4G and 6X4) get their own balance adjustments, and each of the five Gm ranges (different signals) has its own calibration pot, seven trimmers already!

Add to that two adjustments to calibrate heater current test ranges (3A & 300mA), a bridge balance trimmer, internal line calibration, bias calibration, and "shunt adjust" calibration, and you get the picture.

Although Precise 111 has the inbuilt capability to measure anode current and currents of other electrodes, it is much easier to permanently wire in a 10Ω resistor and to install binding posts for that purpose (5).

That way, an external voltmeter can measure the anode current during the Gm test - less switching & time wasting.

BELOW:
1) "Off-Life-On" switch and globe fuse
2) Bias range switch and adjustment
3) "Meter" control: "Line" - "Bias" - "Test" and two filament current positions, 3A and 300mA.
4) "Gm-Em" selector for "Test" mode
5) Added binding posts for anode current measurements
6) Three screen and six plate (anode) voltages, but only in "Em" mode
7) The "Test" switch with momentary (left) and latched (permanent) positions (right), a very good idea
8) "Gas" test switch
9) Plenty of space for additional sockets, more than any other tube tester

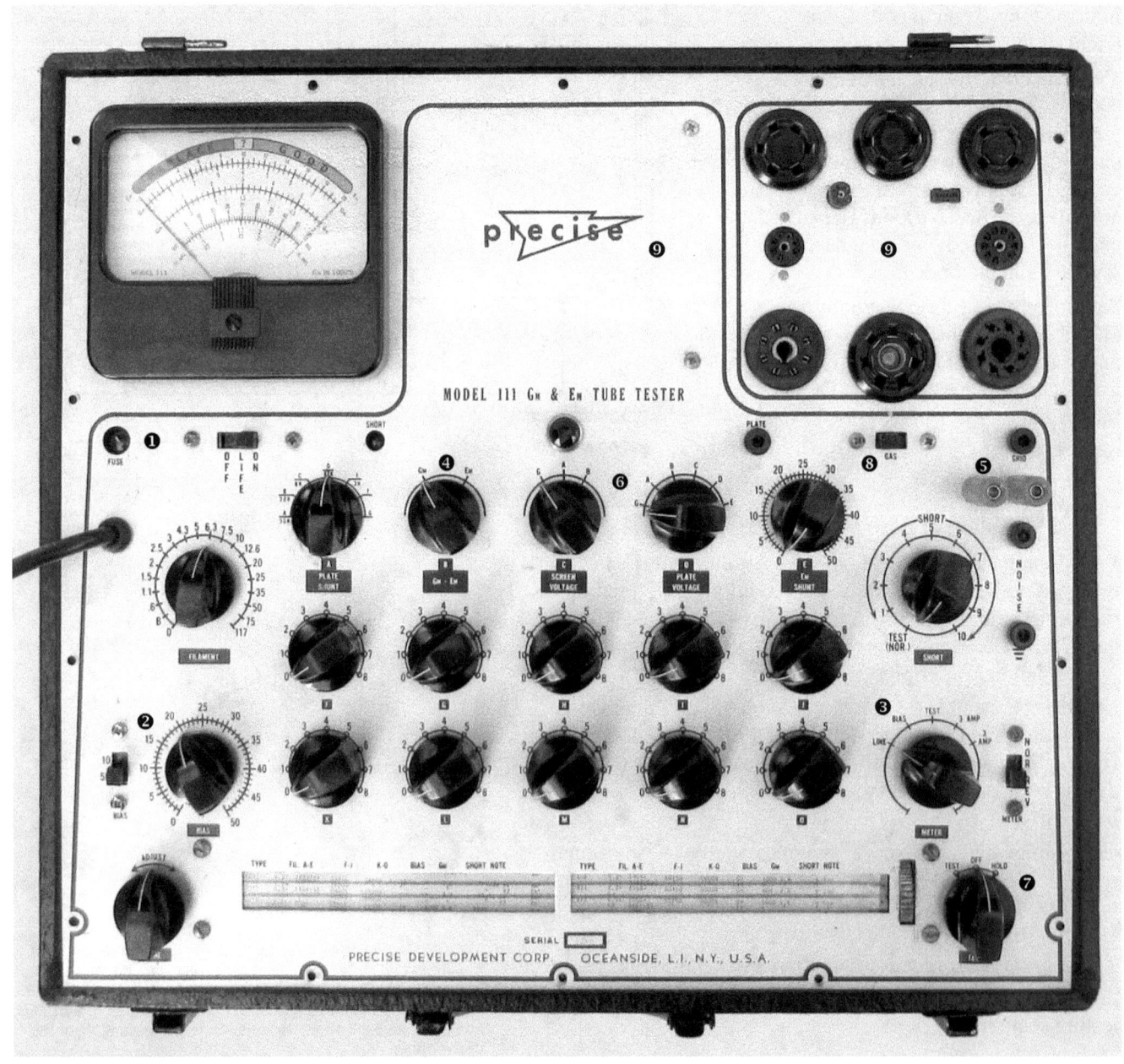

1) "Line Adjust" rheostat
2) 6U4G rectifier tube
3) PT3 - filament supply transformer
4) PT4 - anode/screen/signal transformer

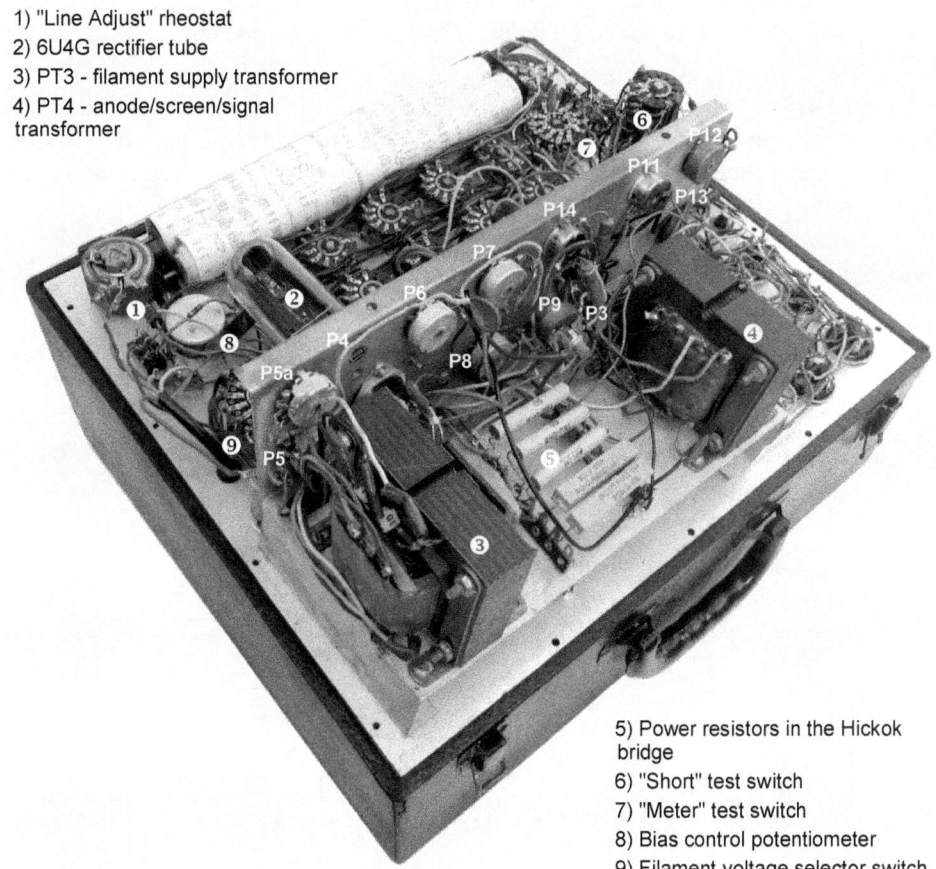

5) Power resistors in the Hickok bridge
6) "Short" test switch
7) "Meter" test switch
8) Bias control potentiometer
9) Filament voltage selector switch

Notice the date on the drawing, March 1954, a full year before the Hickok's patent expired! Precise sensed that being the first to release a Hickok-type tester could be crucial.

Alas, B&K also released their first Hickok-clone (model 500) the same year, and despite B&K 500 being in every sense inferior to Precise 111, B&K testers won in the marketplace.

Perhaps Precise made a mistake by releasing 111 as a kit as well, which could have given the potential buyers the impression of Precise being the "el cheapo" option. Goes to show that buyers' common sense and intelligence should not be overestimated.

Most simply did not and, judging by high prices some mediocre testers fetch on eBay, still do not have the necessary skills for an in-depth comparison of seemingly similar testers.

The four Gm scales (0-3, 0-6, 0-8, and 0-20 mA/V) are welcome, but instead of two scales that are very close (6 and 8mA/V), a better choice would have been 0-3, 0-8, 0-16 and 0-30mA/V.

The bias scale is relatively precise, so the only thing missing is the leakage scale. Alas, Model 111 does not use an analog MΩ-meter to measure leakage but a primitive neon on-off circuit. This is the most serious shortcoming of this otherwise decent tester.

Noise and microphony test

Connect the "Antenna in" terminal of a radio or AM (Amplitude Modulation) receiver to the "Noise" terminal of the tube tester. Connect the GND terminals on the tester and the receiver as well.

Tune the receiver to any position on the dial without a radio station. While performing the "shorts" test (rotating the "shorts" switch through all of its positions), gently tap the heated tube under test at each position. Some microphony is to be expected, but intermittent shorts and internal arcing will clearly be heard.

ABOVE: Four Gm scales obviate the need for any multiplication or guesswork.

BELOW: The "Noise" terminals on any tube tester can be connected to high impedance headphones or to an AM tuner or receiver. Any noise present will be heard in the loudspeaker or the headphones.

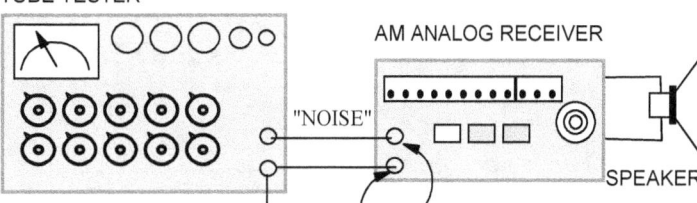

Despite the complexity of the circuit, the schematic isn't that difficult to follow. A directly-heated rectifier with a very high internal resistance (5U4G) is used, so if it's replaced by silicon diodes, the anode & screen test voltages will jump from $160V_{DC}$ to almost $200V_{DC}$!

There are five signal voltages, each with its own calibration trimmer. The "Plate Voltage" switch has six positions, but there is only a single "DC" anode test voltage; the other five are various AC levels for Em (emission) testing. The 6X4 rectifier provides a screen DC supply, and three screen voltages are available. In tetrodes, pentodes, and beam power tubes, screen voltage determines anode current much more than the anode voltage.

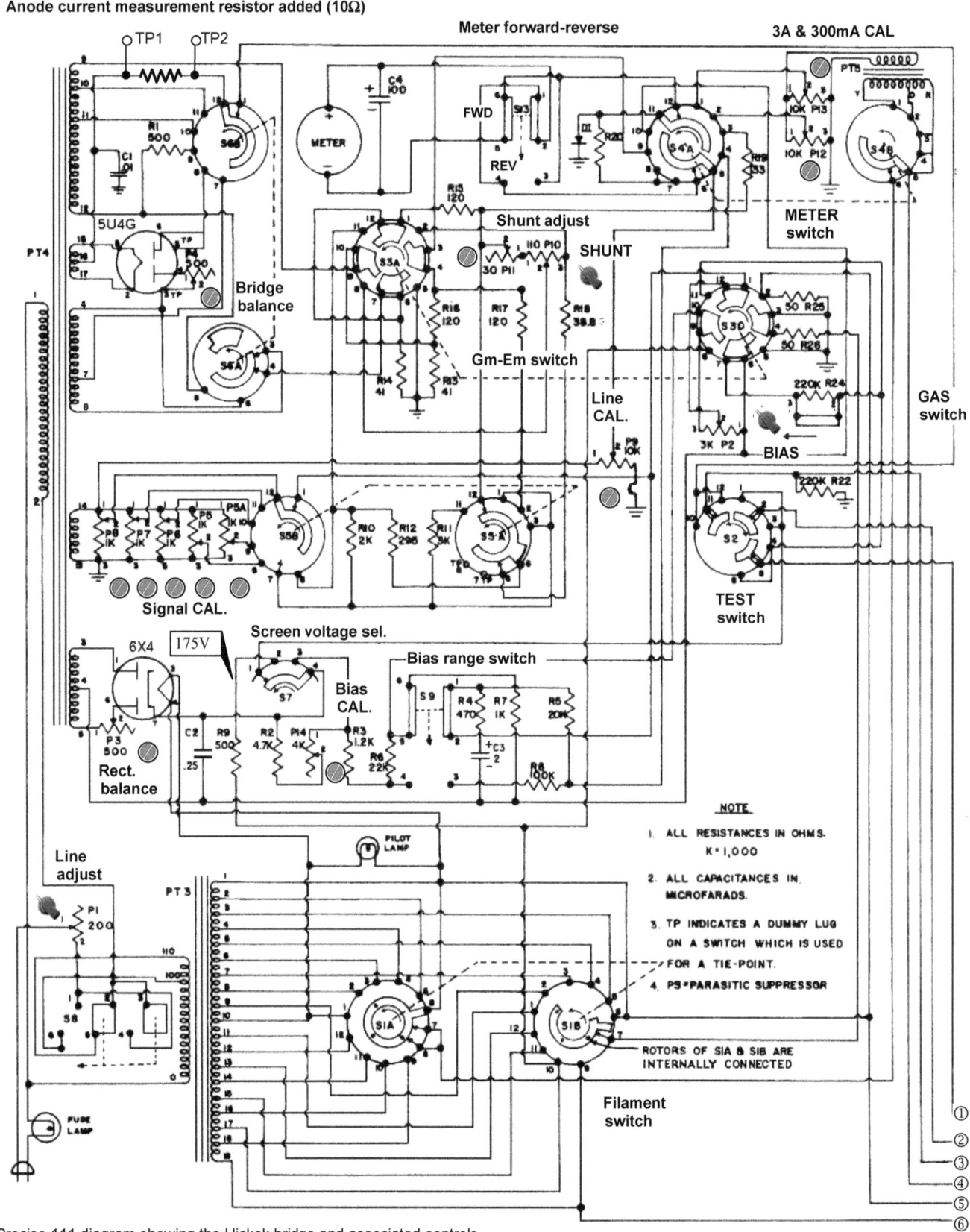

Precise 111 diagram showing the Hickok bridge and associated controls

HICKOK-TYPE TESTERS

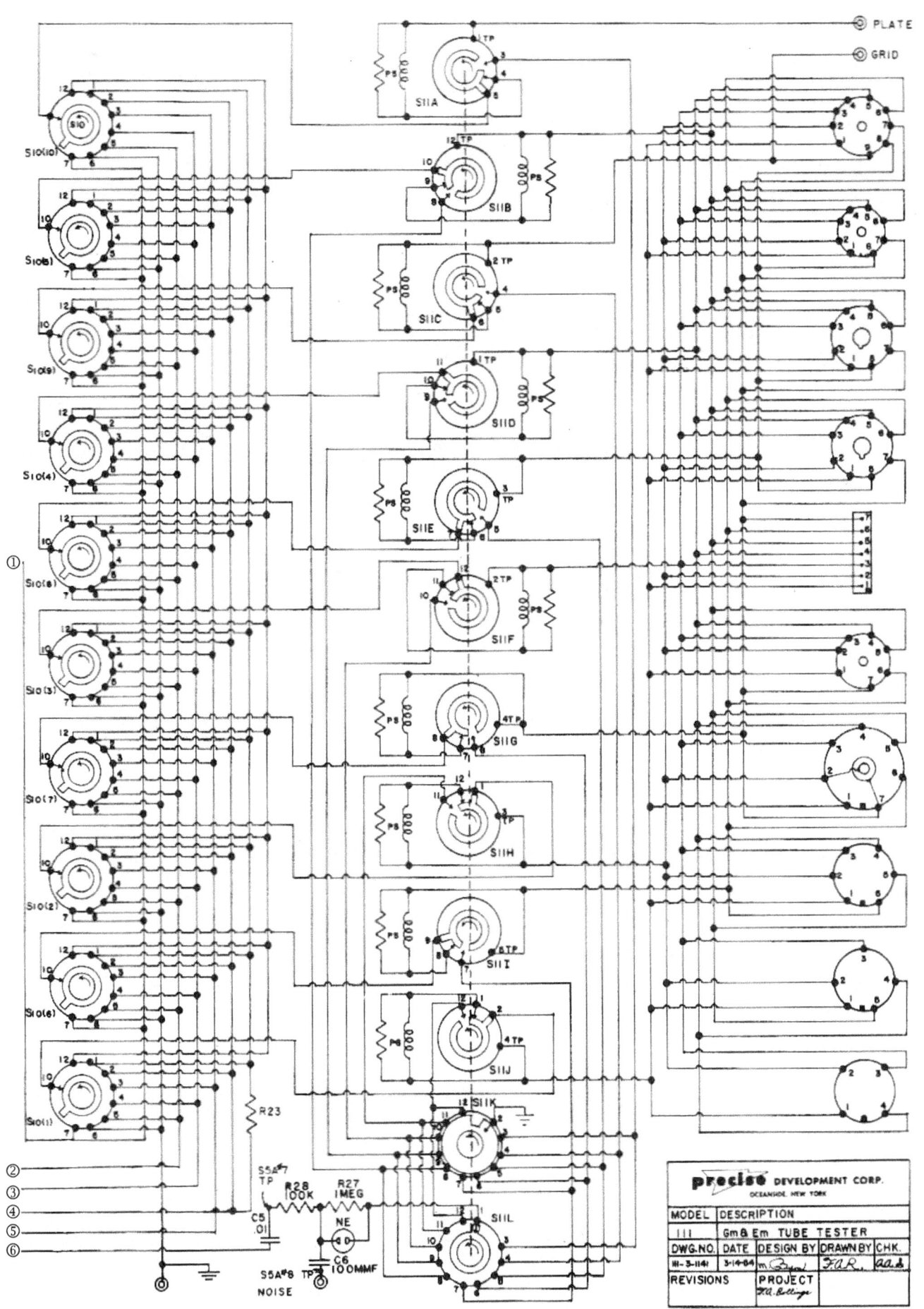

DYNAMATIC DM456

Made by TeleTest Instrument Corp., DM456 is probably the tiniest mutual conductance tester made, much smaller than even Seco 107. Three sockets (Octal, 7-pin mini, and 9-pin Noval), three controls, and one slot for inserting Bakelite cards are all there is to it.

The test settings for 375 tubes are printed on a large cardboard piece, attached to the inside of the cover. There was no user manual or schematics, but some detective thinking will shed light on its inner working.

The expected test results for 12AU7 (4,400 micromhos) and 12AX7 (3,200) are twice as high as on other Gm testers.

Since there is no "Test1- Test2" or "Gm1-Gm2" switch for duo triodes, these are tested in parallel (just as in B&K 500), and hence the test results are a sum of two mutual conductance figures.

Remember, when identical tubes are paralleled, their Gm is doubled and the internal resistances halved, so the amplification factor (μ) remains unchanged.

There are two tubes under the hood, a 6AT6, a triode + duo-diode (1), and a 6AX5 twin rectifier (2), the same tube complement as in B&K500.

There are also two incandescent bulbs (3) in B&K style automatic signal bridge, two selenium rectifiers (4), and a tiny power transformer (5).

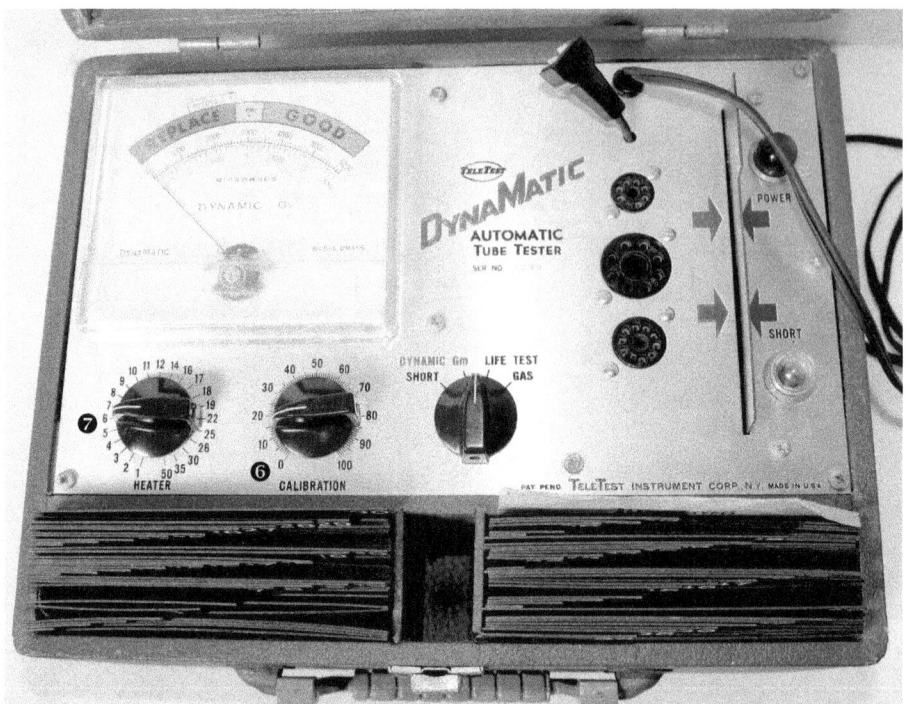

This is a Hickok-bridge-based tester, the test engine of B&K500 combined with a card switching similar to B&K675. The cards are very brittle, but replacements cards made of modern stiff plastic sheets could be made. But then, why go to all that trouble and still have to live with so many limitations of this tester, namely the inability to match twin-triode preamp tubes, a very limited range of tubes that can be tested, and dubious results at low test voltages.

There are only two Gm scales, 0-6,000 and 0-18,000 micromhos, so newer tubes with Gm over 9,000 micromhos (9 mA/V) cannot be tested!

For example, 6V6 beam power tubes are tested with a 1V signal, a grid bias of -7.5V_{DC}, and anode and screen voltages of around +150V_{DC}. During this test, the nominal 6.3V heater voltage sags to an incredibly low 5V_{AC}!

If you live in a shoe-box and every inch of space is at a premium, if you don't need to know if two triodes inside duo-triode tubes are matched, if you don't mind making replacement cards when the originals are lost or broken, if you only test the most common audio tubes (6L6/6V6/12AX7/12SN7 and the like), and if you can buy this tester for under US$50, it may tell you something about the state of your tubes. If you can live with so many "ifs," ...

SIMPLE FIXES & UPGRADES FOR HICKOK-TYPE TESTERS

The earlier the Hickok's model you have, the more you'll have to upgrade. Our model 533A didn't have many improvements included in later models, for instance, in the 539 series. It pays to study circuit diagrams and instruction manuals of later Hickok models and testers that use the same Hickok circuit, such as B&K, Precise, and Mercury. You may get some good ideas.

Fixing the Gm range mismatch

In our Hickok 533A and Weston 798 testers, the Gm readings varied between different positions of the range switch. For instance, if the Gm displayed for a particular tube on a 6,000 micromhos range was 5,000, on the 16,000 range (Hickok 533A), the reading would drop to 4,200 micromhos.

The solution in adjusting the values of resistors that form voltage dividers used to scale the meter readings. In both testers, which date from approximately the same year (around 1949 -1950), wound spool resistors were used.

After determining what resistance values you need, you have two options. You can get rid of those silly spools and replace them with modern precision metal film resistors, or, if you enjoy tedious and unrewarding tasks, you can carefully remove them one by one from the switch they are bolted onto, and start unwinding them until you get to the values you need.

Alas, that only works if you need a lower resistance than the one you have. You are out of luck if you need to add a few turns.

Replacing 83 and 5Y3 tube rectifiers with silicon diodes

5Y3 and 83 tube rectifiers inside earlier Hickok testers can be replaced with solid-state diodes. You'll benefit in many ways. The 83 rectifier tubes are filled with mercury vapors. Mercury is highly toxic and illegal in many countries. Luckily, customs officials are unaware of such hazardous materials in tube testers; otherwise, they would not allow their importation. If caught, you would have to pay for their disposal or even a hefty fine.

If a tube gets broken while in a cold state, mercury would be in a liquid state and its droplets would probably stay contained inside the broken bulb. However, mercury vapors would spread throughout the room if broken while in a hot (operating) state and contaminate it forever!

Also, by not using 83's heaters, you will gain 15 watts of transformer power. Since 533's mains transformer is undersized, these are welcome gains! 5 volts times 3 Amps is 15 watts of power wasted on the rectifier. By replacing 5Y3 with silicon diodes, you'll save another 10 Watts of heater power, for a total saving of 25 watts (out of the 60 or so watts the transformer is rated at!) Your tester will run cooler and consume less power.

The voltage drop across the 83 rectifier is small compared to other, non-mercury types such as 5U4, but it still drops the plate voltage by about 15 volts.

The two diodes inside one rectifier tube age and wear out differently, so you get an unbalanced pulsating voltage, which will affect the Gm readings of your tester. Silicon diodes don't age, and voltage pulses will be equal! See the oscilloscope screenshot (Hickok 533A tester using 83 rectifier tube) on the next page.

There are two ways to perform this upgrade. One involves simply plugging in a solid-state replacement module. You can buy a couple of different models on eBay and online retailers.

Alternatively, you can make a SS module yourself by soldering a couple of diodes and resistors into an octal plug, as per the diagram below. The two resistors should be 10Ω, rated at minimum 1 Watt; diodes can be the garden variety 1N4007 or similar.

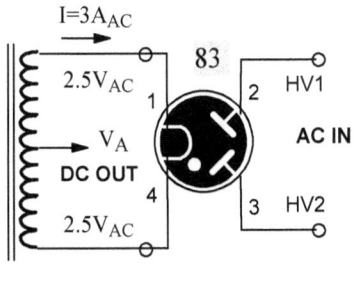

ORIGINAL

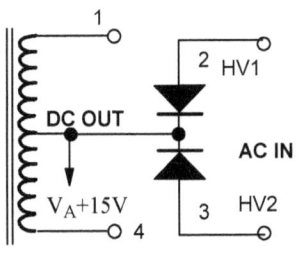

IMPROVED

RIGHT: How to wire a solid state plug in replacement for Type 83 rectifier

LEFT: How to permanently replace #83 mercury vapor rectifier tube with silicon diodes

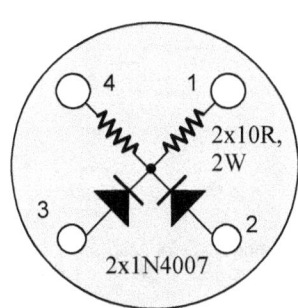

A better alternative involves a bit of surgery inside your tester. Wires from the transformer windings to pins 1 and 4 will be unused. Connect an additional jumper wire from the center tap on the transformer to where two cathodes meet. Ensure you insulate the wires connected to pins 1 & 4, so they don't touch anything and cause a short circuit.

Don't remove them completely, just in case you want to sell your tester later. Some buyers prefer Hickok testers in the original state and don't like solid-state replacements, so you easily re-solder everything back on before the sale.

Adding a rectifier balance control

As illustrated in the photo (below right), the unequal rectified sine pulses introduce an error into Gm readings. Later Hickok testers have a 50-ohm rheostat included for balancing purposes, for instance, R8 on model 539C. You can add one yourself very easily. Connect the fixed ends of the rheostat to pins 1 and 4 of the rectifier tube (the ends of the secondary winding). Disconnect the wire from the CT and connect it to the wiper (slider) of the potentiometer.

Insert a power tube (6L6, for example) and press Gm or "test" button to obtain a meter reading, and you'll get the waveform as pictured. Observe the waveform of the pulsating anode voltage on an oscilloscope. Keep the Gm button pressed and rotate slowly the rheostat one way or the other until both peaks become equal. Be careful not to touch any of the three lugs; they are live and carry high voltage.

ABOVE: SS modules such as TubeDepot SSR are simply an octal plug with two 1N5408 silicon diodes soldered inside.

Another way of achieving balance was used by Precision in their Hickok-bridge testers, such as models 111 and 116. A 500Ω rheostat is included in one leg of the rectifier. The slider is moved one way or the other until both peaks are equal. But then, what if the rheostat is needed in the other leg (depending on which diode is weaker)? The Hickok solution is more elegant and allows for balancing both ways.

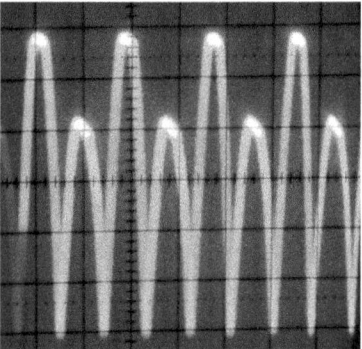

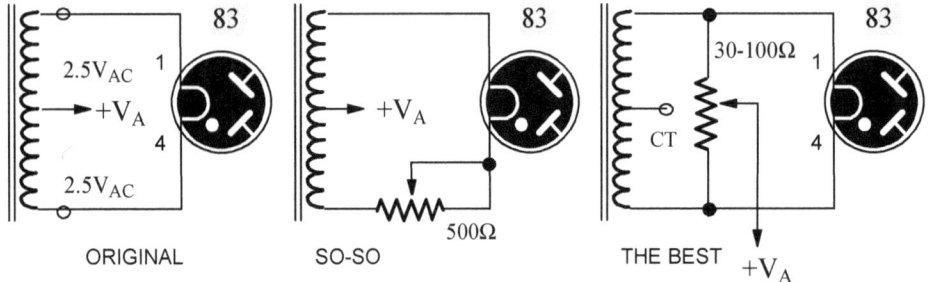

LEFT: Adding a rectifier balance control solves the imbalance in the pulsating anode voltage (Hickok 533A waveform pictured above) and thus greatly improves the accuracy of Hickok-type testers and should be performed before any calibration is attempted!

Adding a signal calibration control

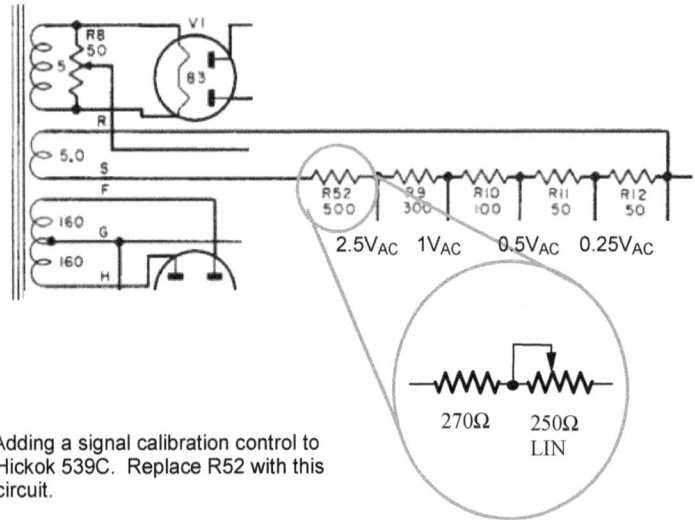

Adding a signal calibration control to Hickok 539C. Replace R52 with this circuit.

While Precise, Mercury, and B&K testers based on the Hickok bridge have an internal signal calibration control, most Hickok testers don't. Calibrating the rest of any Hickok tester without precisely adjusting the test signal amplitude is pointless, so, apart from the addition of the rectifier balance control, this upgrade is the most important of all!

Although the signal circuit will be slightly different, this procedure applies to all Hickok testers. Most use only one signal level, so there is no R9-R12 voltage divider network found in model 539C.

The five resistance values in the voltage divider add up to exactly 1,000Ω, so it is easy to mentally calculate the voltages along the divider string (marked on the diagram).

First, measure the four fixed resistors and replace those whose values had drifted over time with new, precise metal film replacements. Then, replace the 500R resistor with a trimmer pot and a series resistor, as per the diagram. Finally, adjust the 250R trimmer pot until the voltage in point X is exactly 2.5V_{RMS}.

Adding a plate current measurement capability

Even if a tester doesn't have any current measuring feature (and most don't), the anode or even screen currents can be measured via the voltage drop on a precision 1-ohm or 10-ohm resistor inserted in series with those electrodes. The value should not be too high since that would change the tube's operating conditions under test. If a 1Ω resistor is used, 1 mV is equivalent to 1 mA of anode current I_A. With a 10Ω resistor, 1 mA of I_A will produce 10 mV of voltage. The resistor should be inserted in the line that goes to the anodes of tubes under test.

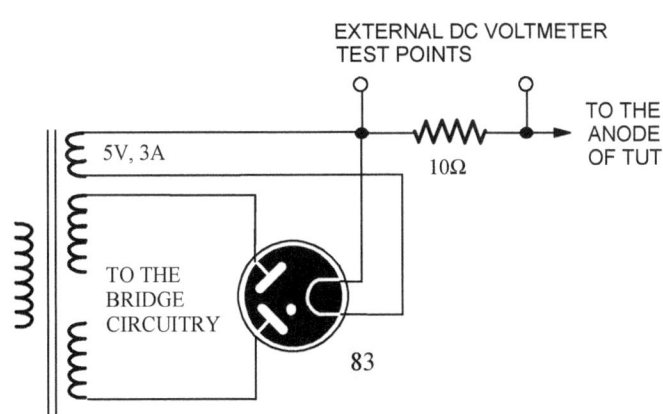

ABOVE: Where to insert the anode current measuring resistor in Hickok-type testers. The rest of the bridge circuit is not shown.

ABOVE: To measure anode or cathode current, add one series resistor and two binding posts and use an external meter. CAUTION: High voltage will be present on the binding posts during testing, so they need to be insulated from the chassis (top metal plate)!

Installing a meter-reverse switch

Hickok-type testers show negative deflection when testing directly-heated triodes such as 45, 2A3, or 300B. Since the heart of those testers is a balanced bridge, there is nothing wrong with such voltage polarity (you cannot damage the tester).

An effective solution that will enable you to test those tubes (although they may not be listed in tube charts) is installing a meter reverse switch. Hickok's own testers all have such a switch, but B&K, Mercury, and Precise Gm testers, which all use the Hickok circuit, don't.

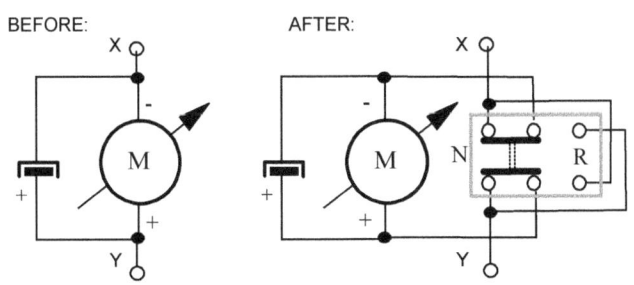

ABOVE: How to install a meter reverse switch

This simple modification using a DPDT (Double Pole Double Throw) switch is illustrated on the left. It can be sliding, rotary, or toggle switch; it makes no difference. If the meter has two anti-parallel diodes for protection, simply leave them in place; they will not be affected.

Strictly speaking, the polarity of the damping elco should be reversed as well, but since the voltages across analog meters are under 0.5V, all elcos will be able to withstand such a small reverse polarization.

Testing full-wave rectifiers

The test data for Hickok 533 and similar testers indicates a couple of issues about testing dual rectifier tubes. Let's use 5AU4 as an example (test data on the next page).

The first figure (1) is a filament voltage of 7.5V_{AC}. However, 5AU4 has a 5V heater! The explanation lies in the relatively high heater current draw of 3.75 Amperes, as seen from its datasheet (2).

The power transformers in Hickok testers (especially older ones dating back to the late 1940s, such as 533) have a very poor regulation due to their small size and low-grade laminations used. With a 3.75A load, the voltage on their nominally 7.5V heater tap will sag (droop) to 5.0V!

The second issue is different "Eng." (for "English") settings for two identical anodes (3), "51" for pin 6 (anode 1), and "46" for anode 2 (pin 4).

With identical "English" or shunt settings, anode 1 at pin 6 would always test stronger, even if the two diodes were perfectly balanced. So, the higher setting lowers the meter reading slightly as a compensation measure.

This problem only applies to filamentary rectifiers, where the filament (heater) is also the cathode, meaning it emits electrons itself. Due to filament voltage distribution (gradient), one anode in a dual rectifier tube is always tested at a higher anode-cathode voltage than the other. The average filament potential for A2 is 3.75V, but for A1 it's only 1.25V, so 2V less. That means that the anode-to-cathode test voltage for A1 is 2V higher, causing the disparity in meter indications.

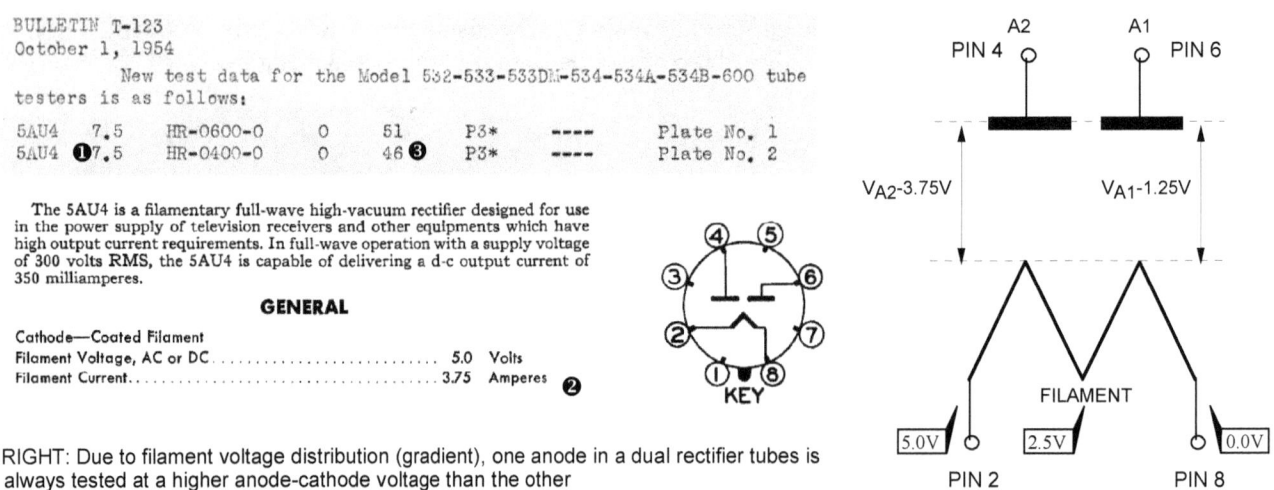

RIGHT: Due to filament voltage distribution (gradient), one anode in a dual rectifier tubes is always tested at a higher anode-cathode voltage than the other

Higher quality testers check rectifiers with 150-300V_{AC} on their anodes (plates). In that case, a couple of volts would make no difference to the readings. However, the Hickok testers in question use a very low test voltage of 35V_{AC}, so one anode is tested at 35-3.75 = 31.25 V and the other at 35-1.25 = 33.75 V. Those 2 volts will make a significant percentual difference and cause even perfectly balanced filamentary rectifiers to test as unbalanced!

Testing rectifiers that normally work with 300-500V using a 35V test voltage is like testing someone's swimming skills in three feet of water. Many rectifiers that test OK at such a ridiculously low test voltage will not work properly at ten to fifteen times higher voltages in amplifier power supplies.

TRUE MUTUAL CONDUCTANCE TESTERS

- HEATHKIT TT-1 AND WESTON 981
- TRIPLETT 3444
- SENCORE MU-140 & MU-150
- RCA WT-110A
- SECO 107, 107-B & 107-C
- MODERN TRANSCONDUCTANCE TESTERS

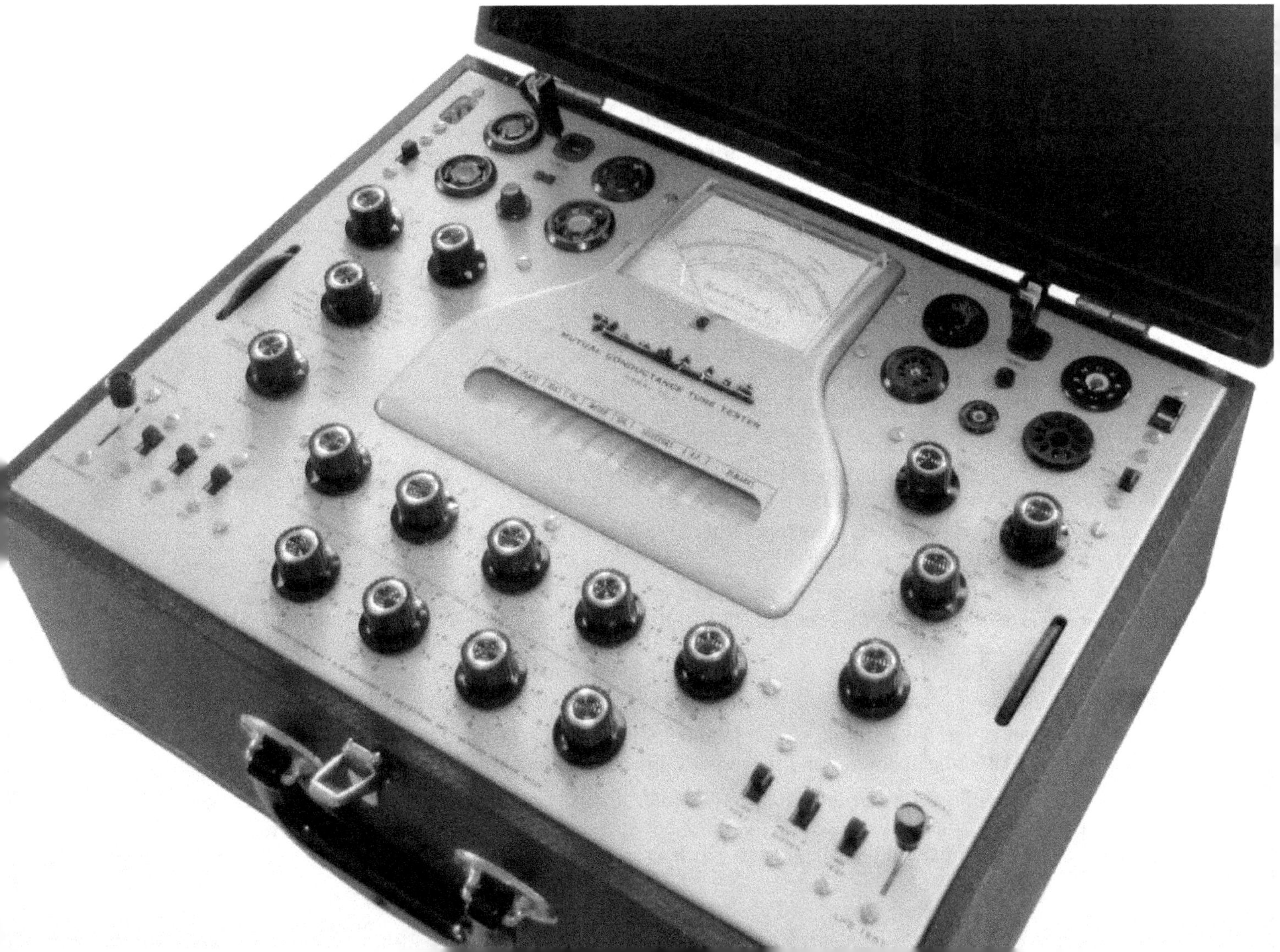

We have now covered two types of Gm testers, proportional mutual conductance testers, which used AC voltages for all electrode supplies and used the tube-under-test itself to rectify them, and Hickok-type Gm testers, which relied on an AC bridge arrangement. The use of AC supplies was common to both types. When these testers were designed and built in the first half of the 20th century, rectifiers and DC power supplies were expensive, so these cost-cutting measures simplified tester design, minimized costs and maximized profits.

In contrast, "true" Gm testers use DC anode, screen, and bias supplies, ideally stabilized (regulated), but none of the USA-made testers went that far, only some rare and expensive laboratory models and a few European testers.

HEATHKIT TT-1 AND WESTON 981

Weston's line of testers moves from 798 through 978 to 981-1, 981-2, and 981-3. While the looks and the switching arrangements are different, there are many similarities under-the-hood.

Heathkit TT-1, a clone of Weston 981, was introduced as a DIY kit in January 1960, priced at $134.95, followed by TT-1A in March 1962, $149.95. An upgrade kit for the older TT-1 (TTA-1-1) was also available for $19.95. That was the upper-mounted panel with five additional sockets (Novar, 5- and 7-pin Nuvistor, Compactron, and 10-in miniature) and three rotary selector switches for pins 10, 11, and 12. It plugged into the main tester control panel. Still, there was no Magnoval socket or European sockets, as was the case with all other American testers.

Heathkit specified the average assembly time for the TT-1 kit as a whopping 40 hours. Considering that, its seemingly moderate price was quite expensive for that time.

Specifications:

- Anode/Screen DC volts: 26, 90, 135, 225, and variable 80- 200V_{DC} for space charge grids
- AC volts: 20, 45, 177
- Low bias 0 to -5V, High bias 0 to -20 V_{DC}
- Signal voltages: 2, 1, 0.5, 0.25 V_{AC} (5kHz)
- Heater voltages: 0.65, 1.1, 1.5, 2, 2.5, 3.3, 5, 6.3, 7.5, 10, 13, 20, 27.5, 35, 47, 70 & 115 V_{AC}.
- Heater current strings: 300, 450, 600 mA
- Gm scale: 0-3 mA/V, Gm range: 0-24 mA/V max.
- Leakage: MΩ-meter, Grid current: ¼ µA sensitivity
- Scales: Gm: 0-3,000 micromhos. VR test volts: 0-200 volts. Leakage: 0-10 Mohms.
- Tubes: 1x3A4 (grid signal oscillator), 1x12AV6 meter driver
- Sockets: 4-pin, 5-pin, 6-pin, 7-pin, combination and pilot light, 7-pin mini, 7-pin mini, 8-pin mini, octal, Loctal, 9-pin mini (Noval).
- Line adjustment: continuous

Gm test circuit

The DC component of the anode current passes through the choke L but cannot pass the capacitor C. The AC component of the plate current, proportional to Gm, splits after the capacitor - part of it is shunted away from the meter by the adjustable internal calibration R, the rest goes through the meter and its series resistor R_S.

The circuit is a high-pass LC filter, tuned to filter out 50 or 60 Hz low frequency (hum). The resonant frequency of the LC circuit is $f_R = 1/(2\pi LC)) = 2.9$ kHz, but that formula is valid only for an ideal LC circuit with no resistance in the circuit.

Here the choke has its own resistance and there is a variable resistor in series with capacitor C, a total of $R\|(R_S+R_M)$.

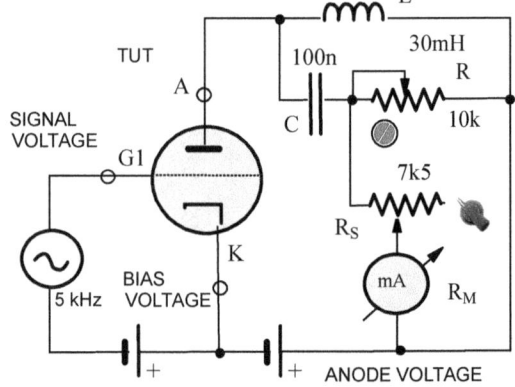

ABOVE: The Gm test circuit of Heathkit TT-1

R_S is an external control marked "Sensitivity" (typically a 7k5 rheostat), and R is the internal scaling shunt (trimmer pot) whose adjustment "calibrates" the metering circuit.

The first problem is that the metering circuit loads the tube and affects the results, so awkward multiplying factors need to be used. This blind copying of a flawed design was a serious mistake by Heathkit.

Adding a cathode follower as an impedance decoupler and a relatively simple tube amplifier to drive the meter would not cost much and would make TT-1 a much better tester. Heathkit used such amplifiers in other instruments (see the VTVM diagram on the next page), so this oversight is mind-boggling.

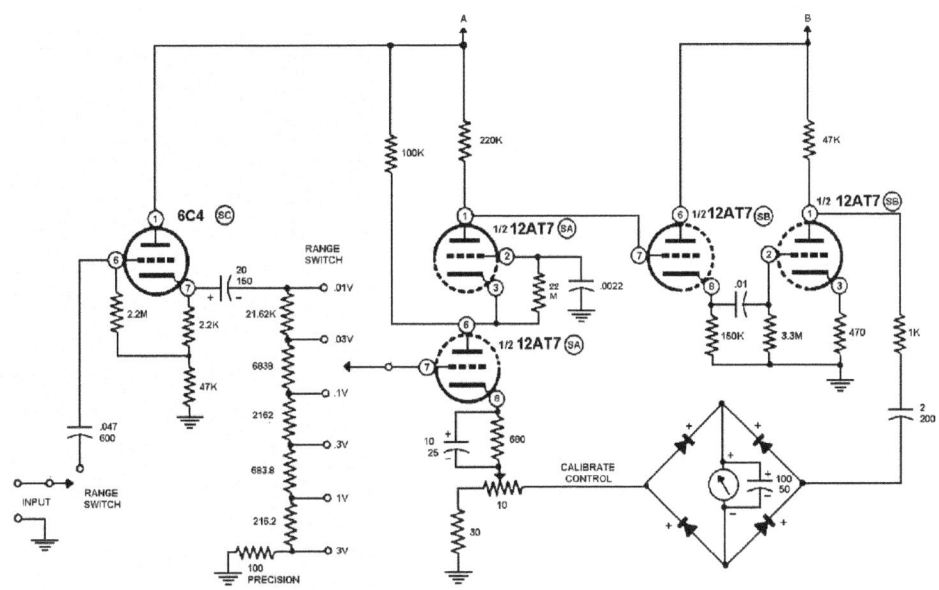

ABOVE: The partial schematics of Heathkit VTVM. Such a metering circuit (even its simplified version) would solve the loading problem of 981 and TT-1 testers.

Simpson 330, a much older Gm tester, also had an LC filter in the anode circuit of the tube-under-test but solved this problem by using a single duo-triode (6SN7) as a metering amplifier. Incredibly, Weston and Heathkit engineers did not learn anything from previous models of other brands!

The second problem is that the meter reading depends on the properties of the choke. A choke is a mechanical component with huge tolerances, so get two TT-1 or 981 testers to test the same tube, and you may get significantly different results!

The issue of testing triodes with low internal resistance

Anode current I_A of a common cathode stage (just as a tube-under-test in tube tester Gm testing circuit) with load R_L is $I_A = \dfrac{\mu \cdot V_G}{R_I + R_L} = \dfrac{V_G}{\dfrac{R_I}{\mu} + \dfrac{R_L}{\mu}} = \dfrac{V_G}{\dfrac{1}{Gm} + \dfrac{R_L}{\mu}}$

R_I is the internal resistance of tube-under-test, and μ is its voltage amplification factor.

Factor R_L/μ is the source of problems. Without it, we would have $I_A=(\mu/R_I) \cdot V_G$, and since $Gm=\mu/R_I$, the meter current with a known grid signal V_G would be proportional to Gm. Now, due to the loading effect of the external metering circuit R_L, it isn't!

For high internal impedance tubes such as pentodes (which also have a high μ), that second factor R_L/μ could be neglected if μ is at least 50-100. For low μ triodes, such as 300B for instance (μ=3.5), with low internal impedance (R_I = 700-800 Ω), that factor is of the same order of magnitude as the first factor, R_I/μ!

So, testers such as Weston 981 and Heathkit TT-1 have their Gm scales calibrated for one specific value μ_0 and one value of $R_L=R_0$. For tubes whose μ is different, the reading must be "fudged" by different multiplication factors so there must be a way of adjusting the external resistance of the metering circuit, so the factor R_L/μ is kept constant and equal to R_L/μ_0.

Weston tube tester patents

The first US patent of interest to us here was No. 1,854,901, awarded to W. N. Goodwin on April 19, 1932. The subsequent patent no. 2,456,833 was awarded to O.J. Morelock on Dec. 21, 1948. Both were assigned to Weston Electrical Instrument Corporation and can be downloaded from online sources.

FURTHER READING

Testing diodes and rectifiers

For each tube, a suitable AC voltage is selected (choice of three), as is a load resistor and a meter shunt. A milliammeter is connected in series with the TUT and measures the current through the diode. This tests the tube's emissive capacity.

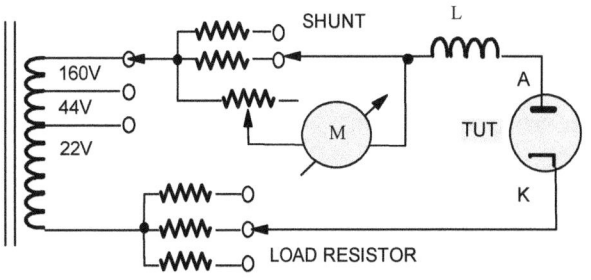

Calibration

TT-1 is possibly the easiest Gm tester to calibrate. The internal "Calibrate" switch has four positions, marked "Bias-Signal-Meter-Operate," and calibration is performed in the same sequence. Before each of the three steps, the "Set Line" adjustment must be made so the meter's needle is right on the "Line check" mark.

With "Bias" set at "20L", trimmer "CD" is adjusted until the meter's indication is again right on the "Line check" mark. The "Set Line" adjustment should be performed again if the meter's needle hasn't remained on the "Line check" mark. The "CD" setting should also be rechecked and finely adjusted if necessary. These controls are interactive.

With zero bias, "Signal" at "1" and "Set Line" adjustment made, the "Gm" switch is turned on, and trimmer "CG" is adjusted until the meter's indication is again on the "Line check" mark.

The purpose of meter circuit calibration is to set the source impedance presented to the metering circuit to the specified value. Remember, this Gm testing method is only accurate for one specific value μ_0 and one value of $R_L=R_0$, and if that value is not set properly, the instrument's indication is meaningless!

Two-point meter calibration is used, first with "Meter" control at max. CW position, trimmer "CF" is adjusted, so the meter indicates Gm=1,100 mmhos. The "Meter" is then adjusted until the meter shows Gm=600 mmhos, at which stage the control knob should be pointing at "40", and if it isn't, the knob is mechanically readjusted to do so.

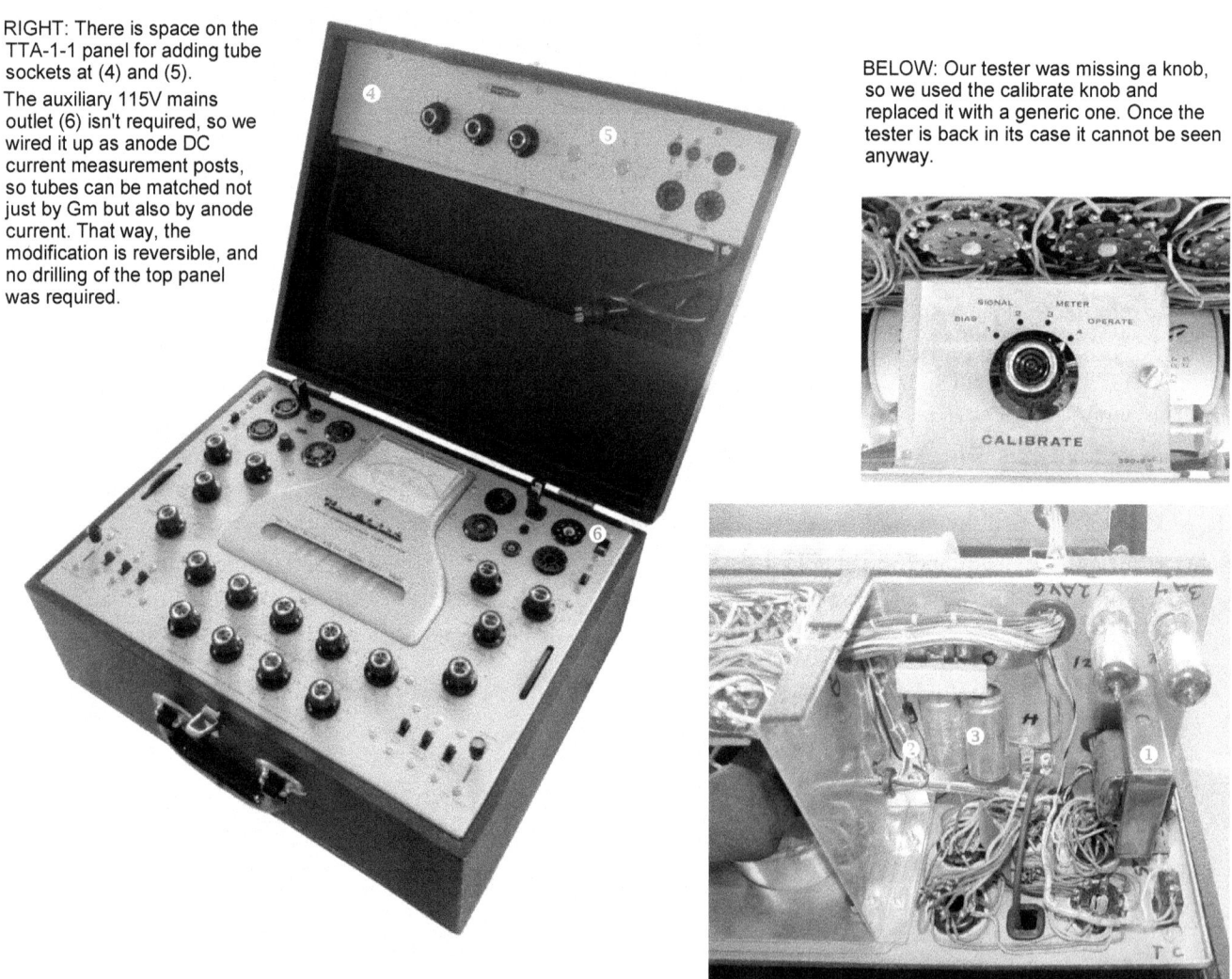

RIGHT: There is space on the TTA-1-1 panel for adding tube sockets at (4) and (5).

The auxiliary 115V mains outlet (6) isn't required, so we wired it up as anode DC current measurement posts, so tubes can be matched not just by Gm but also by anode current. That way, the modification is reversible, and no drilling of the top panel was required.

BELOW: Our tester was missing a knob, so we used the calibrate knob and replaced it with a generic one. Once the tester is back in its case it cannot be seen anyway.

RIGHT: The two tubes, 12AV6 and 3A4, metering anode choke (1), a solid state rectifier diode (2) and the associated filtering elcos (3) in the are one side of the power transformer. This power supply is used only for space charged grids.

Internal construction

Internally, TT-1 is a well-laid-out tester (photo above and on the next page). Five screws (1) hold the chassis to the bottom of the timber cabinet (next page). The roll chart mechanism is spring-loaded (2), very smooth, the best of all testers we have seen so far! Most of the power supply components are on the top "shelf," easily accessible (3), as are the three calibration pots, two mounted on one and the third one (4) on the other side "lip."

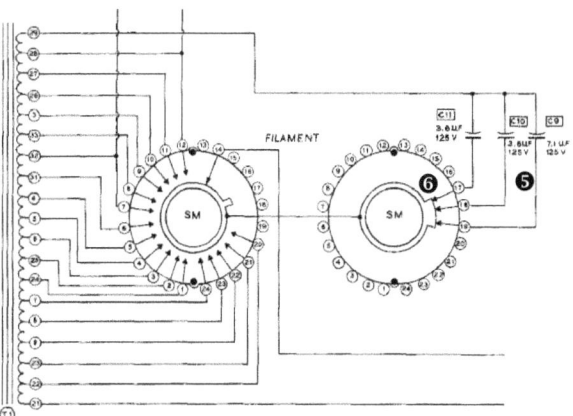

BELOW: The 'Filament' switch is shown in 600mA position (fully CCW). As the contact (6) moves clockwise, in the next position (450mA), C11 is disconnected, and only C9 & C10 are in the circuit. Finally, in the 300mA position, only C9 is in the circuit.

The triple oil capacitor (C9, C10 & C11) is on the other vertical shelf (5). One, two, or all three are used (in parallel) as ballast capacitors in the primitive "constant" current heater supply. 115V tap supplies the heater string through these caps, which act as simple current regulators.

Our tester was missing the five screws, so the chassis wasn't attached to the case. This is usually considered a warning sign that somebody opened the tester up to try to fix it and didn't bother to return them, meaning the tester was beyond repair. However, in reality, that could also mean the "fixer" did not have the knowledge or skills to proceed with the repair. For that reason, that eBay auction had only one bid - ours and sold for the starting price. The tester arrived undamaged, and all internal components were original; nobody had messed around, and that is always the best news. The tester worked well and was close to perfect calibration.

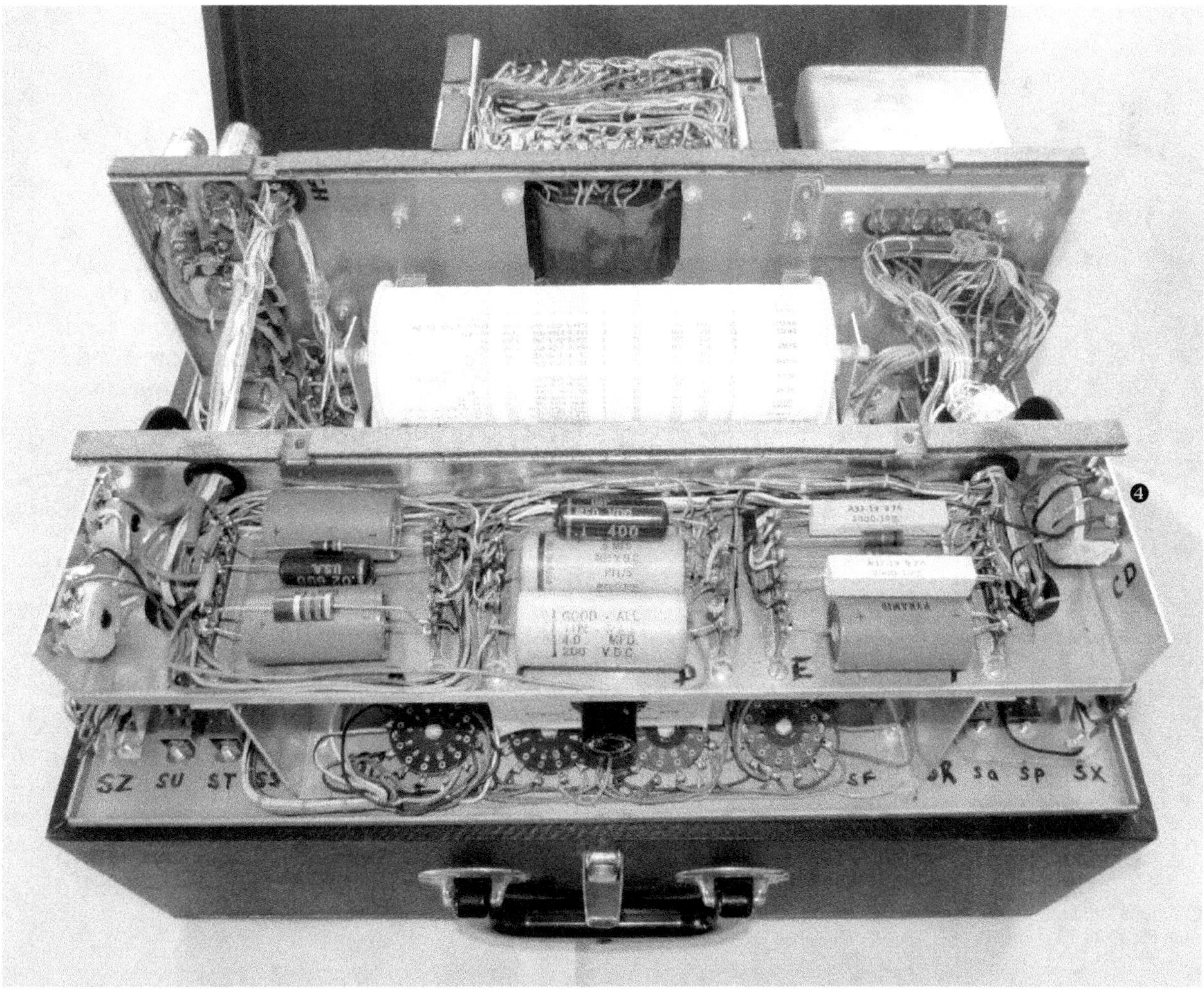

Constant current (kind of) heater supply

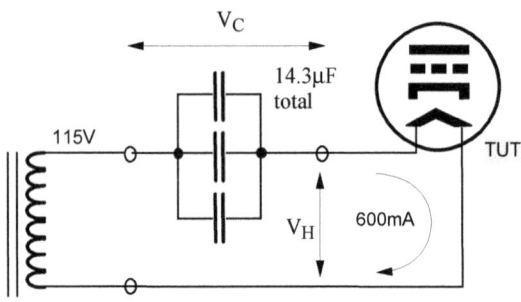

Switch position	"Selector" switch connects the pin to
0	OFF (pin not connected)
1	K (cathode bus)
2	G3 (suppressor grid)
3	Anode (plate) bus
4	G2 (screen grid)
5	CG1 (control grid tube 1)
6	Filament positive (+)
7	Filament negative (-)
8	Space charge grid bus
9	K
10	P
11	G

Switch position	"PLATE" switch positions
0	26 V_{DC}
A	20 V_{DC}
B	45 V_{DC}
C	90 V_{DC}
D	135 V_{DC}
E	177 V_{DC}
F	225 V_{DC}
G	Variable DC for VR tubes

Since the voltage drop on a very low resistance (under 1Ω) filament is much lower than across the ballast cap, the capacitor's impedance determines the AC current in this simple series circuit, so it is assumed that V_H is zero and V_C is 115V. Let's see if it's correct.

For 600mA current loop, all three caps are in parallel for a total of 14.3μF, and at 60Hz mains frequency, its reactance is $X_C=1/(2\pi fC) = 185.5\Omega$, and the current will be I=115V/185.5Ω = 620 mA, close enough.

However, on our 50Hz mains frequency in Australia, the reactance will be higher ($X_C=222.6\Omega$), and the current will be lower (I=517mA). This can be rectified by adding small film caps in parallel with the three existing ones. Start by calculating the needed capacitance for C9 (300mA), subtract 7.1μF, which is the add-on cap for C9, then calculate the required value for 450mA and determine the required add-on capacitance for C1, and so on. The film caps you add must be rated at 125V_{AC} or higher!

Adding a plate-cathode or screen current measurement capability

DIY PROJECT

Incredibly, TT-1 does not have a current measurement capability, but adding a 1Ω or 10Ω resistor to enable such measurements by an external multimeter (on DC Volts) is easy. Insert it into line (1), as shown, or line (2) to measure anode current. For screen current, insert it into line (3) or (4), and for cathode current into line (5).

With 6L6 and zero bias, the anode current is around 75mA, and with a maximum negative bias, it drops down to 25mA. This is another shortcoming of this tester, -20V bias is not nearly negative enough for many tubes, and transfer curves and cutoff points cannot be measured or plotted!

Also, in TT-1, anode & screen currents flow as soon as the settings are made, and the tube is inserted in its socket, even when "Gm Test" is not pressed! Another dubious design decision!

ABOVE: Where to insert the anode, cathode and screen current measuring resistors in Heathkit TT-1

TRUE MUTUAL CONDUCTANCE TESTERS

TRIPLETT 3444

Triplett 3444 (also made as Westmore 501) was released in 1959 and superseded in 1968 by model 3444-A, a solid-state version with a few badly needed functional improvements. The power transformer was bigger, had better regulation, and the power supply could supply up to 150mA of anode/screen current.

A few more socket types were added, and the anode current measuring capacity was increased from 50mA to a maximum of 150mA. However, this rare model comes up for sale on eBay USA perhaps once or twice a year, so we will limit our analysis to model 3444.

There are many positives but also some negatives. It is a great idea that the roll chart is illuminated and housed in its separate compartment. The 200mA meter is also illuminated. The tester is well laid out and a pleasure to use. The combination socket (4-, 5- and 6-pin) is a great space-saving idea. To our knowledge, the only other testers that used such a socket were Conar models 221, 223, and 224.

The range of other sockets is very limited, which could be considered the first limitation of this tester. There are no European sockets of any kind or more modern sockets such as Magnoval, Novar, and Compactron.

1) The three most commonly used sockets (7-pin mini, Noval and Octal) are recessed and socket saver module plugs into them
2) Two pin straighteners can be removed and additional sockets installed there
3) A bank of nine 12-position selector switches, each connects one tube pin to various test points
4) "Line test" push button and "Noise" terminals for headphones
5) "Short-Leakage" test switch
6) Although marked "Plate voltage", the "C" switch also selects screen and signal voltage levels
7) "Gas-Value" switch and "P1-P2" switch for testing dual tubes
8) "D" switch selects one of four Gm ranges, two anode current ranges, a Thyratron or rectifier testing function
9) Bias control potentiometer

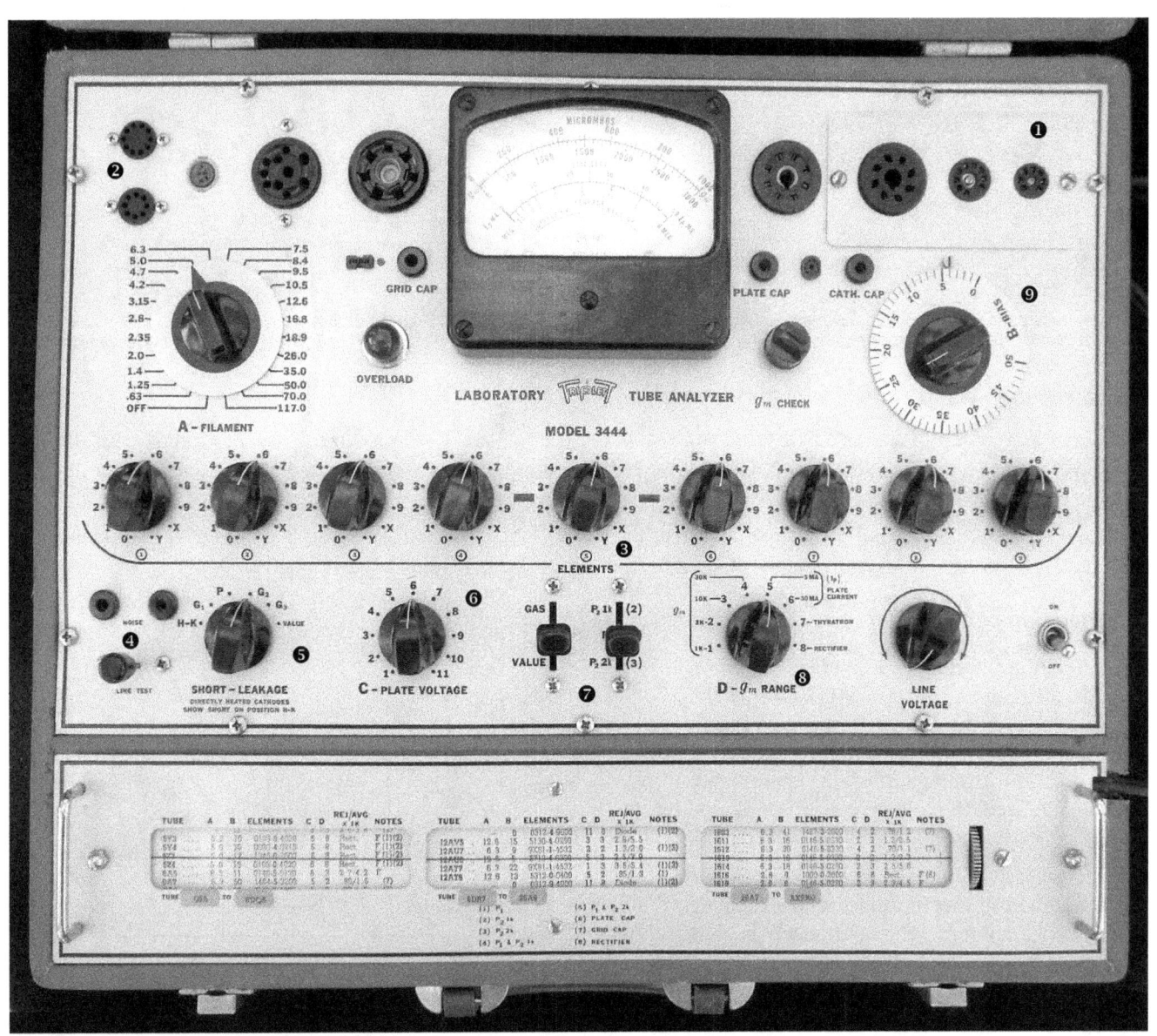

Internal layout

Internally, the tester is relatively well laid out; most components are easily accessible. However, the componentry around the meter amplifier (3) and oscillator (4) is crowded and difficult to access and replace.

The second major limitation is obvious from the photo: the single smallish transformer (1) that provides all test voltages. Many cheaper and less capable testers have two transformers, one to supply heater voltages only, so when a heater circuit is loaded with say 2A of heater current, such a load does not drag the anode and screen voltages down as well. Alas, that happens on 3444. These DC voltages may be filtered, but they aren't regulated in any way! The same shortcoming applies to the grid bias DC voltage.

The transformer is small, and its laminations are made of ordinary silicon steel, resulting in poor voltage regulation and significant sag under load.

1) Power transformer   2) Step-down autotransformer (240-117V), added by us   3) 12AU7 tube (meter driver amplifier)   4) 6C4 tube (5kHz oscillator)   5) Oscillator transformer   6) Line adjust rheostat   7) "D" switch   8) "C" switch (Plate voltage)   9) "Short-Leakage" switch

DC power supplies

The bias supply circuit features a half-wave solid-state rectifier X3 and a simple capacitive filter (30µF), resulting in a $-80V_{DC}$ negative voltage. Rectifier X4 and its capacitive filter C14 (also 30µF) provide a half-wave rectified DC voltage of around 145V, used as anode supply for the metering amplifier and for the "line test" circuit.

The high voltage power supply was drawn all over the place (schematics on the next page) and is hard to identify at first. The full-wave rectified pulsating voltage after the bridge X2 is filtered by C7 (50µF).

The metering circuit

The metering circuit is a 2-stage common cathode amplifier using a 12AU7 duo-triode. The output of the first stage (higher gain due to 200k anode resistor) is capacitively coupled to the 2nd stage (lower gain, 47k in the anode), which then feeds a full-wave rectifier bridge with the 200µA moving coil DC meter.

The meter is actually in the feedback loop back to the cathode of the 1st stage, whose cathode resistor (a trimmer potentiometer) is left un-bypassed, providing additional local NFB.

The 750pF bypass cap at the input shunts high frequencies and interference signals to GND.

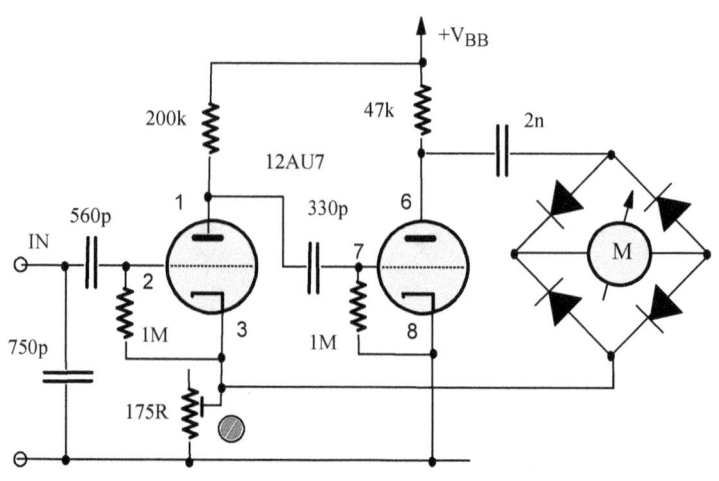

The two-stage meter amplifier of Triplett 3444.

TRUE MUTUAL CONDUCTANCE TESTERS

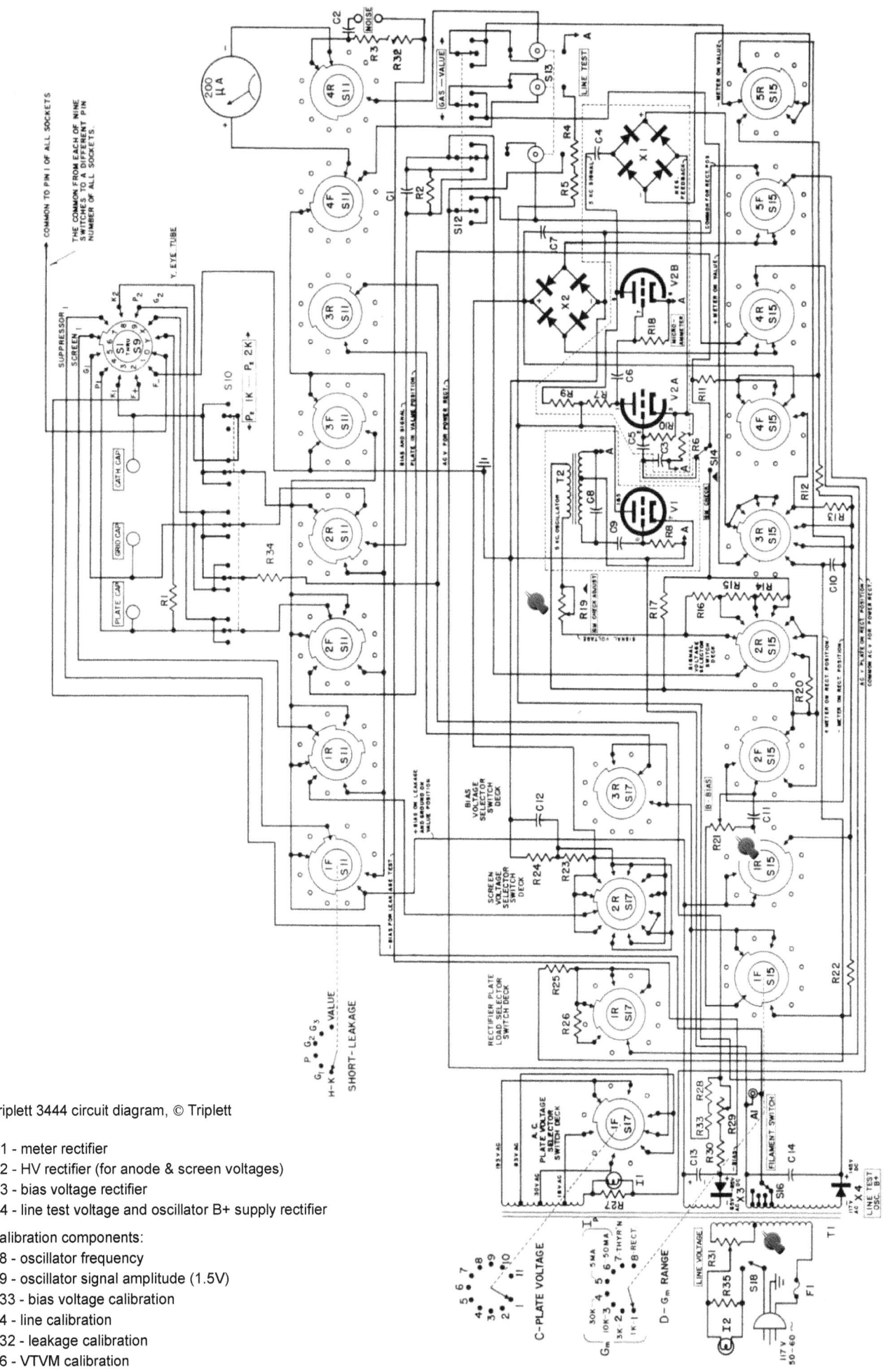

Triplett 3444 circuit diagram, © Triplett

X1 - meter rectifier
X2 - HV rectifier (for anode & screen voltages)
X3 - bias voltage rectifier
X4 - line test voltage and oscillator B+ supply rectifier

Calibration components:
C8 - oscillator frequency
C9 - oscillator signal amplitude (1.5V)
R33 - bias voltage calibration
R4 - line calibration
R32 - leakage calibration
R6 - VTVM calibration

Switch C position	Plate V_{DC}	Screen V_{DC}	Bias range V_{DC}		Switch position	Connection to
1	250	250	0-5		0	OFF (pin not connected)
2	250	250	0-50		1	HEATER (filament) negative (-)
3	250	100	0-50		2	HEATER (filament) positive (+)
4	250	100	0-5		3	K1 (cathode tube 1)
5	100	100	0-5		4	P1 (plate tube 1)
6	100	100	0-50		5	CG1 (control grid tube 1)
7	100	45	0-50		6	G2 (screen grid)
8	100	45	0-5		7	G3 (suppressor grid)
9	30	30	0-5		8	K2 (cathode tube 2)
10	30	12	0-5		9	P2 (plate tube 2)
11	12	12	0-5		X	CG2 (control grid tube 2)
					Y	Anode (plate) for "magic eye" tubes

FAR LEFT: The 11 possible combinations of the 3 factors, the bias range, the anode and the screen voltage, selectable by switch "C" on Triplett 3444 tube tester.

LEFT: The meaning behind "Elements" switches' positions.

Triplett 3444 uses four grid signal levels, 33mV, 100mV, 0.333V, and 1.0 V. The test frequency is 4 kHz.

There are four anode (12, 30, 100, & 250V) and five screen (12, 30, 45, 100, & 250V) DC voltages, but there is no range selector switch. The two bias ranges (0 to -5V and 0 to -50V) are automatically selected depending on the position of the "Plate voltage" switch ("C"). The table above outlines the 11 voltage combinations available.

Since the bias DC voltage is continuously adjustable, 0 to -10V or 0 to -50V, most tubes can be tested in any point along their transfer curve.However, the 11 test positions are this tester's most serious limitation. If anode and screen voltages aren't continuously adjustable, it could not be called a "Laboratory tube analyzer"!

The 50mA limit was deliberately introduced by the designers due to the limited current supply capabilities of its selenium rectifier and transformer. Once we installed a silicon rectifier and doubled the current measuring capability to 100mA, the "Overload" light glowed at 65mA (full brightness at 80mA), the DC current through a 6L6 tube with zero bias. However, 100mA was supplied with no problems for 5-10 seconds, enough to test higher power tubes.

Manufacturers of tube testers were very quick to give them names they didn't deserve, such as "laboratory tube analyzer" in this case, yet very slow to improve their circuits (it took Triplett nine years to come up with model 3444-A!) and justify the use of such epithets. They could argue that even the most primitive tester could be used in a lab, thus deserving such a title. So, the ambiguity of language and the lack of strict and meaningful official definitions and regulations allowed the makers of tube testers to call them whatever they wanted!

How to test an unlisted tube on your Gm tester: EC8010 example on Triplett 3444

Since EC8010 is a rare European tube, it is not listed on Triplett 3444 tube analyzer's chart. Even the updated supplement does not mention it or its close American relative, the 8556. So, a bit of detective work was required. The exact procedure for determining the settings for an unlisted tube will, of course, depend heavily on a particular tube tester, but the methodology outlined here can be used on most other Gm testers. Here are the major steps:

1. TUBE PINOUT: You need a tube pinout and typical operating condition, meaning you need its datasheet.

2. YOU NEED A SUITABLE SOCKET ON YOUR TESTER: If your tester uses prewired sockets (no switches for tube pins), such as the B&K Dyna-Quik series or Precise 116, you need to find a tube that is listed on your tester's chart that has an identical pinout. That sounds easy but can be a long and frustrating exercise, and, in the end, you may conclude that there is no prewired socket for such a tube.

In that case, you need to choose one socket that you don't need (for some obsolete TV or RF tube not used in audio) and rewire it to suit the tube you want to test. Make sure you mark such a change in the tester's manual and its faceplate.

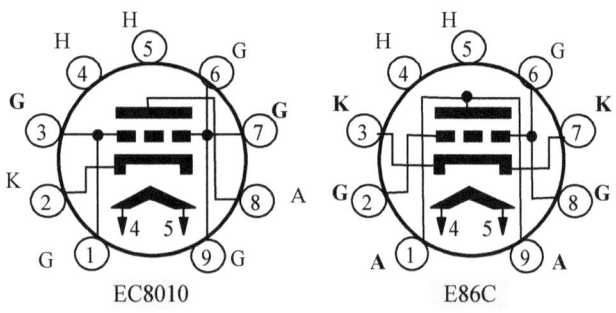

EC8010 is quite similar to E86C, another lovely tube, and with a bit of rewiring, it can be used as its substitute in audio amplifiers. Let's say you have a tester with a socket prewired to test E86C or 6CM4. Three pins are OK, 4 and 5 heater pins and 6, the grid. The others need to be rewired; the exact mod will depend on how the tester was wired in the first place. Alternatively, you can make a socket adapter.

3. If your tester has an array of switches for its 9 or 10 (older models) or 12 pins (more modern ones), you need to understand what each switch does and what each position means. Triplett 3444 has nine switches, marked 1 to 9, one for each tube pin.

For EC8010, pins 1,3,6,7&9 are all internally connected to the control grid. Connecting only one of those would be enough for testing (and leaving the other pin switches in position 0 - off), but we may as well connect them all. If the tube socket on your tester is a bit loose, as they all get with frequent plugging and unplugging of tubes, and that particular pin loses contact, you'd lose the grid bias, and the tube could overheat and get damaged or destroyed! If two or more pins are wired together, that is unlikely to happen.

The heater ends are pins 4 & 5; we can put switch 4 in position 1 and switch 5 in position 2 or the other way around. On Triplett 3444, heater pins are marked + and - for some reason, but AC heating is used, so both settings would work. The cathode is pin 2, so switch two goes into position 3, and the anode is pin 8, so switch #8 must be in position 4! Thus, one of the few possible variations of the setting is (SW1-9): 5351-2-5545

4. HEATER VOLTAGE SETTING: This is usually obvious (except for dual heater tubes), which can operate on 6.3V and 12.6V. In this case, the setting "A" for EC8010 is 6.3V.

5. GRID BIAS, PLATE, AND SCREEN VOLTAGES: Unfortunately, the anode (and screen) test voltage on Triplett 3444 cannot be varied continuously (can be achieved by a suitable modification) as the bias voltage can, so there are only two switchable choices, $100V_{DC}$ and $250V_{DC}$. Consulting the anode curves below, at $100V_{DC}$, we can test this tube at any point along the marked range depending on the setting of the bias potentiometer. The anode current will vary from 2mA to 42mA, which is perfect for Triplett 3444, so we could match tubes not just in terms of Gm (mutual conductance) but also anode current!

We could choose a $250V_{DC}$ anode test voltage. For instance, the -4V bias setting would test the tubes in Q3, but the allowable bias range would be very narrow, so we could easily underbias the tube! The anode current would shoot up, and the tube's anode power rating would be exceeded. The maximum allowed anode voltage for EC8010 is only V_{AMAX}= 200V, so we should not use the $250V_{DC}$ setting.

The grid bias setting should be between 0 and -4V, so we need to select one of the 0-5V bias ranges on the tester. From the decoding table (previous page), it is obvious that positions 5, 6, 7, and 8 are the candidates. The screen grid voltage settings are irrelevant in this case; EC8010 is a triode and has no screen grid. Positions 5 and 8 use the 0-5V bias range, so either of those can be used with EC8010.

6. The Gm RANGE: The data sheet for EC8010 lists nominal mutual conductance at around 28 mA/V (28,000 micromhos). Older testers, such as most Hickoks, cannot go that high and cannot test such high Gm tubes.

Luckily the highest Gm range on Triplett 3444 is 30mA/V, but even then, some very strong tubes may go over the scale, especially at higher anode currents where Gm increases (test point Q2).

To prevent that from happening, we should test these tubes at lower anode currents, in test point Q1, for instance, or even lower, below 10mA of anode current.

In that region, the slope of the anode curves (mutual conductance) is lower, from 20-25 mA/V (draw a tangent in a particular point and see for yourself), so that will ensure that even the strongest tubes will not go off the 30mA/V top end of the tester's scale.

With a bias set at 8 (-0.8V), the results for our pair of tubes were Gm1=23.5mA/V and I_{A1}=15mA, and Gm2=22.5mA/V and I_{A2}= 17mA, so the tubes were quite well matched.

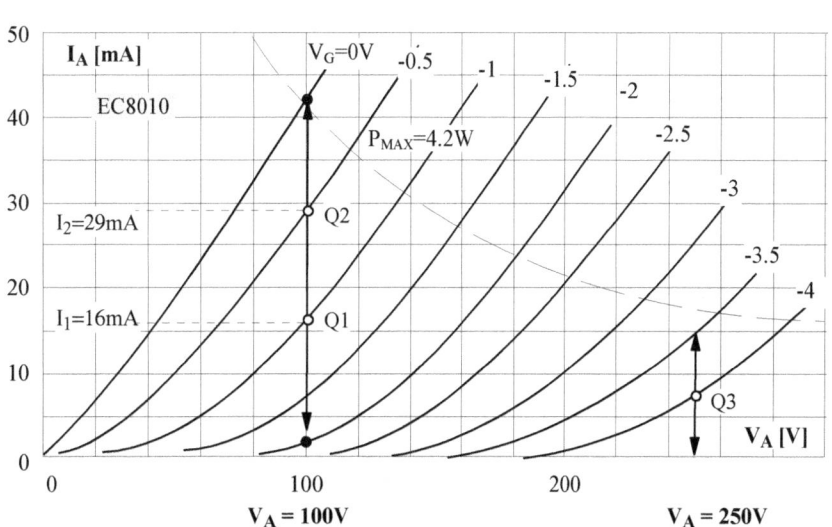

ABOVE: There are only two "high" anode test voltages on Triplett 3444, $100V_{DC}$ and $250V_{DC}$, so EC8010 can only be tested along the two vertical lines marked.

Increasing the anode current meter range on Triplett 3444

IMPROVEMENT

Doubling the anode current measurement range from 50mA to 100mA is not difficult on Triplett 3444. Even if you don't want or cannot follow the messy circuit diagram, there is a short-cut trick. A similar strategy can be used on other Gm testers, so this procedure is of universal value!

Positions 5 & 6 of the "D" switch are 5mA and 50mA, respectively. Although the switch has nine poles (not all are shown in the detail below), only one is of interest, wafer "3R" (1). The drawing shows the "D" switch in position #1, and we count the following positions in a clockwise manner. The first four are Gm ranges, and wafer 3R shows all four bridged together (1). Wafer 2R shows precision resistors R14-R15-R16 connected between the four Gm contacts. We are interested in contacts 5&6 (3), and we find R13(18.84Ω) connected between them and R12 (2Ω) connected between the other end of R13 and "something else."

When you see resistors specified as "non-inductive," of very precise values (18.84Ω) and very tight tolerances (0.25%), you can bet they are shunt resistors for the meter. In such instruments, the two shunt resistors are usually connected in series on one (5mA) range, and only the lower value (2W) resistor is the shunt for the other range (50mA). 18.84+2=20.84Ω.

We know that the meter's FSD (Full-Scale Deflection) is 200mA or 0.2mA.

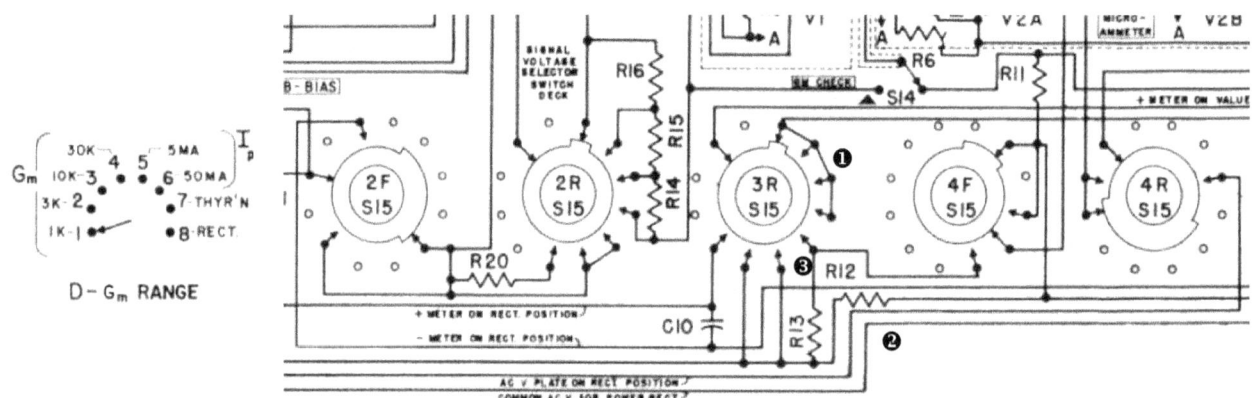

At full scale the voltage drop across the meter (and shunt in parallel) would be V=20.84Ω *4.8mA = 0.1V. At 50mA flowing through the parallel shunt and the meter (0.2mA through the meter and 49.8mA through the shunt resistor), the voltage drop must stay the same, so we can calculate R_{SH50mA} = V/I_{SH} = 0.1V/0.0498A = 2.0Ω! Our theory has therefore been confirmed.

Conclusion: the 50mA range can be doubled by simply switching another 2Ω resistor in parallel to get a total shunt resistance of 1Ω! Check: R_{SH}=0.1V/99.8mA=1.002Ω!

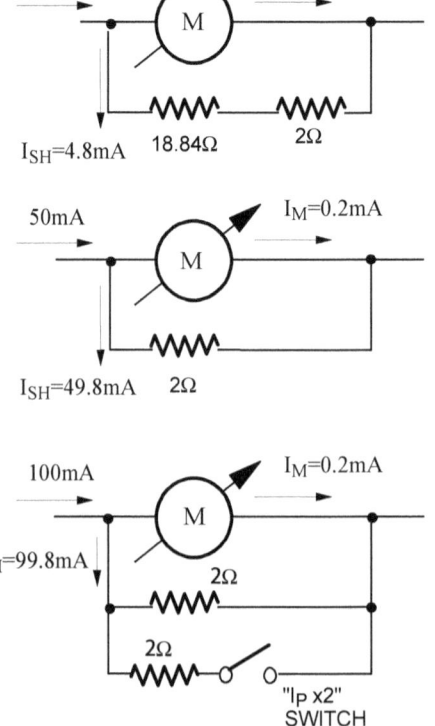

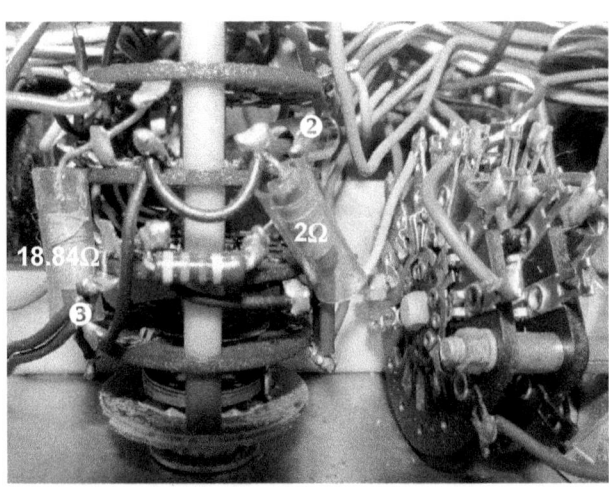

LEFT: "D" switch with two current shunt resistors, R12 (2) in the background and R13 (3) in the foreground on the left.

TRUE MUTUAL CONDUCTANCE TESTERS

SENCORE MU-140 & MU-150

Any tester with the unfortunate "Sensitivity," "Meter," "Shunt," or "Calibration" control potentiometer is seriously compromised. Instead of changing test ranges and meter's full scale by a switch, these testers use a single scale and vary the reading for each tube type by this potentiometer. This affects the results since the accuracy and repeatability depend heavily on the accuracy and the setting of this imprecise potentiometer.

Just when you thought that the absence of such control made Sencore MU140 and MU150 into serious testing machines, you notice the unusually labeled "Signal" potentiometer, also with 0-100 graduations. Instead of shunting the meter, Sencore made the amplitude of the signal fed to the control grid continuously variable.

An adjustable test signal may be a good idea when testing low bias tubes such as 12AX7 and 6DJ8, but from the accuracy standpoint, it suffers from the same drawbacks as shunting the meter, compromising the accuracy, and repeatability of tests.

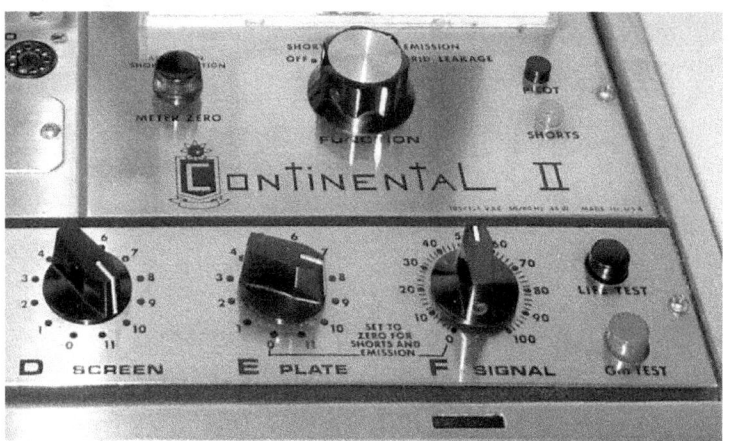

RIGHT: The controls of MU150 ("Continental II") are simple and well laid out.
BELOW: MU-140 circuit diagram, © Sencore

1) Heater transformer
2) Anode/screen and misc. power supply transformer
3) +/-210 V_{DC} rectifiers
4) Tube voltage regulator
5) Square wave oscillator (astable multivibrator) for grid signal during Gm tests
6) Anode AC signal amplifier and voltage doubler rectifier
7) Neon "Shorts" indicator
8) Each of the three feed-through selector switches picks one pin for anode, screen and control grid, and passes the rest through
9) VTVM, a differential amplifier driving the analog meter

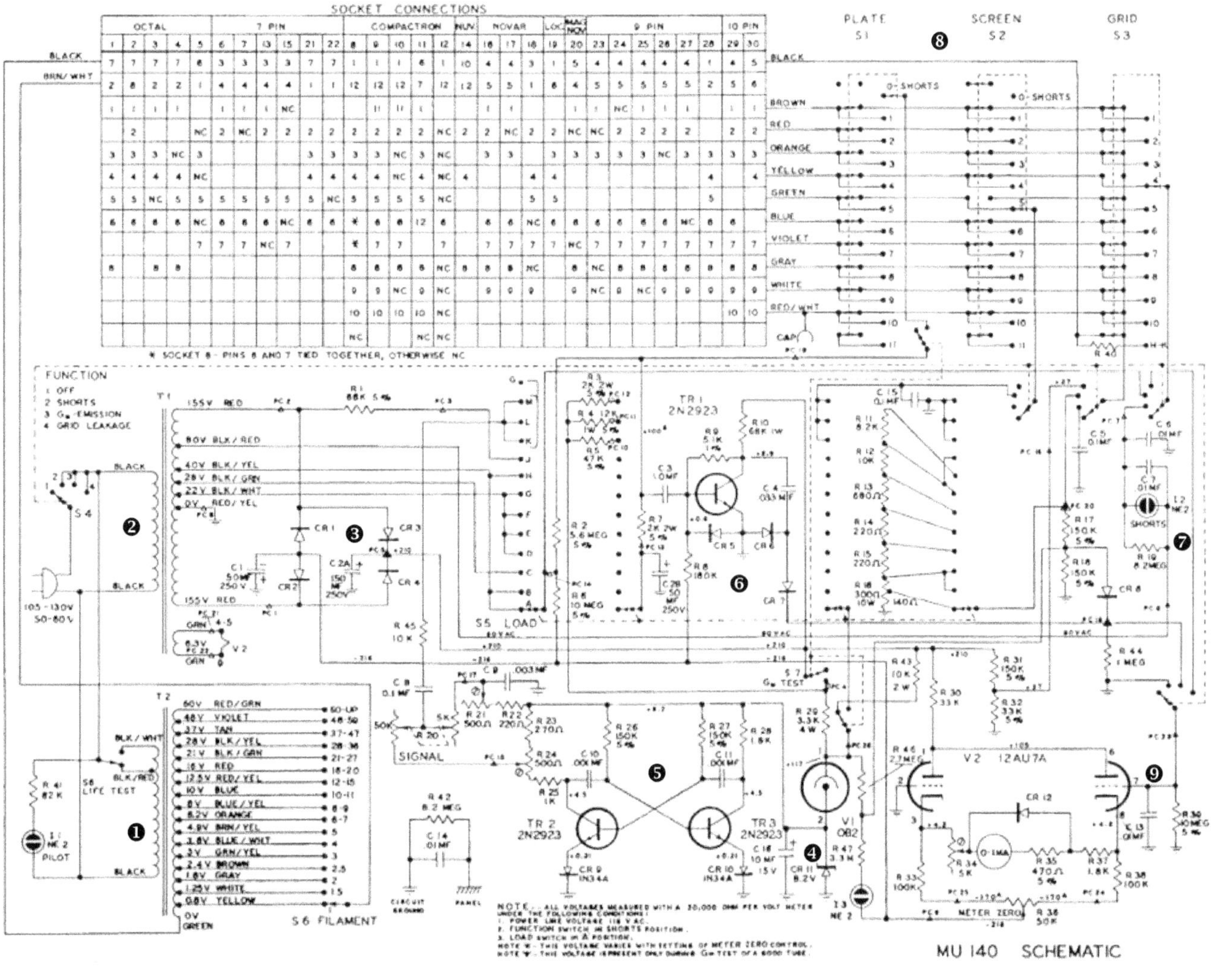

MU 140 SCHEMATIC

The automatic biasing circuit

Let's assume a power tube such as 6L6, tested at 25mA of anode current. Apart from the fixed anode load resistor R7 (2k), the "Load" switch selects one of three additional resistors, R3 (2k) for 25mA test, R4 (12k) for 7mA, and R5 (47k) for 2mA bias.

In this case, we have a total of 4k anode load, and since 25mA flows through it, the voltage drop on R3 and R7 together is exactly $100V_{DC}$. The anode voltage is thus $+210-100 = +110V$. The control grid is biased through the R2-R6 resistive voltage divider, and the DC voltage on the grid is $(+110+216)*10/(10+5.6) = 326*10/15.6 = +209V$ above the referent voltage (-216V), so the grid is at $-216+209 = -7V$ above the ground level (zero DC voltage).

If, for some reason, the anode current tries to increase to 28mA, the voltage drop across R3+R7 would increase to $4*28 = 112V$ (+2V), and the voltage across the R2-R6 divider would decrease by 2V to 324V.

The grid is now $324*10/15.6 = +207.7V$ above the referent voltage (-216V) or at $-216+207.7 = -8.3V$ above the ground level (zero DC voltage). The grid bias has increased from -7.0 to $-8.3V_{DC}$, and such increased negative bias will reduce the anode current. This is a typical negative feedback mechanism based on the feedback resistor R2 between the tube's anode and control grid.

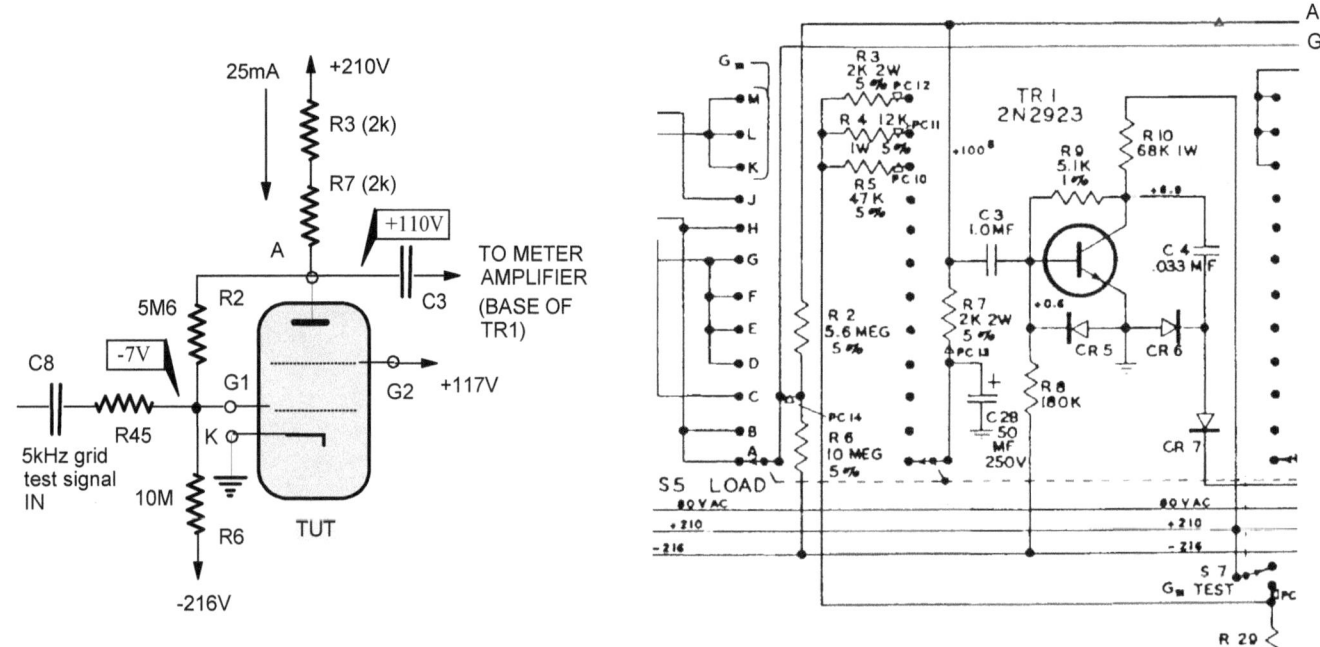

ABOVE RIGHT: Part of the circuit diagram showing the "LOAD" switch, associated resistors and the simple bipolar transistor gain stage that amplifies the AC component of anode current on its base © Sencore

ABOVE LEFT: Simplified diagram illustrating the automatic bias control mechanism.

The final verdict

On the positive side, two power transformers were a wise choice. The life test was properly implemented by lowering heater voltages only. However, there was no provision for line adjustment. Because the screen DC voltage is stabilized and the automatic biasing regime always tests each tube type at the same point, one could argue that line adjustment isn't required, but heater voltage levels will also vary with line variations and affect test results.

On the cosmetic level, most MU140 and 150 testers are now 50+ years old, and their cases suffer from serious corrosion (pitting). Also, the screen-printed markings on the control panel fall off so easily that even mild wiping with alcohol or a similar cleaning agent removes them in a matter of seconds. We had one of these units in for repair. The resistors that Sencore used have drifted significantly, so many have to be replaced, and due to its crowded PCB, that is a tedious task. Due to many connections and bundled wires crisscrossing the tester's interior, if you are not extremely careful and possess a high level of manual dexterity, you may fix one problem but create two or more new bad connections.

As for test conditions, the screen voltage is around $+117V_{DC}$ (from pin 1 of the voltage regulator tube OB2) while the anode voltage is somewhat lower, at $+100V_{DC}$, which is OK for checking tubes but too low for serious testing.

There is no choice of a test point; each tube is tested according to a fixed current level, at 2, 7, or 25 mA. The meter lacks minor graduations and does not display Gm in real units, but in % only. Thus, this tester is not the best choice for tube matching.

Calibration modules need to be made to simulate a tube, and the amplitude of the square wave grid signal needs to be accurately adjusted while being monitored on an oscilloscope.

RCA WT-110A

A very fast a simple tester to use, RCA model WT-110A uses DC plate & screen voltages and a 4 kHz grid test signal, which makes it a formidable testing platform, but the rest is sacrificed to obtain a speedy reading.

There is only one (fixed) test point, the meter scale is relative (0-10, not in terms of transconductance units) and very imprecise (no minor subdivisions), there are only four tube socket types (Octal, Loktal, Noval, and miniature 7-pin), and, of course, the major dislike factor, the use of flimsy cardstock test cards.

The functionality is weird, to say the least. The tester measures Gm before checking for shorts! This should never be the case; the shorts check should always be performed first. Inside, the wiring is messy, especially around the test lever switch, and the component layout is suboptimal.

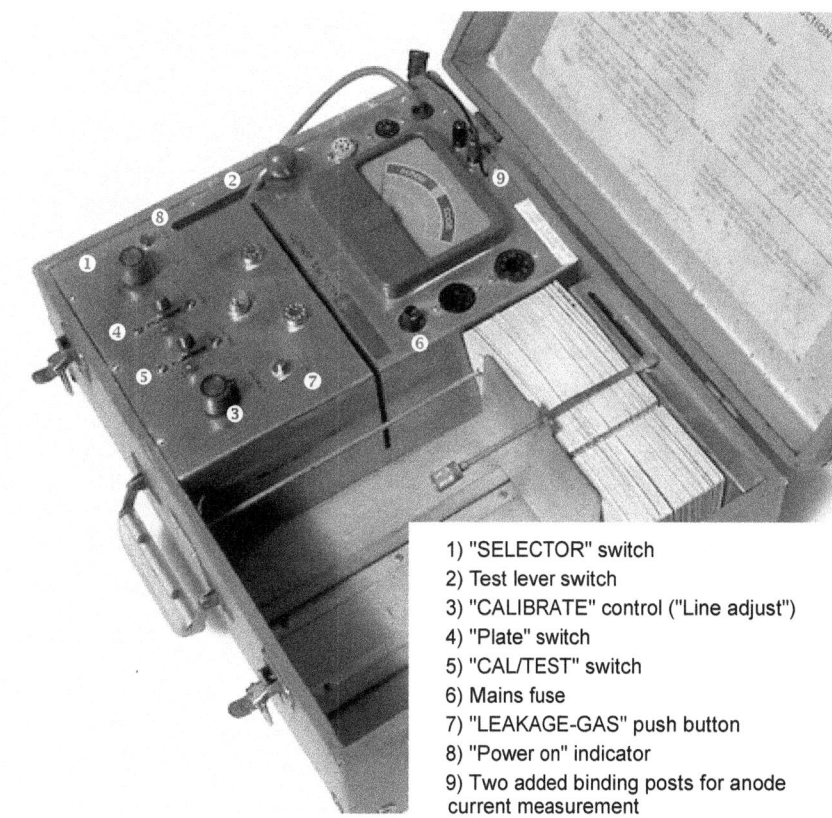

1) "SELECTOR" switch
2) Test lever switch
3) "CALIBRATE" control ("Line adjust")
4) "Plate" switch
5) "CAL/TEST" switch
6) Mains fuse
7) "LEAKAGE-GAS" push button
8) "Power on" indicator
9) Two added binding posts for anode current measurement

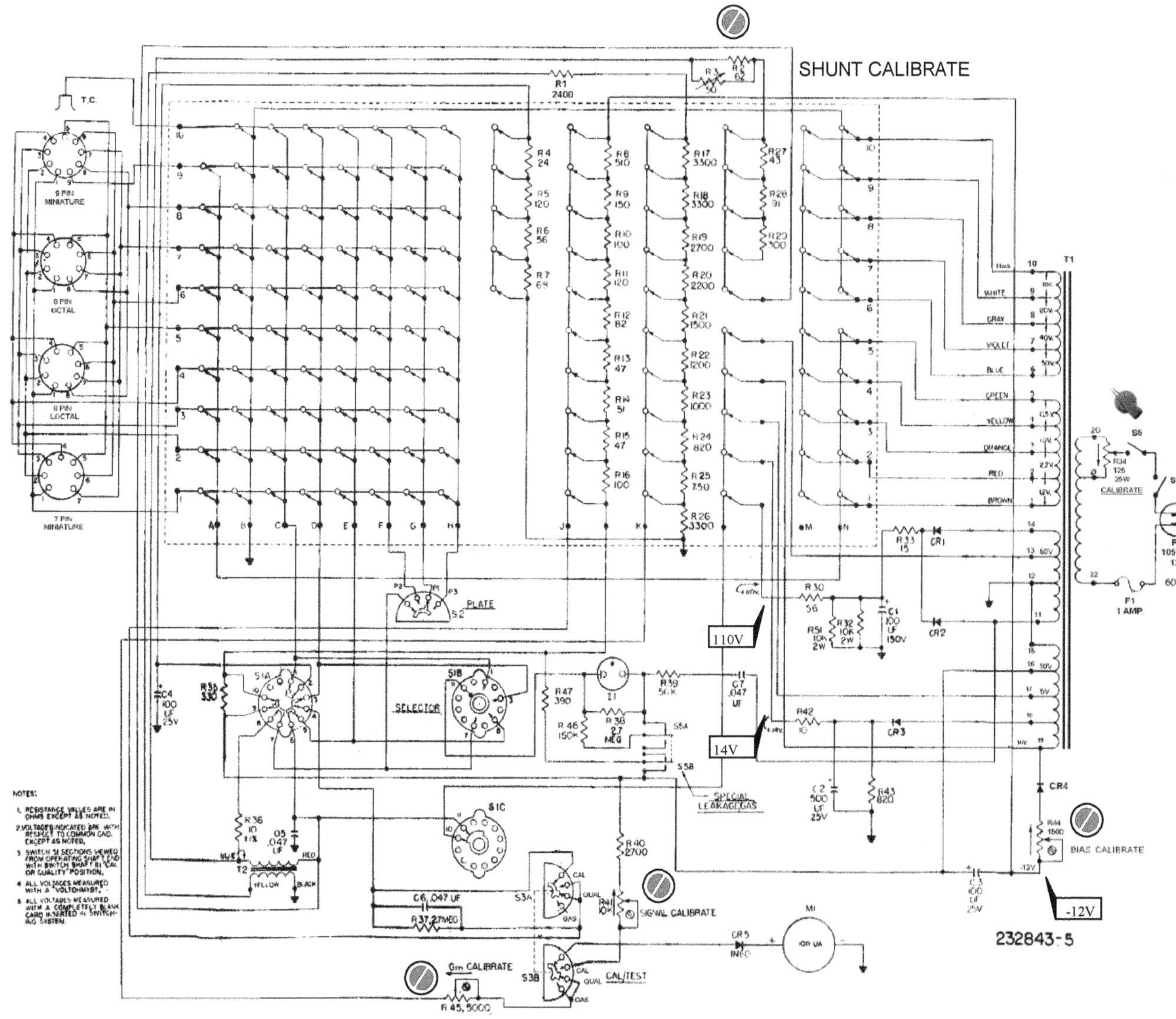

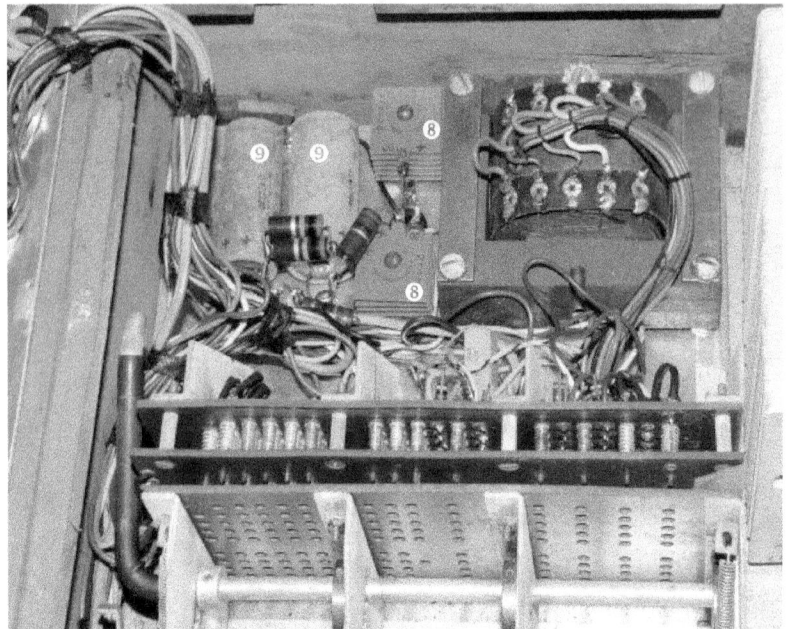

1) "Selector" switch
2) Signal oscillator transformer
3) "Calibrate" control ("Line adjust")
4) Two replacement capacitors
5) "Signal calibrate" adjustment
6) "Gm calibrate" adjustment
7) "Shunt calibrate" adjustment
8) Two original selenium rectifier diodes to be replaced
9) Two original HV elcos to be replaced
NOTE: The "Bias calibrate" adjustment is on the other side of the chassis

What is wrong with RCA WT-110A ?

1. The anode voltage is very low, about 110 V_{DC}.
2. Mains frequency voltage is used for grid signal. Tube hum (if present) would impact Gm readings.
3. Primitive neon-based shorts- and leakage-test.
4. The pin selection mechanism has no modern replacement. If faulty, it would have to be repaired or rebuilt, which would be a nightmare.
5. Very limited socket range. No older (4-, 5-, 6-pin) or later sockets (Nuvistor, Magnoval, Compactron).
6. The meter does not display true Gm, just the so-called "English" scale of "The Good, The Bad and The Suspect"! If you want to match tubes, look elsewhere!
7. Without its required card, you cannot test a tube (unless you figure out the required settings and punch a replacement one yourself).

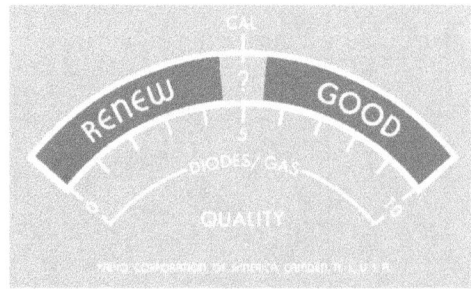

ABOVE: RCA WT-110A competes with B&K700 scale for the title of "The most useless tube tester meter scale of all times"!

The switching matrix and operating conditions

To understand the operation and to be able to make your own cards for newer tubes, we must understand the switching matrix. The pins (connections) on the card (next page) correspond to the matrix pins on the circuit diagram reproduced on the previous page. We have drawn some connections and resistors here on the sample card for easier reference.

Three AC voltages are used for emission checks, 5V, 18V & 50V, two DC voltages for Gm testing, 110V and 14V, and ten DC bias voltages (0, -1V, -1.5V, -2V, -2.5V, -3.5V, -4.5V, -5.5V, -7V & -12V).

The cathode resistance can be any combination of four resistors, 24Ω, 56Ω, 68Ω, and 120Ω.

For example, connections for 6L6 tube are marked as filled dots, -7V_{DC} bias and 110V_{DC} plate voltage are used, and all four cathode resistors are connected in series, a total of 268Ω. All tubes are tested with 1V_{RMS} AC signal.

Notice a peculiar way of obtaining various heater (filament) voltages. Various taps of the power transformer's secondary are brought out to lines P1 to P10, the voltages between adjacent taps are marked (10V_{AC}, 20V_{AC}, etc., from top to bottom).

TRUE MUTUAL CONDUCTANCE TESTERS

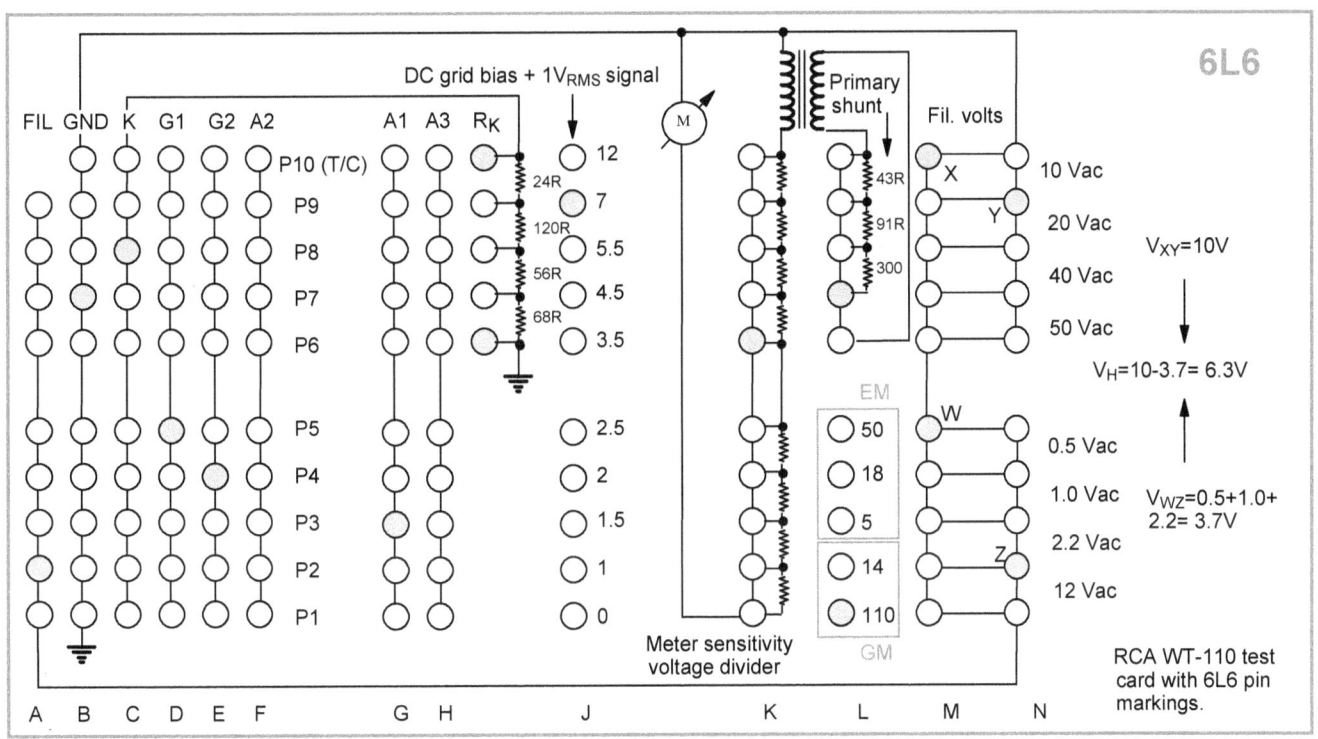

All heater terminals in the matrix are connected horizontally and vertically, as indicated by solid lines. Since pins X and Y make contact through their respective holes in the card, and so do pins W and Z, the AC voltage XY is 10V, and the AC voltage WZ is a sum of three voltages, 0.5+1.0+ 2.2 = 3.7V! However, starting with the top heater terminal connected to GND, we have VYX + VXZ. Since VXY and VWZ are in phase, VYX and VWZ are of opposite phases, so their voltages subtract, and we end up with 10-3.7= 6.3V for a 6L6 heater!

SECO 107, 107-B & 107-C

In 1959, when model 107 cost US$139.95, Hickok 752 was selling for US$298, and Triplett 3444, a much better tester than either of those, cost just over a hundred bucks more than Seco (cheaper than 752).

The On-Off switch is part of the "Eye Adjust" control potentiometer (1). Line adjustment works in a standard way by shunting part of the primary winding of the mains transformer by a rheostat (2) and thus changing the AC and DC voltages inside the instrument.

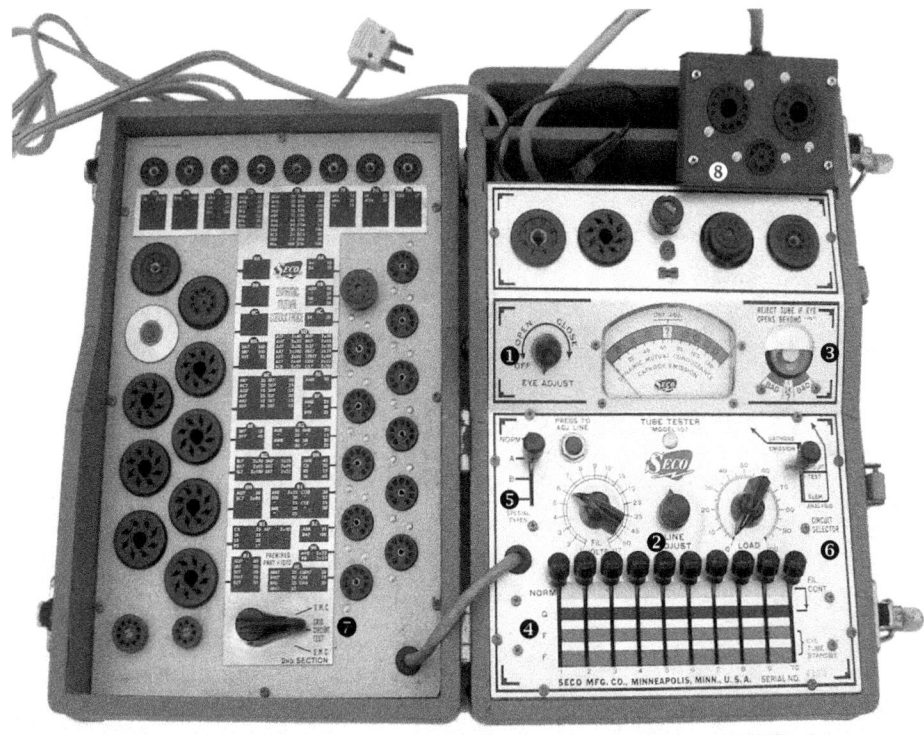

There are no internal calibrations, so that is the only way to center the meter to the "Line Adj." mark in the center of the tester's scale.

Next comes "zeroing" the 6AF6G "magic eye" tube (3) by adjusting (1) until the lit segment at the bottom shrinks into a narrow slit. This magic eye tube indicates gas, grid leakage, or inter-electrode shorts.

It is quite a clever way of performing all those tests. However, there is no numerical indication as on an MΩ-calibrated megohmmeter scale. The wider the lit segment opens up, the lower the insulation resistance and the more serious the short is.

The ten lever switches (nine pins plus top cap) have only four positions (4). "F1" and "F" select heater pins, and "Q" selects the control grid of the tested tube. Cathodes on the prewired and main chassis, plus all unused pins, are left in the "Normal" position, which is grounded by one contact of the "Circuit Selector" switch during CE or cathode emission tests (see diagram below). Therefore, the cathode emission is not a full emission indicated as current flow between the cathode and the anode, but only a low-level current between the cathode and the control grid, just as with all other Sico grid testers such as models 78, 88, and 98!

The "Special Types" switch (5) changes some of the test voltages for certain tubes. Setting "A" simply patches through pin 9, which is left open in all positions of the "ST" switch. Position "D" is used with rectifier tubes.

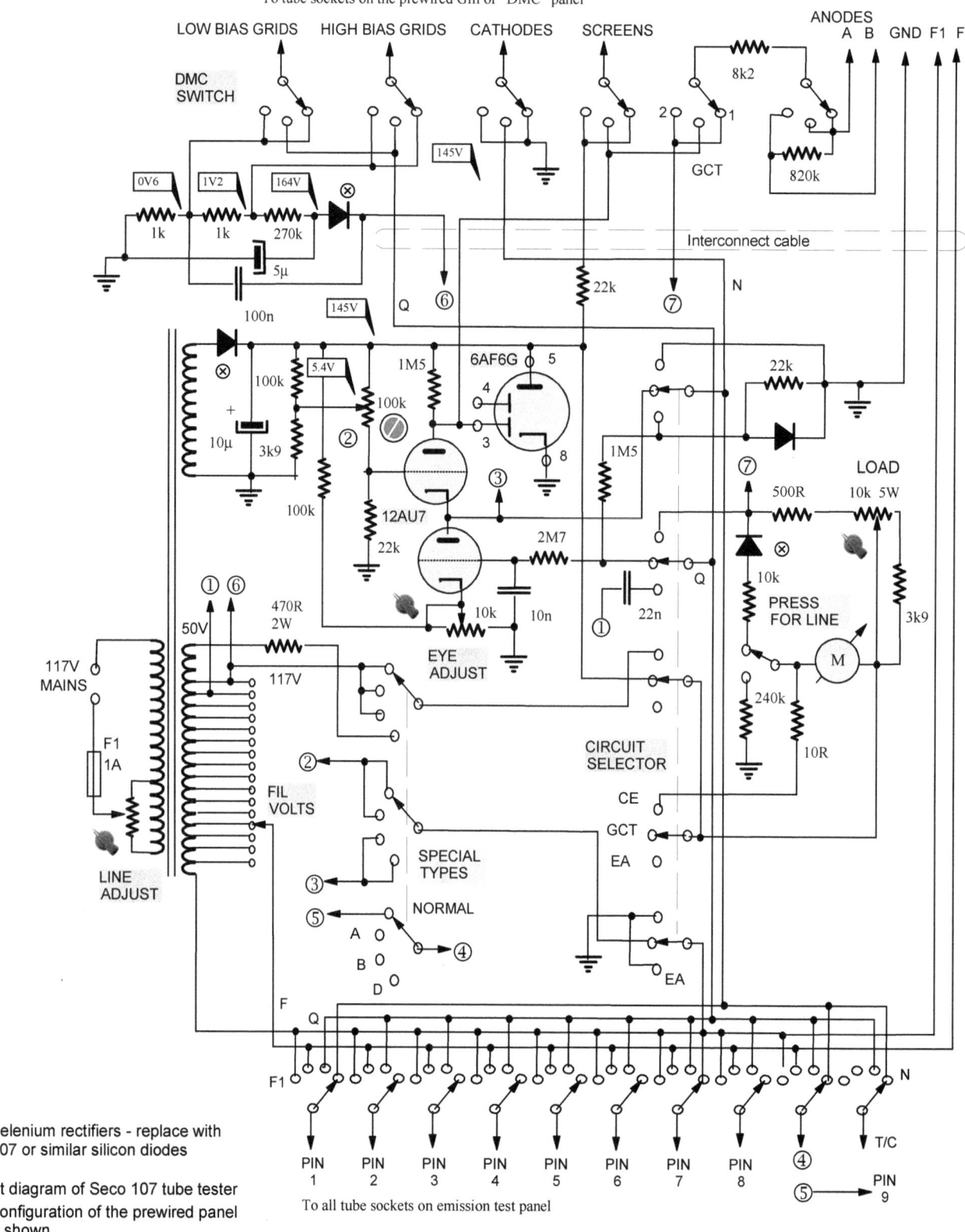

⊗ Selenium rectifiers - replace with 1N4007 or similar silicon diodes

Circuit diagram of Seco 107 tube tester
The configuration of the prewired panel is not shown.

The "Elementary Analysis" of the "Circuit Selector" switch (6) is a fancy name for the basic short test between electrodes. The middle position, "Grid Circuit Test" or GCT, enables gas, grid current, and leakage detection (indicated on the magic eye) on both the main and prewired panel.

The test switch on the prewired panel (7) selects the grids, screens, and anodes of dual voltage amplifying tubes and grounds their cathodes.

Setting "A" simply patches through pin 9, which is left disconnected in all positions of the "Special Types" switch. Position "D" is used with rectifier tubes.

A modernization panel with three more sockets (8) was soon added to model 107, but a more versatile model 107B soon replaced it.

Small signal triodes are tested with almost $100V_{DC}$ on the anodes, and due to relatively low bias, high currents flow, almost 5mA in case of 12AU7, which is good news, meaning these tubes are tested at current levels they would normally operate at in amplifiers. As a comparison, since the same $-0.7V_{DC}$ bias is used, 12AX7 lovelies draw almost 2.5mA, which is a high current for those tubes.

The anode voltage for power tubes such as 6L6 dips to the very low level of around 25V, but the screens are at 75V. In pentodes and beam power tubes, the screen voltage determines anode current much more than the anode voltage. Still, 6L6 is tested at a very low current level of 10-11mA! The grid signal is way too high, a whopping $2.3V_{RMS}$, so this tester suffers from the same issue as B&K testers when testing low bias tubes such as 12AX7 and 6DJ8! Test voltages and resulting anode currents for a couple of common tubes are in the table below.

	PLATE VOLTS (V_{DC})	SCREEN VOLTS (V_{DC})	GRID BIAS (V_{DC})	SIGNAL (V_{AC})	ANODE CURRENT (mA_{DC})
12AU7	95	n/a	-0.7	2.3	4.9
6L6	25	75	-0.7	2.3	10.5

Seco 107-B and 107-C

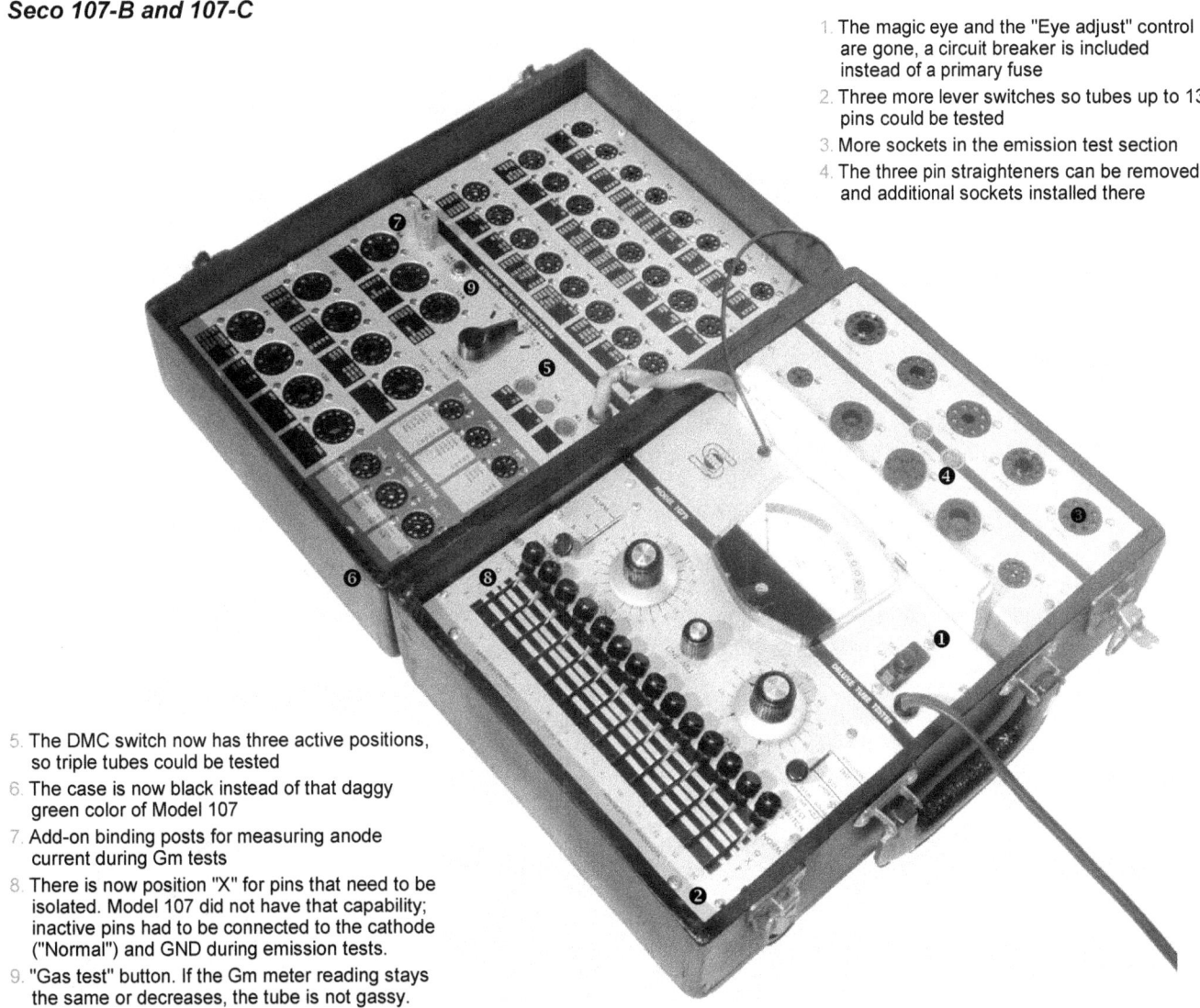

1. The magic eye and the "Eye adjust" control are gone, a circuit breaker is included instead of a primary fuse
2. Three more lever switches so tubes up to 13 pins could be tested
3. More sockets in the emission test section
4. The three pin straighteners can be removed and additional sockets installed there
5. The DMC switch now has three active positions, so triple tubes could be tested
6. The case is now black instead of that daggy green color of Model 107
7. Add-on binding posts for measuring anode current during Gm tests
8. There is now position "X" for pins that need to be isolated. Model 107 did not have that capability; inactive pins had to be connected to the cathode ("Normal") and GND during emission tests.
9. "Gas test" button. If the Gm meter reading stays the same or decreases, the tube is not gassy.

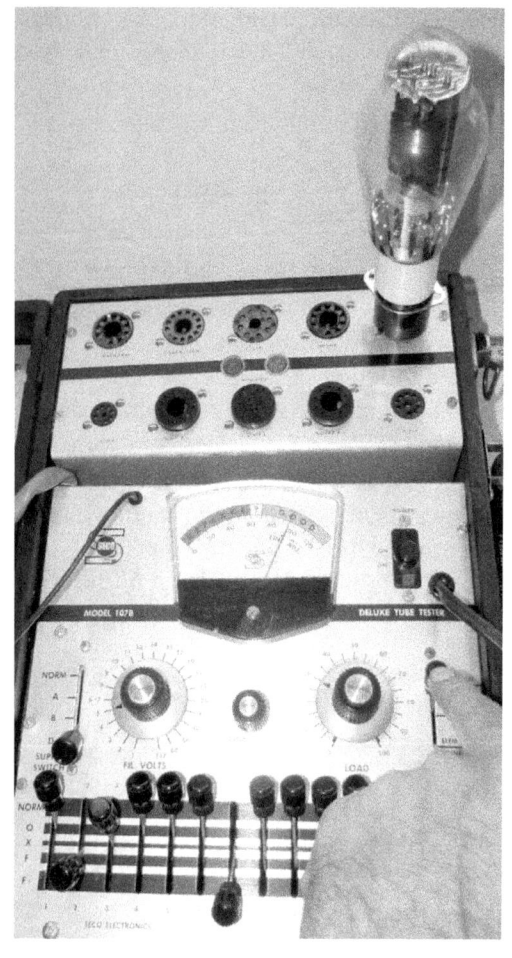

ABOVE: The prewired DMC panel is much nicer and has more sockets than that of the older model 107.

LEFT: Testing a 300B triode using an octal adapter. The "Load" setting of 30 was found experimentally so that 107B's 90% emission reading was in agreement with Triplett 3444 tester's Gm reading of 90% for the same tube. Subsequently, the test results of other 300B tubes we checked followed 3444 results very closely.

Leakage indication via the "magic eye" was gone; 13 lever switches and more prewired sockets on the Gm panel made the tester "future proof."

For some reason, model 107C, almost identical in other ways to 107B, had seen the return of the magic eye, so either Seco designers had a change of heart on their own, or the users of model 107B expressed their displeasure to Seco about that issue and Seco's management listened.

MODERN TRANSCONDUCTANCE TESTERS

Orange DIVO VT1000

Three sockets, three buttons, three "Test Status" LEDs ("Good-Worn-Fail"), and 15 LEDs for tube-type selection by the user and indication of tube's numerical rating. That is the summary of VT1000's control panel.

This compact, easy-to-use, and relatively affordable (US$499.-) valve tester is certainly an original product. How does it work, and how well?

The range of tubes that can be tested is very limited. The octal socket is for power tubes such as 6L6, 6V6, KT66, EL34 (6CA7), KT77, and KT88 (6550). Interestingly, EL34L and 5881 are treated as distinctively different tubes from their EL34 and 6L6 brethren. One Noval socket is for EL84 (6BQ5) only, the other for duo triodes such as ECC81 (12AT7), ECC82 (12AU7), ECC83 (12AX7), ECC99, and 12BH7.

Some of its limitations are obvious. It does not test common audiophile tubes (such as 6DJ8, 6SN7, or 6SL7), or pentodes (such as EF86 or 6AU6), not to mention older 4-pin triodes such as 2A3 & 300B, so cherished by audiophiles.

It does not test rectifier tubes at all! It seems to be geared more towards guitar amp users (it is an Orange product, after all).

Test points (test voltages) are not controllable by the user, so VT1000 belongs to the family of quick "Yes-Maybe-No" checkers. From that perspective, its cost may be hard to justify by prospective users.

Finally, when a fault is indicated ("Fail" LED lit), the user has no idea what kind of fault or failure it is.

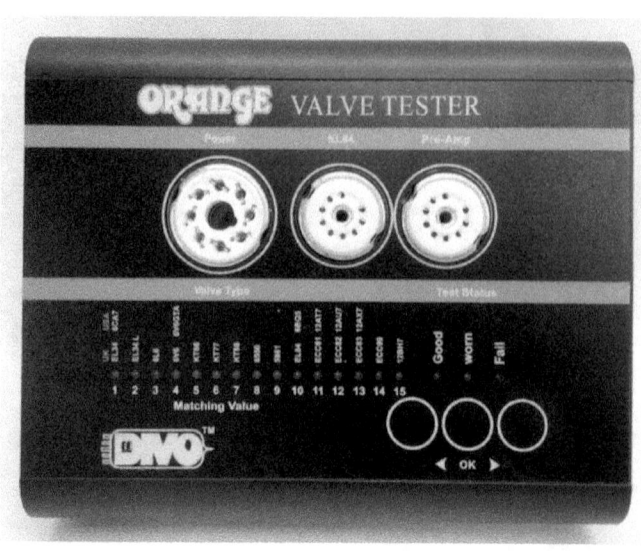

> **"Thermionic Valve Tester" patent application**  FURTHER READING
>
> US patent application US 2015/0042345 A1 by Orange Music Electronic Company Ltd. and KBO Dynamics International Ltd. is available online and explains a few aspects of VT1000's operation. It references patent GB 2462368, but we were unable to find such a patent online.

A few user reports of VT1000 passing tubes that subsequently did not work properly (or at all!) in a guitar amp can be found online, plus a few cases where power and preamp tubes reported as "matched" by the tester were anything but. However, such cases are to be expected, no tube tester is 100% accurate and a complete substitute for the actual amplifier a tube will be used in.

Like all testers, VT1000 "matches" tubes in one point only, and if their operating point in an amplifier is different (which is almost certain!), the two tubes aren't likely to be matched anymore.

Since we haven't yet had the pleasure of properly evaluating this tester, we need to look for clues elsewhere. Before we return to the patent application, the Owner's Manual will shed some light on a few burning questions.

There is next to nothing in the patent application about the actual algorithm used to grade and match tubes. The only hints are from its Owner's Manual, which mentions matching being "based on a summation of the many results obtained during the test and has been specifically designed to reflect the operation of the valve, according to its normal function in an audio amplifier. For example, power valves are graded with their emission and control grid performance as primary factors, whereas preamp valves are graded with different parameters to reflect their role in signal amplification and phase splitting applications."

It is unclear if "emission" means cathode DC current and what "control grid performance" means, transconductance or voltage gain of the tube.

"RED - If a red LED (Light Emitting Diode) is illuminated, this indicates that the valve has failed and should be replaced immediately. Note: Some specific faults found by the VT1000 may be either transitory or not immediately obvious when used in normal operation. However, they could manifest themselves when the valve comes under stress or heavy load during a performance."

This is a de facto admission that some tubes will be declared as "Fail" but will still work fine in an amplifier. In other words, the design suffers from false positives (just like in medicine, testing "positive" for a problem or ailment that does not exist) declaring a tube as faulty when it still may work *in a particular (less demanding) application*!

"After the valve has been tested, the tester will assign a matching number ranging from 1 to 15. Basically: the higher the value, the higher the gain of the valve." This seems to indicate that "matching" is by the mju factor only, not by mutual conductance, internal resistance or anode/cathode current.

"In order to keep the total test time to a minimum, the unit is calibrated at manufacture to test cold valves. If a valve is subsequently re-tested while still warm, then a slightly different reading may be observed." This is understandable; no user wants to wait 10-15 minutes for a tube-under-test to fully warm up and stabilize.

Patents and patent applications are notoriously boring and frustrating to read since there is so much repetition of the obvious and very little illumination or clarification of the critically important. Perhaps that is being done deliberately, so such vagueness enables as many variations and options to be caught under its provisions. Thus, we have distilled this info here into a more concise summary.

The power supply consists of three DC-DC converters. The first boosts the $19V_{DC}$ input voltage to an adjustable anode and screen grid voltage supply of between +100 and $+400V_{DC}$, the second generates negative grid bias voltages up to $-85V_{DC}$, and the third supplies the required heater voltage in the 4-12.6 V_{DC} range.

A 1Ω cathode resistor between each of the three sockets' cathode pins and GND enables the corresponding DC voltage to be monitored by the microcontroller. It also monitors all three electrode voltage levels and provides PWM (Pulse-Width Modulation) control pulses to the FET switches inside voltage regulators (converters).

After the user inputs the tube type to be tested and plugs such a tube in, the first test performed is with cold heaters, $+385V_{DC}$ is applied to "valve electrodes" and $-85V_{DC}$ to its control grid. This "arc test" checks for flash-overs between electrodes. This accounts for the first 20 seconds of the 2 minute test time.

The tests are performed much quicker, but there is a 30 second or 60 second delay between the completion of the test and the display of its results. It was assumed that a user would try to pull out the still hot tested tube immediately upon learning about the test result, so this delay allows TUT to cool down enough to be safe to touch.

Since this is a digital tester, normal parameters of healthy, marginal, and faulty tubes are stored in the microcontroller's memory. Test results are compared to those figures, and the algorithm determines the final diagnosis on that basis.

Finally, the only parameter test that the patent application explains in some detail is the Gm measurement. Control grid voltage (bias) is adjusted until a preprogrammed level of cathode current (for that particular tube type) is reached. Then the microcontroller adjusts the grid bias again until a different preprogrammed level of cathode current is achieved ("typically 1 to 10 mA difference"). The mutual conductance is then calculated by dividing the difference in cathode currents by the difference in grid bias voltages, as $Gm=\Delta I_K/\Delta V_G$!

While this isn't an issue for triodes, strictly speaking $Gm=\Delta I_A/\Delta V_G$, unless anode and cathode currents are equal (no screen grid current!) or tetrodes and pentodes are tested as triodes, this Gm determination would be erroneous for such tubes. It is not clear how VT1000's designers solved this problem. In conclusion, VT1000 uses a grid-shift method to calculate transconductance.

Maxi Matcher II Digital Tube Tester

MaxiTest LLC, based in Seattle, USA, offers a few tube matchers. Octal power tubes (6L6, 5881, EL34, 6V6, 6550, 6CA7, 7581, KT88, KT90, KT100) can be tested and matched on Maxi Matcher II (US$697 in 2017). EL84/6BQ5, EL509, 7027, 7591, 7868, 807 and directly-heated triodes such as 45, 2A3, and 300B require an adapter.

Two anode voltages are available, 325 and 400 V_{DC}, and five fixed bias voltages (-60, -48, -36, -24, and -14 V_{DC}), but only one heater voltage, 6.3V_{AC} @ 6A maximum. Tubes are tested using a 1kHz grid test signal for two parameters, anode current (0-120 mA_{DC}) and mutual conductance (0-19.99 mA/V), both at ± 1.5 % accuracy.

Maxi Preamp Digital Tube Tester

It tests 12AX7, 12AU7, 12AT7, 12AY7, 12DW7, 12BH7, 7025, 5751, 5814 and other Noval preamp tubes with the identical pinout. Duo-triodes with different pinouts, such as 6DJ8, 6CW7, 6AQ8, 6BQ7, 6GM8, 6EU7, 6CG7, 6FQ7, 6N1P and 5687 can be tested but require an adapter.

It measures Gm (0-5 mA/V), voltage gain µ (0-120) and noise (0 to -92 dBV). High µ triodes are tested for gain (µ) at 1 mA cathode current, medium µ triodes at I_K=2 mA and low µ triodes at I_K=4 mA. See the test table for full details of testing regimes.

Up to four tubes can be plugged in and preheated simultaneously. In 2017 its retail price was US$989.-

	Gain I_A	Gain bias	Gm V_A	Gm R_K	Noise R_A	Noise R_K
High µ	1 mA	-1V	200V_{DC}	1k	100k	1k
Med µ	2 mA	-3V	200V_{DC}	1k5	100k	1k5
Low µ	4 mA	-6V	200V_{DC}	1k5	100k	1k5

MaxiBurn

This tube burn-in station can accommodate up to 30 octal power tubes (6L6, 5881, EL34, 6V6, 6550, 6CA7, 7581, KT88, KT90, KT100, and 7027). It heats their filaments, supplies an anode voltage of 100V_{DC}, and "drives a nominal plate current through each tube." EL84, EL509, 6BQ5, 7027, 7189, 7591, and 7868 power tubes can also be used via adapters.

A quintet or five identical tubes must be plugged in simultaneously (or in multiples of five, 10, 15, 20, etc.) because the heater voltage is 35V_{AC}, so 35/5= 7V_{AC} per tube, which will sag under load towards 6.3-6.6V, depending on how many quintets of tubes are plugged in. Tubes must be identical because the same current flows through all five heaters in series.

MaxiBurn is not cheap; in 2017, it retailed for US$1,450.-!

Nominal currents through common power tubes with 100V_{DC} on anodes   INVESTIGATION

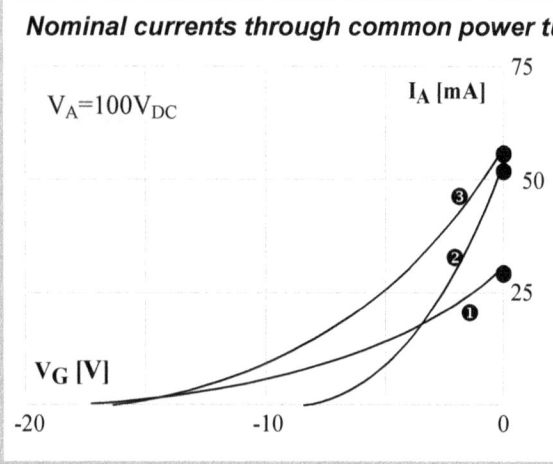

Since there was no mention of any bias voltages in the MaxiBurn unit, the currents that flow with 100V_{DC} anode voltage can be read from each tube's transfer curve for that voltage and zero bias (V_G=0V).

Three such transfer curves were compiled and placed on the same graph (left). TX curve (1) is for 6V6. With zero grid bias I_A will be around 27-28 mA. TX curve (2) is for 6BQ5 (EL84), which under the same conditions will pass around 52 mA, while the zero-bias anode current of 6L6 (TX curve #3) will be slightly higher, around 54 mA.

REPAIRING & UPGRADING VINTAGE TUBE TESTERS

- CHOOSING & BUYING A TUBE TESTER
- SAFETY RULES & PRECAUTIONS
- TYPES OF FAULTS, COMMON CAUSES AND FAULT LOCATION METHODS
- COMPONENT TESTING
- ANALOG (MOVING COIL) METERS
- A CRASH COURSE ON PROPER SOLDERING
- SOCKET ADAPTERS

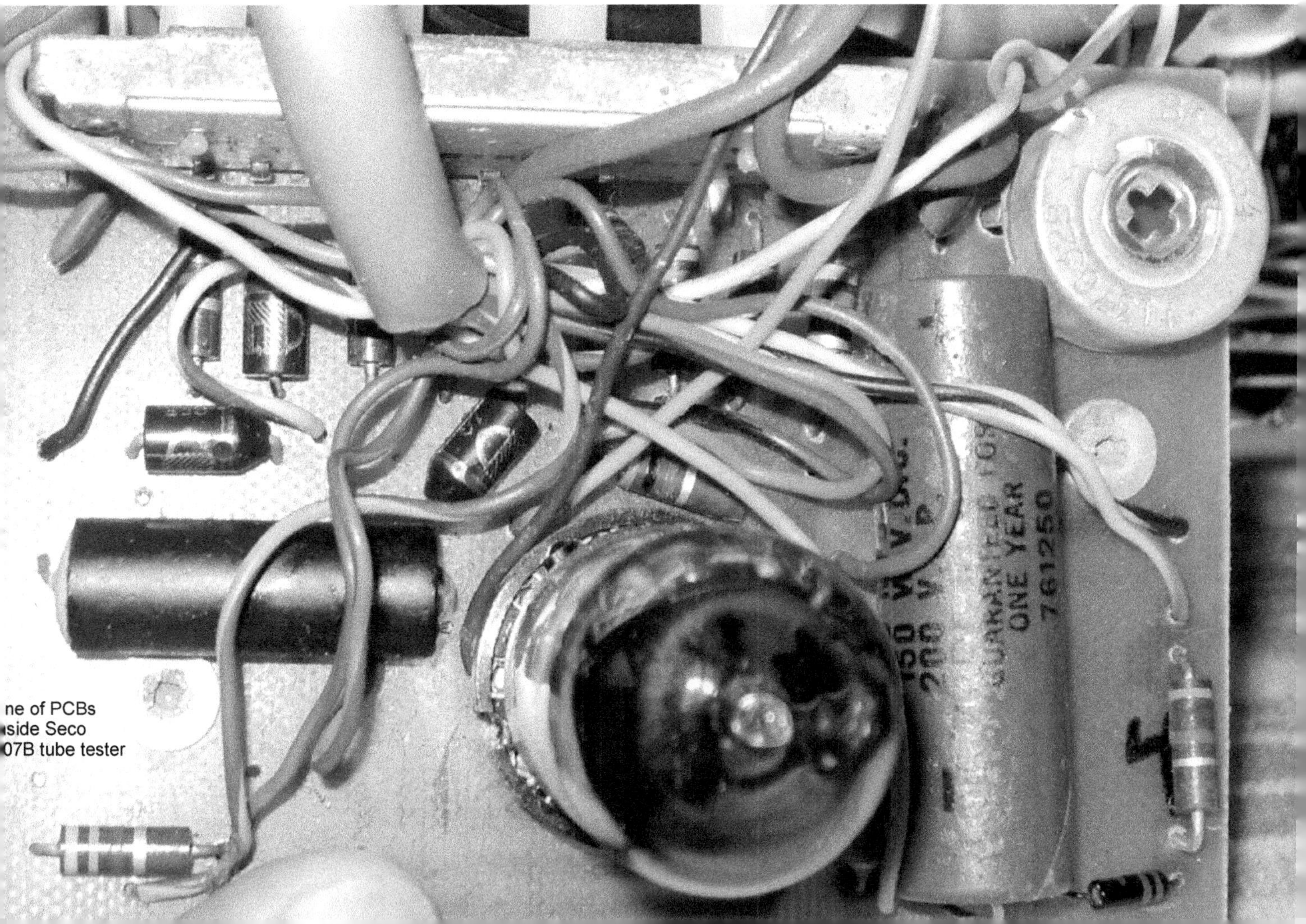

One of PCBs inside Seco 107B tube tester

CHOOSING & BUYING A TUBE TESTER

Reading discussions in online fora, I've noticed many myths and misconceptions that plague potential users of tube testers. Many express a reluctance to buy a certain tester due to one or more of the reasons outlined in the table below. Others follow trends and buy a tester that is well regarded by many but is wrong for them - the tester is not an optimal choice when their needs and other situational factors are taken into account.

Many compromises were made in the design and manufacture of vintage tube testers, and all testers have strengths and weaknesses. You have to decide which of those weaknesses you can live with and which are a deal-breaker for you. I've divided factors into five major areas of concern in the checklist below. They are not in any particular order, certainly not in the order of importance. As with most decisions in life, a compromise is needed.

For instance, if you are a guitar player who suspects a tube or two to be approaching the end of their life in a guitar amp that "doesn't sound right" anymore, and you don't want to lug your amp to a repair technician and pay a fortune for tube testing and "matching," you may only want a grid circuit or emission tester that will give you a rough idea of tube's emission capabilities and gas/grid current condition.

And, since only a limited range of preamp and power tubes is used in guitar amps (mostly 12AX7, occasionally 12AT7 and 12AU7 in preamp/driver stages, and 6V6, 6L6, EL34, and 6BQ5 in output stages), you don't need testers with 4- and 5-pin or Magnoval, Nuvistor and Novar sockets, all you need are Noval (9-pin), octal (8-pin) and perhaps a 7-pin miniature sockets. Except for the early (pre-WWII) models, all tube testers have those.

As with cars, some testers hold their value much better than others. Many even increase in value with passing years, so you may sell them for much more than what you paid for them a few years back.

Factors to consider before buying a tube tester — CHECKLIST

- *Your needs:* quick yes/no checks - tube matching for own use - tube matching for commercial sales
- *Technical/Design/Construction issues:* testing method (principle) - accuracy - sensitivity - reliability - repeatability - test conditions (voltages, currents, frequencies) - ease of calibration - how often is calibration needed (stability and drift)
- *User-friendliness:* complexity - ease of use - speed of testing - mobility (transportability) - likability (enjoyment factor)
- *Versatility*: range of sockets & tubes tested - parameters and tube conditions tested or checked
- *Financial factors:* acquisition cost - repair and calibration cost - resale value

---

REASON FOR NOT BUYING A PARTICULAR TUBE TESTER (TT)	VALID?	REMEDY?
No specific socket (e.g. 4-pin socket for 300B)	NO	Add such a socket or replace an existing yet obsolete or rarely used socket (e.g. Acorn with Magnoval)
Tubes you want to test are not listed in tester's book or rollchart	NO	A common issue with older testers (40's & 50's) and newer tubes (1960s). Consult datasheet for those tubes, learn how you tester works and figure out the required settings!
TT is not a well regarded model	YES	If you want to make a quick buck reselling the tester you recently bought
	NO	You will get it cheap(er) and it may serve you well once you figure out how to use it. For instance, Sylvania 219, Weston 789, Seco 107 and Precise 116 can often be bought cheaper than many basic emission testers, yet are decent and competent testers!
TT test only emission, but you want to test Gm	YES	Gm tests carry more weight (authority) if you want to test and match tubes for sale
	NO	An emission test is fine for most purposes, good enough to weed out 90+ % of bad tubes
TT is in poor condition	YES	If wire insulation has deteriorated, if the transformer has burned out, or if contacts are badly worn or oxidized, stay away!
	NO	Cosmetic issues can easily be fixed, burned out analog meters can be replaced, as can knobs, lamps, resistors and capacitors
The seller of the TT lives in another country	NO	You may buy testers cheaper from overseas, even when transport costs are factored in. For instance, tube testers are rare in Australia and prices are higher. We bought all our testers from USA, all arrived without any problems and most worked well
	YES	If unhappy with your purchase, it is harder and more expensive to return the tester and get your money back.
Your friends or forum members tell you to stay away from a certain model	YES	If your friends are experts in tube electronics and electronic measurements, than you should listen to what they have to say.
	NO	If you improve your knowledge of tube testers you will be more likely to make the best decision. Many forum members don't know what they are talking about, are biased or even deliberately spread misinformation. Why would you trust someone else's judgment better than your own?

Some online resellers of tubes post photos showing their expensive and rare tubes (Ameprex Mullard, Telefunken, and such) being tested on cheap and dangerous emission checkers. Dangerous because many of those simple checkers use test voltages and currents that can damage or destroy sensitive preamp tubes.

Also, such tests carry very little weight with buyers who expect rare & expensive tubes to be tested for mutual conductance and anode (plate) current to see how closely matched they are and how they compare to the standard or "bogey" value published by tube manufacturers. You'd imagine such sellers would invest in a decent and capable tube tester that would give credibility to their claims about the condition of tubes they are selling.

If you are a serious audiophile with an expensive collection of tubes and buy such tubes online often, you need a mutual conductance or at least a dynamic conductance tester that can also test for gas, leakage, and grid current problems. Many supposedly matched pairs and quads of tubes we bought over the years on eBay from sellers in China and Hong Kong were not matched at all, let alone closely matched as those sellers claimed.

A certain emission tube checker tested our newly-bought "matched" pair of 2A3 tubes as "91" and '92", which could be construed as a well-matched pair considering the capabilities of such an instrument. However, when tested on Triplett 3444 tube analyzer, one tube's anode current was 36mA, while the other passed 47mA with the same anode and bias voltages. Testing these power tubes as diodes, with a low AC voltage of 30 or so Volts, as emission testers do, clearly does not indicate their true condition and *cannot be used for tube matching*!

The socket trouble

TIPS& TRICKS

When buying a tube tester, pay attention to its tube sockets. Most factory-wired units had riveted sockets. If a socket looks different or is attached with screws, you know that it had been replaced in the past.

On prewired testers such as B&K 700/707 (pictured), the octal socket for tubes such as 6L6, 6V6, and EL34 (#29) is the most likely to be troublesome due to its constant use (1). The same applies to the Noval socket for testing 12A*7 series, socket #8 (2). The Noval socket wired for 6BQ5 (#23) is also likely to have been used a lot (3).

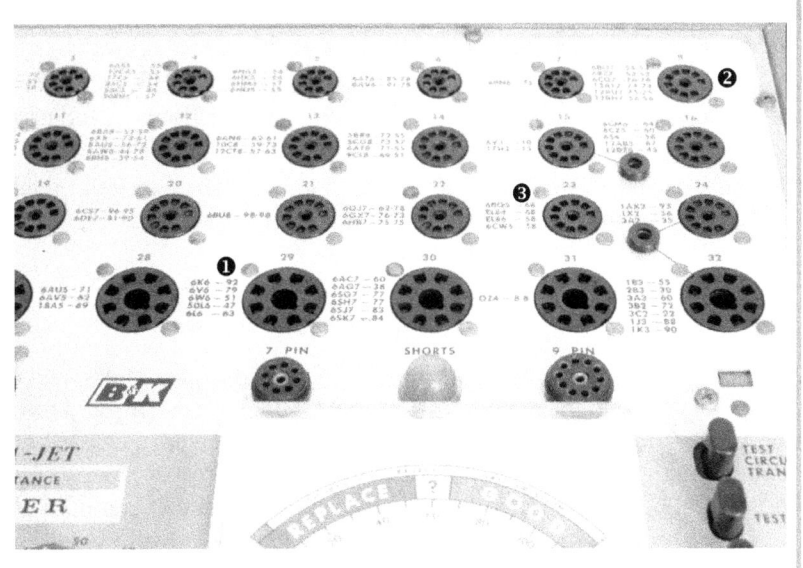

SAFETY RULES & PRECAUTIONS

Electric shock - how it happens and how to avoid it

Just as people can drown in 10cm of water and 10m deep water, they can get a lethal electric shock from 100V, which can be as deadly as 1,000V! The severity of an electric shock depends on many factors, such as the age, gender, and physical condition of the victim. However, the main factor is the level of current flowing through the body.

The threshold of perception is around 1mA. Currents up to 5mA produce tingling but not severe pain. Muscular contractions start at around 10mA, while 100mA of current would start the fibrillation of the heart muscle, preventing it from pumping blood and causing death unless the fibrillation is stopped.

Above 300mA, the heart's contractions are so severe that fibrillation is prevented, so normal heart rhythm will probably resume if the electric shock is halted quickly. Thus, 100-300mA is the most dangerous range.

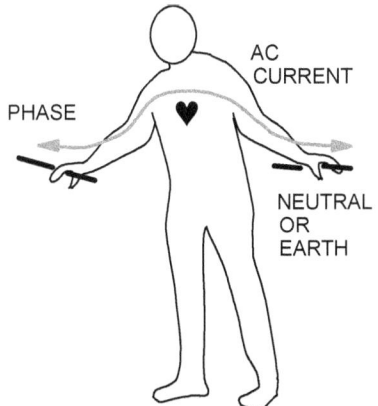

The worst situation: current flowing through the upper body (torso) and the heart

The lethal voltage level depends on the resistance of the current-conducting path through the body, which in turn depends on skin resistance and how the contact is made. Skin resistance, in turn, depends on its moisture level. Dry skin can have as much as 500kΩ of resistance, while the wet skin's resistance can be as low as 1kΩ!

As electric current flows, it punctures and breaks down the outer layer of the skin, and skin resistance falls rapidly. This is why it is paramount to break the contact with the live conductor as quickly as possible. The most dangerous situation involves a voltage between two hands or arms, as illustrated when current flows through the upper torso where the heart is located. That is why you should only use one hand when measuring voltages. The other hand should not be touching anything - keep it in your pocket!

Electrical hazards and required safety measures

CHECKLIST

- Tube testers involve high voltages that may be lethal, so numerous precautions must be taken to mitigate the risk of an electric shock when powering up, testing, and working on "live" circuits.
- Install a device called ELCB (Earth Leakage Circuit Breaker), ground fault interrupter, or RCD (Residual Current Device) in your house's or workshop's switchboard. These are now mandatory by law in Australia, but electrical laws in many other countries are much more lax. These protective devices may save your or your children's lives (if they poke a knife into a toaster, for instance).
- Unplug the tube tester from the wall outlet while working on it. Once unplugged, discharge the power supply capacitors before doing any work. Don't just use a straight wire for discharging; the spark and the high discharge current may damage the capacitor. Make a discharge cable with insulated crocodile clips at both ends and a 1kΩ 2W resistor in series to limit the discharge current.
- When working on a "live" instrument use only one hand. Don't touch the chassis or any other part of the tester with the other hand. If practical, wear thin cotton gloves.
- Wear shoes with rubber soles. Never stand barefoot on a bare concrete floor while working with electrical appliances or amplifiers. Use a rubber mat. Timber flooring and carpet (wool or polypropylene) are also good insulators.
- Use tools with insulated grips, and don't touch their exposed metal parts. Use an isolation transformer. That will minimize the probability of an electric shock.

A critical safety upgrade for most vintage tube testers

Most amplifiers, tube testers, capacitor checkers, signal generators, and other audio or test equipment built before the 1970s in the USA used a two-pin mains plug. There was no earth pin, so these units were not earthed (or grounded) at all. Furthermore, these 2-prong plugs are reversible, so instead of the neutral being connected to the metal chassis, you can end up with the live 115 V_{AC} on it. In many cases, the fuse and the power switch are in different branches of the mains circuit, and by current standards, they must be in the same "live" circuit. Many cheaper testers had no fuse at all!

In many cases, there are also film capacitors connected between the phase or neutral and the chassis. Once these caps become leaky, they pass the mains voltage onto the metal chassis.

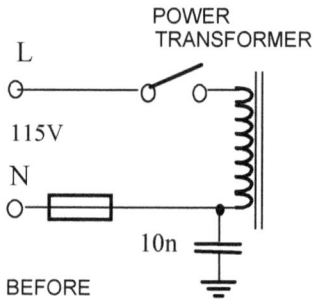

BEFORE

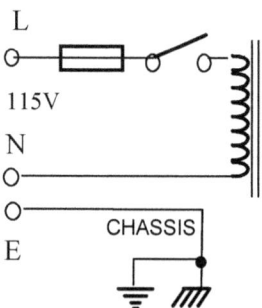

ABOVE: The fuse and the On-Off switch must be in the L (Live) line!

LEFT: How the mains (power) circuit should be wired up

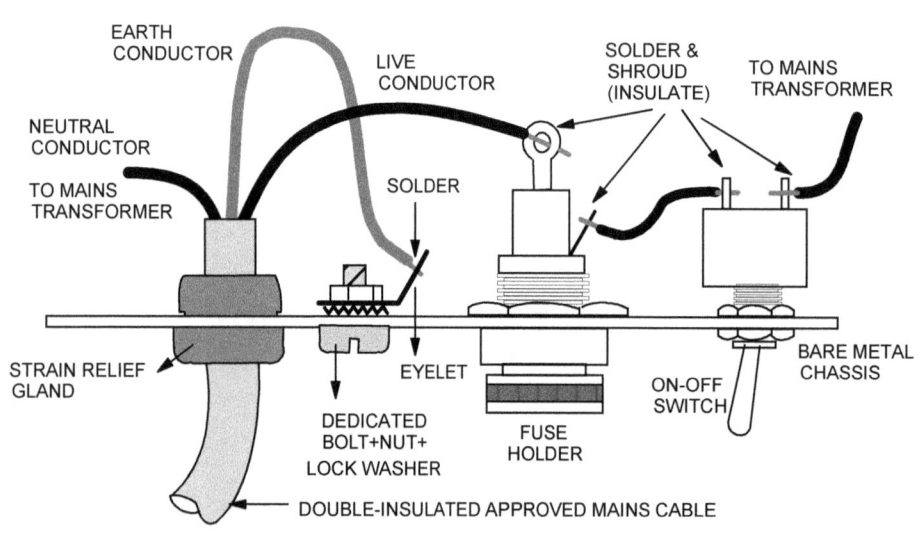

These days only X2-rated and approved capacitors can be used in such critical positions (the short circuits cannot develop), but such technology and approval did not exist 50 or 60 years ago when these amps were made. As such, these capacitors (even if not leaky), often called "widowmakers" for obvious reasons, are a serious hazard and must be removed.

Before you start repairing or upgrading anything in a vintage tube tester, your first job is to perform this safety conversion. Bin the two-core mains cable and the two-pin plug, and replace them with an approved 3-core mains cable and plug.

Remove any capacitors and resistors connected between phase, neutral, and ground. Connect the earth lug via an insulated green or green-yellow wire to a dedicated bolt with good galvanic contact with the metal chassis.

TYPES OF FAULTS, COMMON CAUSES AND FAULT LOCATION METHODS

The first step towards locating and fixing faults is understanding their causes and types. Some faults have a singular cause; two or more contributing factors cause others. Often one such factor would not be sufficient to cause trouble, but two factors certainly do.

The prime suspects amongst the components

As a "rule of thumb," the main suspects among the components are tubes and capacitors, especially electrolytic capacitors, which age and dry out, becoming leaky and losing capacitance. Due to their nature (moving parts and poor contacts due to dirt, grime, and oxidization), switches are also a common cause of problems.

Many current made-in-China tube sockets are of poor design or construction. Socket pins lose contact with tube pins, break off or even fall out. Corrosion and oxidization seriously affect even the best sockets of yesteryears.

While checking DC voltages or AC signals, if you get intermittent readings, consider replacing a suspect tube socket straight away; otherwise, you could be wasting lots of time trying to find the fault around the socket connections.

Resistors and transformers are less likely to be the source of problems, although old, carbon composition resistors (molded type) drift significantly in value, and some circuits that require precise resistance values (such as resistors in Gm, signal, or current shunts) may not operate properly and cause inaccurate readings.

However, the title of the Prime Suspect goes to wires, sockets, and contacts. Let's look at them in more detail.

Contact issues and intermittent troubles

Intermittent faults are the most frustrating of all. They suddenly appear for a while, a moment, a few seconds or minutes, and then all is well again. That unpredictability makes them difficult to trace and locate. The cause behind all intermittent faults is contact loss between two points in the circuit that should be connected or unwanted contact or short between points that should not be connected together!

Quite a few factors can cause the loss of contact. Bad or "cold" solder joints are probably the most common issue, especially in poorly constructed DIY equipment and vintage kits. Although a joint of two or more wires passes the visual inspection, it is uncertain what is happening inside the joint.

Mechanical stress or corrosion caused a break in a wire or the solder joint itself. If in doubt, re-solder all the joints in the affected area, for instance, around a tube stage that does not work properly. This is often faster than signal troubleshooting.

In amplifiers and tube testers that use printed circuit boards, especially double-sided, there could be a poor solder joint or a break in the metal rivet that joins two sides via the plated-through hole. A quick ohmmeter check between two copper layers would confirm such an issue, illustrated in a).

Illustration b) shows a situation which also passes a purely visual inspection. The hookup wire has an invisible break under the insulation. That is the reason why visual inspection should always be performed together with the mechanical check.

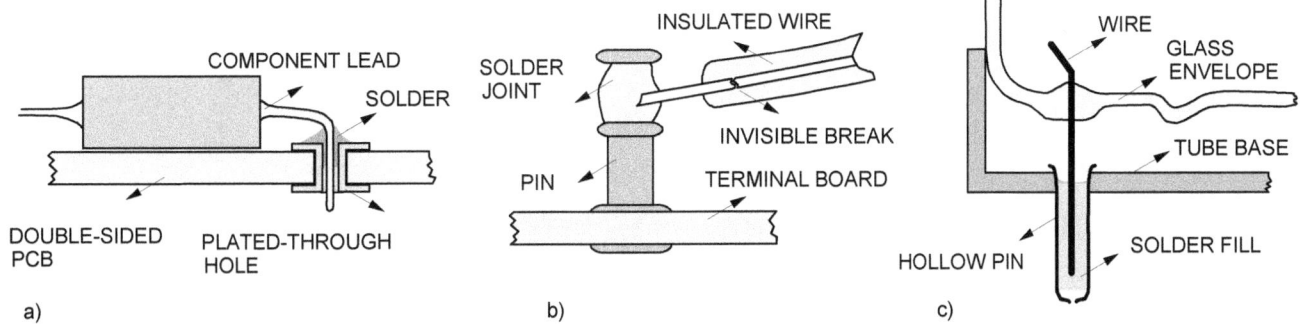

Use a small insulated screwdriver, tweezers, or a dental probe to jerk wires and component leads around. 3-piece dental tool kits are cheap, around $5-6 online, including a probe, tweezers, and a mirror. A mirror and a magnifying glass are valuable to see what is happening in hard-to-reach places!

This issue usually goes back to when wire-strippers or cutters were used to strip the insulation off the wire before soldering. If too much force was used, the wire would be partially cut-through and, if subjected to vibration or mechanical stress, could break off days or even years later.

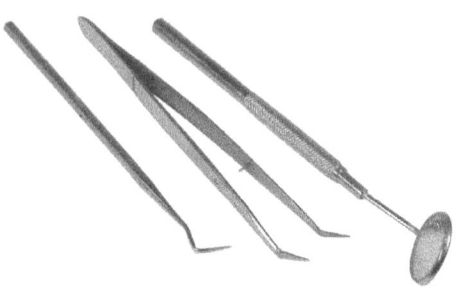

These cheap basic dental kits are useful in identifying "cavities" in solder joints and other contact problems.

A contact defect can also happen inside a tube. If a loss of contact happens inside a glass envelope, nothing can be done, and such a tube should be discarded. A common problem, illustrated in c), is poor solder joints inside the tube pins, especially in the current Chinese-produced tubes.

In octal and other larger tube sockets, the wires that bring tube elements out to the base must be soldered inside the hollow pins. Due to mechanical stresses and thermal expansion and contraction, a poorly soldered contact will soon be lost. So, if a tube tests "bad" or dead, don't throw it away in haste; always re-solder all the pins first and then test it again. It could be just a poor contact, and a tube could be perfectly healthy otherwise.

How to correctly pull tubes out of their sockets

Be extremely careful when pulling tubes out of their usually very tight sockets. Never jerk a tube sideways. This can bend or break delicate pins and loosen the socket contacts. Some tubes have a bottom glass pinch, which can be snapped off, and even the glass envelope can be cracked due to excessive force. Both situations would render a tube useless. Always pull the tube straight up, with minimal lateral movement!

Checking order

The checking order is always the same.

1. VISUAL INSPECTION: Inspect the internals for any overheating, burned-out, missing, broken, loose or disconnected parts, links, and components. Identify non-original parts and previous repairs and modifications.
2. MECHANICAL CHECKS: Using a dental probe or a similar tool, jerk wires, and component leads and check soldered connections for any loose or cold joints, broken or detached wires, and components, anything that could not be identified by a visual inspection only.
3. COLD & COMPONENT CHECKS: With an ohmmeter and LCR meter, check connections, continuity, and component values (inductors, capacitors, resistors, chokes, transformers) without powering the tester up.
4. HOT CHECKS: Check AC and DC voltages in critical points, such as electrodes of internal tubes (if any), AC voltage levels on transformer taps, outputs of bias, anode and screen power supplies, etc.

COMPONENT TESTING

Testing solid state diodes

Since analog ohmmeters use their internal DC battery to push current through an unknown resistor, a diode should conduct in one direction only when the + of the battery is connected to its anode. The measured resistance should be very low, 20-50 Ω.

When the + of the battery is connected to the cathode, the diode should not conduct at all, and the ohmmeter should indicate a very high, almost infinite resistance. The cathode is the negative electrode, marked with a band or a line on one side of the diode's body, corresponding to the line on the symbol.

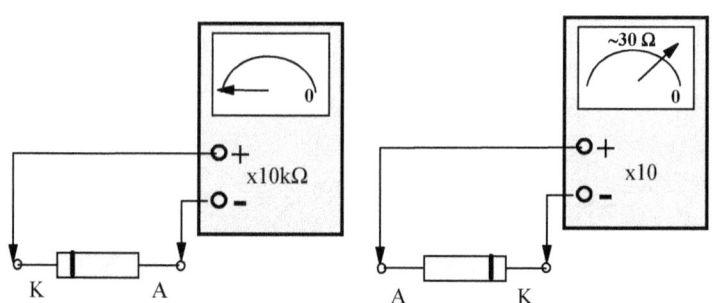

ABOVE: Ohmmeter diode check must show very low resistance one way and a very high resistance the other way.

In some multimeters, the + lead is connected to the - (negative) pole of the battery, so you need to mentally reverse the connections when testing with such ohmmeters. Its - (negative) lead will have a positive DC test voltage on it, and its + lead will be the negative side.

In this case, it seems that everything is the opposite, that a diode under test conducts when it's reversely polarized and doesn't conduct when it's directly polarized, which would not make sense unless you understand the internal battery connection.

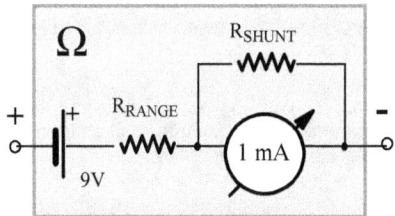

ABOVE: The internal circuit of some analog ohmmeters. Notice the - end of the battery marked as + end on the outside of the meter!

Testing bipolar transistors

A bipolar transistor is made of two PN junctions (P stands for "Positive," N for "Negative") or "diodes." The first is the base-collector junction (test #1), and the other one is the base-emitter junction (test #2).

Using an analog ohmmeter, you should get a low resistance one way (base-collector junction is positively or forward polarized so the diode is conducting and its internal resistance is small), and when we swap the two test leads, we should get a high resistance because the diode is not conducting and its internal resistance is high. These readings will be the opposite for a PNP transistor.

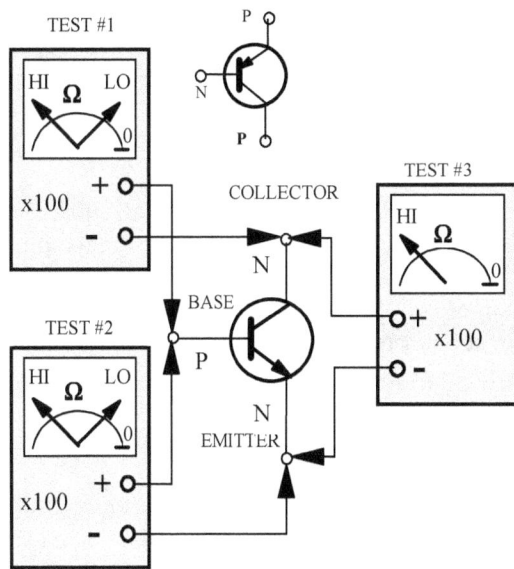

Since the emitter-collector path comprises two "opposing" PN junctions (one forward- and the other reverse-polarized), the resistance reading must always be high for both NPN and PNP transistors. What low and high means will depend on the transistor; power transistor resistances will generally be lower than those of low power transistors.

However, these ohmmeter tests are not conclusive. Even if a transistor passes these rough tests, it may still be faulty. This is called a false negative. Transistors that don't pass this quick check (test "positive") are definitely bad - there are no false positives. It is not possible for transistors that show "bad" on this test to be good and work in a circuit.

Testing Zener diodes

Zener diodes work in the reverse quadrant, not in the forward-biasing arrangement as ordinary diodes. This means that the cathode of the Zener diode needs to be more positive than its + electrode (anode).

To determine the Zener voltage V_Z of a diode, you'll need an adjustable DC power supply and a series resistor R_S. If your DC power supply (or a battery) is not adjustable (most are), connect a 1kΩ potentiometer across it so you can adjust the test voltage. Increase the voltage until there is a sudden current jump, meaning you have reached the V_Z point (breakdown when the diode suddenly starts conducting current in reverse), and read that voltage V_Z on the voltmeter V.

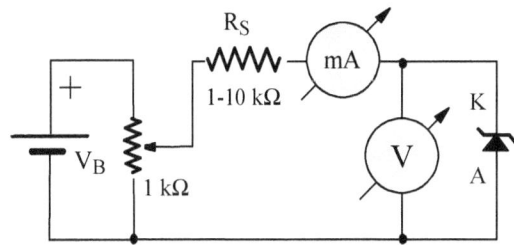

ABOVE: How to find out the Zener voltage of a Zener diode and check its operation at the same time

Testing electrolytic capacitors

To test electrolytic capacitors, use an analog ohmmeter on x1kΩ (x 1,000) range. If the capacitor is open circuit, there will be no meter indicator movement. For a good capacitor, the meter needle should move quickly towards the low resistance side of the scale as the battery inside the ohmmeter charges the capacitor, and then it should slowly drop back towards infinity. The final resting point of the needle will indicate a very high resistance, the insulation resistance of the capacitor.

If the capacitor is short-circuited, the needle will move towards zero Ω and stay there.

RIGHT: If ohmmeter meter needle moves quickly towards the low resistance side of the scale (a) and then drops back towards the very high resistance (b), the elco is most likely good.

BELOW RIGHT: Turn the potentiometer all the way CW or CCW, connect an analog ohmmeter and slowly turn its shaft from one end to the other. The needle should travel slowly and smoothly.

Testing potentiometers and rheostats

Analog ohmmeters or multimeters are the best for this task. The ohmmeter needle should move smoothly, without jumps or breaks, as you measure the resistance between one end and the center tap (wiper or slider).

Digital auto-ranging meters jump when moving from one range to another, which could be falsely interpreted as a break in the variable resistor.

How to make your own test adapters

These test jigs or similar adapters make it quick & easy to measure DC voltages and AC signals on tube pins without opening the amplifier or tube tester up and exposing yourself to dangerously high voltages and hot tubes. Each pin on the tube socket is brought out to its own test point, so measuring the voltage between pins or each pin and GND is a breeze. Simply plug the male adapter into the tube socket, plug the tube into the jig, and you have access to all pins.

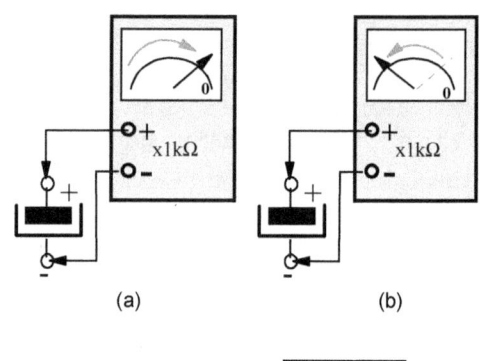

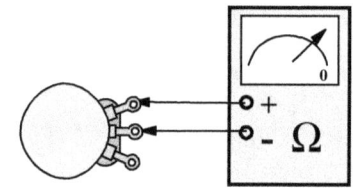

RIGHT: These Noval (left) and Octal (right) DIY test sockets are the most useful tools of all! Build one for all commonly used socket types (7-pin mini, Noval, Octal, Magnoval, RimLock, etc..) Pin 1 is marked with a plastic sticker.

BELOW RIGHT: Testing 6L6 through the Octal test jig on Heathkit TT-1. The multimeter is measuring voltage between pins 5 & 8, which is control grid DC bias voltage, around -12.3 V_{DC} in this case.

Replacing selenium rectifiers

Selenium rectifiers age, regardless of their use. So, even NOS (New Old Stock) rectifiers sitting on a shelf for forty to sixty years have aged and should not be used as replacements. Gradually, their forward voltage drop increases, and the reverse leakage current increases. Their DC rating reduces as they age; they run hotter and may even catch fire. The smoke is highly toxic.

The bigger the size of the plates, the larger the current capacity of the selenium rectifier. The more plates in a stack, the higher the rated voltage of the selenium stack. Each plate can take 30V, so by counting the number of plates you can determine the voltage rating of the selenium diode!

When you replace selenium rectifiers with silicon diodes, the DC voltage will increase by 5 to 20 Volts. The forward voltage drop of silicon diodes is much lower, only around 0.6 Volts. This increase may be useful or may present a problem, depending on the circuit and the application.

In the bias circuit mentioned, the new voltage will be higher (more negative), and the anode current through the output tubes will be lower, so you will have to re-bias the output stage.

If such a voltage rise is unwanted, simply add a suitably sized (in terms of power rating and resistance) resistor in series with the silicon diode(s).

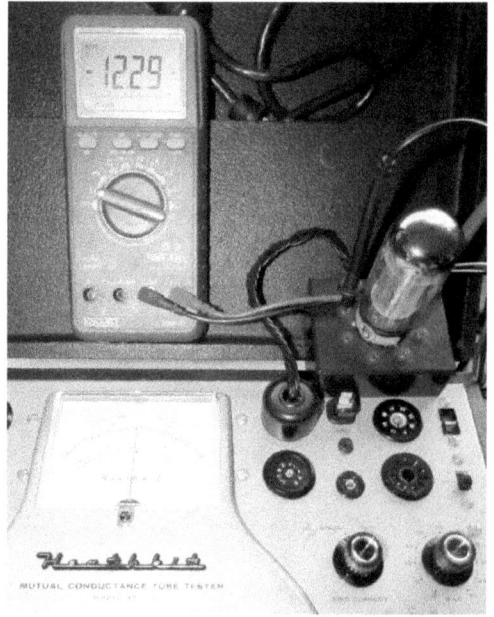

ANALOG (MOVING COIL) METERS

Replacing analog meters

If you have a tube tester or any other instrument whose meter is sticking or has a burned-out coil, you can replace it not just with an identical meter but also with a meter of different (higher) sensitivity and lower internal resistance. As a rule, the replacement meter must be of equal or lower FSD (Full-Scale Deflection) and equal or lower internal resistance, but cannot be of higher FSD or higher internal resistance!

Say you need to replace the analog meter on the Hickok KS15560 tube tester. The meter has an FSD of 200mA and internal resistance of 2,365Ω. You have a 100mA FSD meter from a dead Hickok 600A tester. That meter has an internal resistance of 1,165Ω. Calculate the required values of the shunt R_S and series resistor R_A to be added.

Step #1: Determine the current to be diverted through the shunt resistor. In this case, I_2=100mA will go through the new meter (M2) at full-scale deflection, the old meter (M1) had I_1=200mA through it, so the shunt has to divert I_S = I_1-I_2 = 100µA.

Step #2: determine R_S. Since half the old current has to go through the shunt, R_S has to have the same value as the new meter's resistance, 1,165Ω. Each will take half of the 200µA!

Generally, you will have to calculate the R_S from the formula for the voltage drop, $I_2*R_{M2} = I_S*R_S$. You know all three, except R_S, so you can calculate it easily.

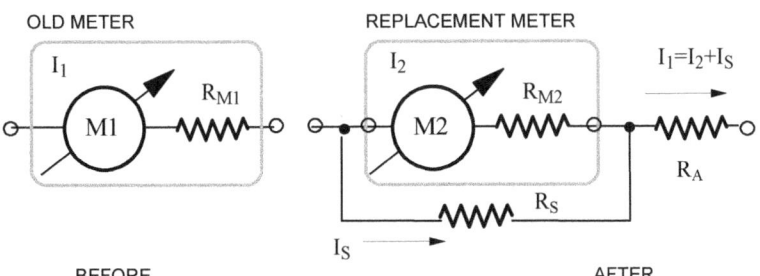

Step #3: Determine R_A. Calculate the total resistance of the new meter and R_S in parallel ($R_{M2}\|R_S$). Then, subtract that figure from the original meter's resistance, which is the required R_A value.

We have 1,165Ω and 1,165Ω in parallel, so the total resistance is half of 1,165 or 582Ω. The old meter had R_{M1}=2,365Ω, so we need R_A= 2,365 - 582 = 1,783 Ω in series!

A standard value of 1k8 (1,800Ω) should be sufficiently close.

A quick primer on shunts and scales

This nice & large moving coil meter was salvaged (among other useful parts such as rheostats, precision resistors, and similar meters) from a vintage Japanese-made TEC transistor tester. The meter was made in the USA by Precision Meter Co., Inc.

The front markings of interest are "accuracy 2% FS" (1), where FS stands for "Full Scale" and "F.S. 5mA" (2).

The mounting board at the meter's back carries three shunt resistors, starting from the meter's positive terminal (3); their values are 0.2Ω (4), 1.8Ω (5), and 18.0Ω (6). A hookup wire was soldered to each of the junctions, four in total, meaning the transistor tester had three current ranges. How did this circuit work, and what were the instrument's ranges?

To perform this analysis, we are only missing one figure: the meter's internal resistance. So, before we proceed with the analysis/reverse engineering of the shunt circuit, we'll have to determine it experimentally.

Since I have only two hands, I hooked up the digital multimeter to the analog meter with short test leads terminated with crocodile clips to take the photo while measuring the meter's internal resistance. The figure measured was 20.5Ω (next page).

Before doing this kind of low resistance measurement, make sure your multimeter or ohmmeter displays 0 ohms when you short its test probes. If it doesn't (many cheaper meters read 0.2-0.6 Ω instead of zero), you'll have to subtract that figure from all test results.

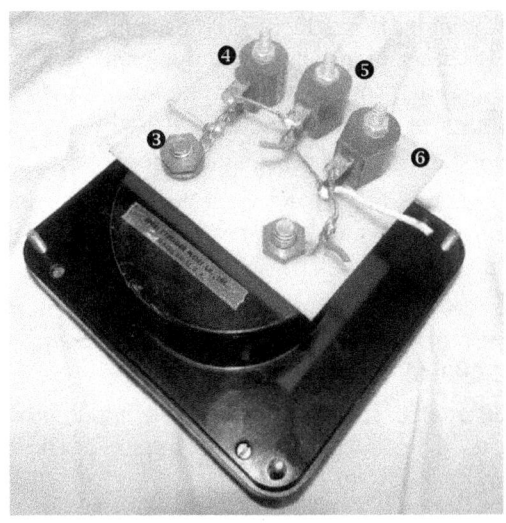

Thanks to our Taiwanese friends that designed and made our Escort multimeter, the short circuit resistance measures 0.0 Ω, no calibration or "zeroing" needed! Don't you wish vintage tube testers were that precise and stable?

I didn't spent more than four years of my life studying electronics at university without learning a trick or two. One of these tricks is never to use any extension test leads for low resistance measurements. Sure enough, when I measured the meter's resistance directly, the result was 19.9Ω!

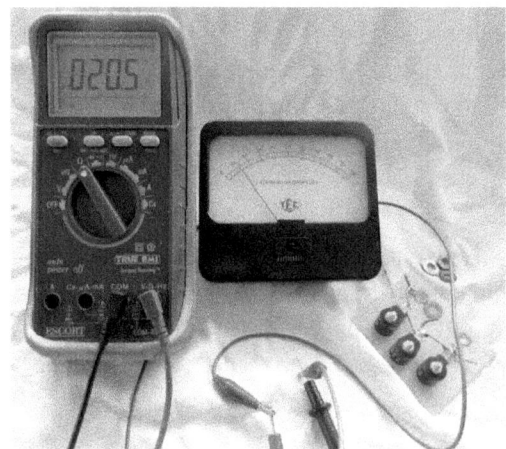
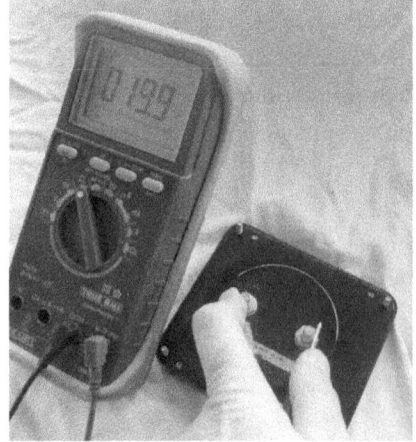

RIGHT: Using "universal" leads and crocodile clips the internal resistance of the meter was measured as 20.5 ohms.

FAR RIGHT: When measured directly at meter terminals, the internal resistance was accurately measured as 19.9 ohms!

Let's start with all three resistors in the circuit. In all cases, we look at full-scale currents and voltages. There is 5mA flowing through the meter whose resistance is 19.9Ω, so the voltage drop between points A and B (across the meter and the three shunt resistors in series) is $V_M = 0.005A * 19.9Ω = 0.0995V$ or 99.5mV (DC, of course).

The total shunt resistance is 0.2+1.8+18 = 20Ω. Since 19.9Ω is the same as 20Ω (within the margin of measurement error) for practical purposes, we can conclude that an identical current of 5mA will flow through the shunt.

So, at full scale, the meter will indicate 10mA (the user will have to divide the readings on a 0-100mA scale by 10). If a user flicks the "Range" switch into the next position, the 18R resistor will be short-circuited, and we get a simplified drawing on the right.

Now we need the current between points A and D. We have a 20Ω meter and a 2Ω shunt. Intuitively, you should feel that a 10-time lower shunt resistance will divert ten times higher current away from the meter so that 50mA will flow through the shunt, or 55 mA in total for that range.

However, the transistor tester's faceplate was marked 0-10mA and 0-50mA, not 0-55mA! The shunt current needs to be 45mA, and since the voltage across the meter and the shunt is the same, 99.5mV, the total shunt resistance needs to be 99.5/45 = 2.21Ω instead of 2.0Ω. So much for (the myth of) Japanese precision and attention to detail!

Finally, with the "Range" switch into the third position, both 18R and 1R8 resistor will be short-circuited, and we get a simplified drawing so we can determine the current between points A and C.

Again, 0.2R is 100X lower resistance than 20R, so shunt current will be 100 times higher than meter current, or 500mA. Actually, the range was 0-500mA, so the shunt current needed to be 500-5=495mA, meaning the shunt resistance of 99.5/495=0.201W was needed. We probably had it, but a multimeter cannot measure those 0.001 of an ohm or 1mΩ!

If we build our own tube tester using this meter to display anode (plate) current, we could use it as it is, shunts and all! The three current ranges suit tubes as well.

The 0-10mA range is perfect for most preamp tubes, 0-50mA for high current preamp tubes and small output triodes and pentodes, and 0-500mA for large power tubes.

Of course, it is assumed that the anode power supply of our tube tester would be able to supply 1/2 Ampere of DC current, but that is a story for another day (or chapter) ...

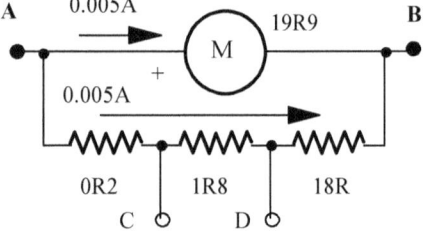
ABOVE: 0-10mA range on a 0-5mA meter

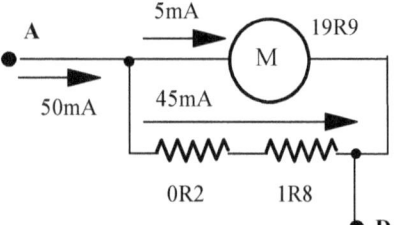
ABOVE: 0-50mA range on a 0-5mA meter

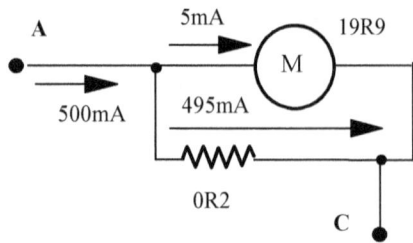
ABOVE: 0-500mA range on a 0-5mA meter

> **Analog instrument symbols**
>
>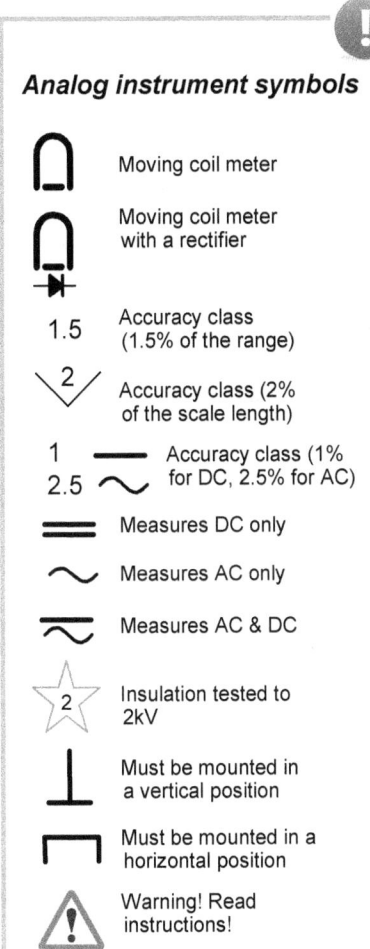
>
> - Moving coil meter
> - Moving coil meter with a rectifier
> - 1.5 — Accuracy class (1.5% of the range)
> - 2 — Accuracy class (2% of the scale length)
> - 1 / 2.5 — Accuracy class (1% for DC, 2.5% for AC)
> - Measures DC only
> - Measures AC only
> - Measures AC & DC
> - Insulation tested to 2kV
> - Must be mounted in a vertical position
> - Must be mounted in a horizontal position
> - Warning! Read instructions!

> **INVESTIGATION**
>
> This enlarged detail from Philips Fluke 2505 analog multimeter illustrates a few important specifications deduced from the printed symbols.
>
> The internal resistance of the multimeter is 10 MΩ, and the frequency range of the signals it can measure (within the specified accuracy) is 10 Hz to 30 kHz. The measurement range extends to 70 kHz, but the accuracy is not guaranteed above 30 kHz.
>
>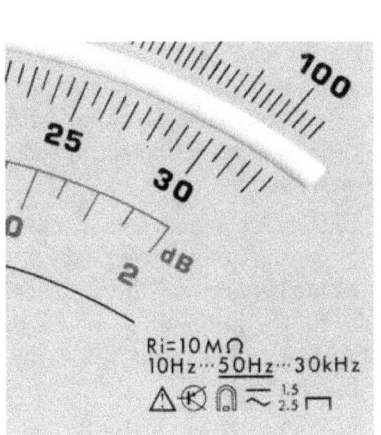
>
> The underlined 50Hz frequency means that its accuracy is higher for 50Hz signals (in this case, 2.5%, compared to 5% on other frequencies).
>
> The transistor symbol indicates an active instrument with bipolar transistors (usually in differential amplifier circuits).
>
> We are already familiar with the other symbols. Notice a higher DC accuracy (1.5%) than 2.5% for AC signals. This is the accuracy of the moving coil meter itself, not the whole multimeter, and it's lower for AC signals due to the non-linearity of the solid-state rectifier (AC to DC converter).

A CRASH COURSE ON PROPER SOLDERING

Soldering irons

Many DIY constructors and repairers ask me if they should buy an expensive Weller soldering station or a cheaper alternative? We have never had a Weller, so we cannot comment on their quality or longevity.

This WER 927D soldering station (pictured) and similar budget stations served us well. You even get a set of interchangeable tips of various shapes and sizes, a very handy bonus (1).

The sponge (for cleaning the tip) is about 1mm thick, an utter joke (2), but everything else seems fine. The only irritating aspect is the fact that the plastic holder (3) is too light, so whenever you put the iron back into its "holster" (4), the whole thing falls off. A heavy metal holder would be much better. Alternatively, bolt it down to the main unit's top (5).

For large or rough desoldering jobs for which 40-60 Watt regulated soldering stations would be too good or not powerful enough, you will need a high power soldering iron, 100-120 Watts.

One such job is unsoldering components from a metal chassis, which is such a large heatsink that even 80 Watts of heat would not be enough, and it would take a few minutes to get the solder hot enough to remove such components. Automotive parts retailers and craft shops also sell these high-power soldering irons.

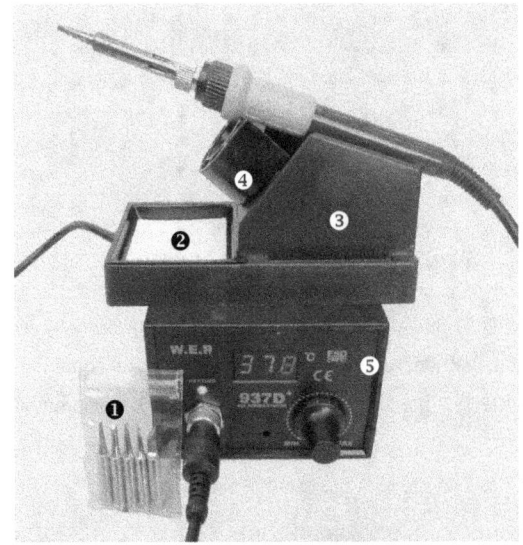

ABOVE: Temperature-regulated soldering iron
BELOW: High power soldering iron

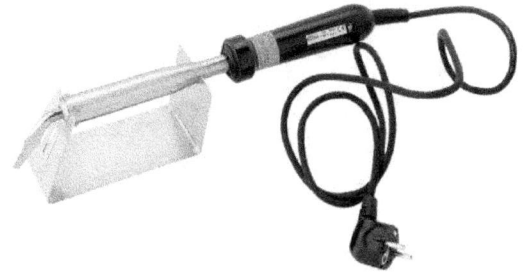

How to solder

A brand new soldering tip must be tinned first, i.e., covered with a thin coat of solder. During use, the soldering iron tip will get dirty. Gently clean it by rubbing the tip on a wetted sponge. Press the tip against the wires/contacts to be soldered together for a few seconds. Then touch the hot joint with the solder. The solder should melt upon contact. If it doesn't, the contact hasn't been heated enough.

You should never place solder onto the tip of the soldering iron, only on the joint to be soldered! The more component leads-wires and the larger they are, the longer it will take for the joint to heat up enough to take on the solder and melt it. Knowing how long to heat the joint comes with practice.

The solder should flow smoothly and cover the whole joint. If it '"blobs" and falls off, the surfaces are not clean enough; they are dirty, greasy, or oxidized and must be cleaned before attempting to solder the joint again. The tip should be touching the component lead or wire and the terminal lug or pin simultaneously, as illustrated.

Large joints with many wires/leads may need to be heated from various sides, not just at one point. Don't just heat one lead entering the joint; try to touch two or three together; otherwise, one will be hot, and others will be cold and will not accept the solder as well.

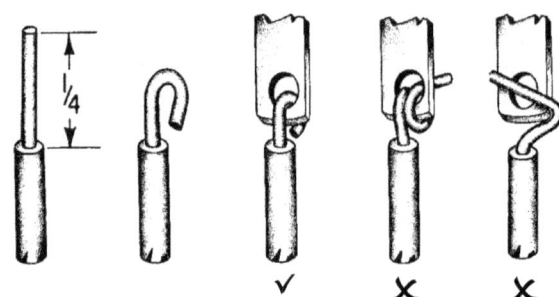

ABOVE: Three steps involved in preparing a wire for soldering onto a terminal or tube socket lug. The copper lead should only be looped once through the hole, so future unsoldering is fast and easy. The two rightmost examples are wrong, one being a double loop, the other an improper one.

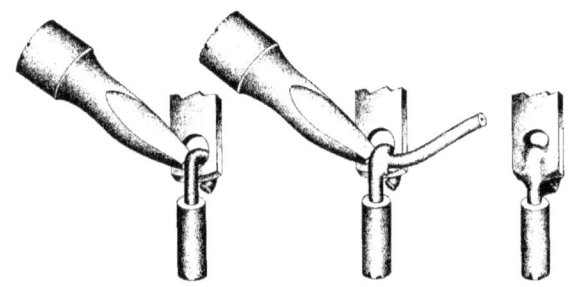

ABOVE: Once the copper lead is looped through the hole, the tip of the soldering iron should be touching both the lead and the lug, thus heating both together. The solder is then brought against the lead and the lug; it melts to cover both without even touching the soldering iron's tip.

SOCKET ADAPTERS

ADAPTER PLUG	CABLE	10 PIN MIN	NOVAR	NUVISTOR 5 PIN	NUVISTOR 7 PIN	COMPACTRON
1	brown	1	1	2	1	1
2	red	2	2	4	3	2
3	orange	3	3	8	5	3
4	yellow	4	4	10	10	4
5	green	5	5	12	12	5
6	blue	6	6		6	6
7	violet	7	7		7	7
8	orange-wht	8	8			8
9	yellow-wht	9	9			9
Grid Cap	white	10				12

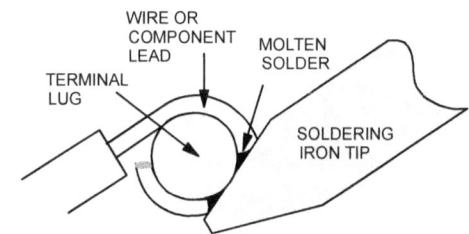

ABOVE: The tip should be touching both the lead and the lug, thus heating both together.

ABOVE: The pin wiring of Precision G-140 adaptor

BELOW: Mercury AD-4, a switchable 4-socket adaptor

RIGHT: Precision G-140, a fixed pinout with 5 different sockets

Buying a vintage adapter is usually faster and cheaper than building your own. Sure, you will never get exactly the sockets you need, but it is possible to remove a socket or two that you don't need and replace them with the socket(s) you will use.

Eico 610, 610A, and Coletronics B-16 (identical to Eico 610a) are most commonly offered for sale.

Most tester manufacturers released at least one adapter, as Mercury had done with their AD-4 and Precision with G-140, pictured on the previous page.

Conar 6AD adapter is identical to Precision G-140.

Mercury AD-4 used an octal plug and had four sockets (10-pin mini, 5-pin Nuvistor, Novar and Compactron) and two 12-position switches. G-140 had an additional 7-pin Nuvistor socket but used fixed wiring, outlined in the table (previous page), and a 9-pin (Noval) plug.

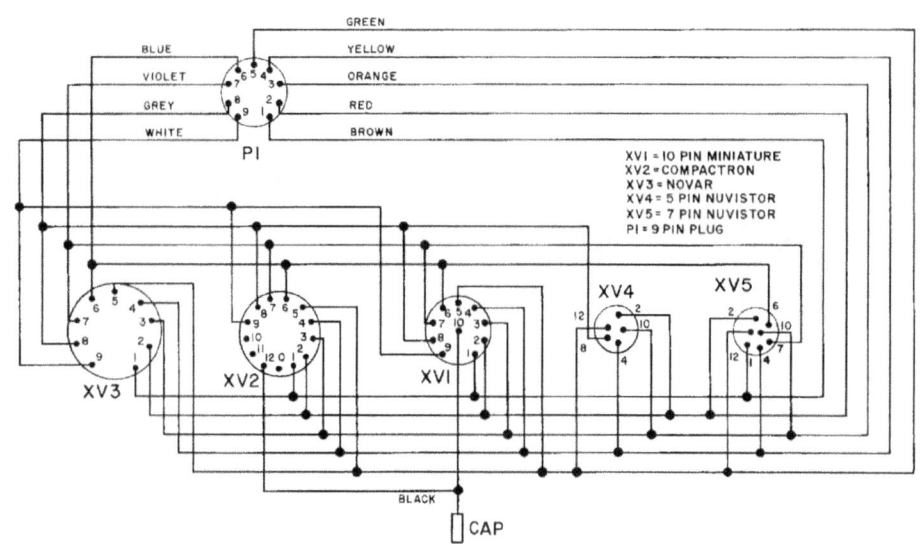

ABOVE: Eico 610 adapter
BELOW: Eico 610A adapter is identical to Coletronics B-16!

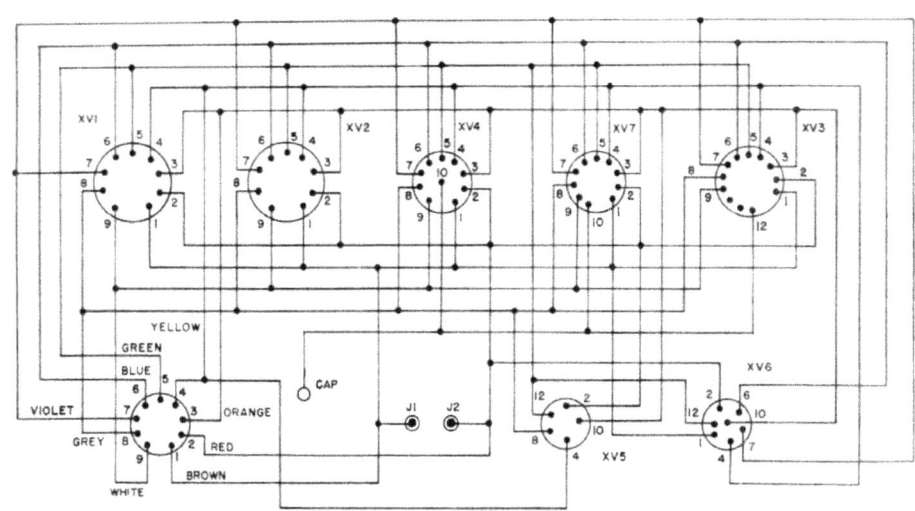

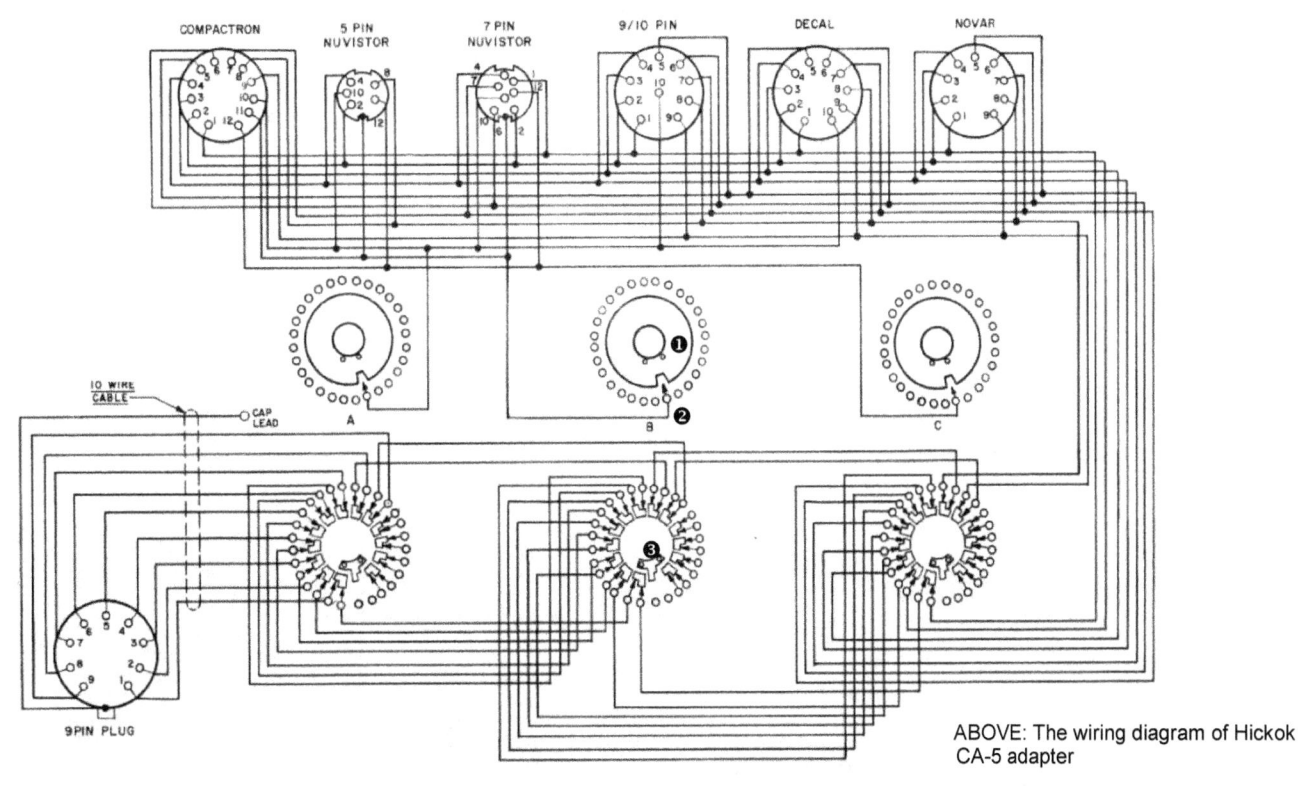

ABOVE: The wiring diagram of Hickok CA-5 adapter

Hickok CA-4 used an octal plug and had eight rotary switches. By using a 9-pin (Noval) plug in model CA-5, Hickok managed to achieve the same with only 3 switches! 14-position switches were used, but only ten positions, marked 0-9 were active. Each switch connected one of the three additional pins to one of the pins (1 to 9). In position "0" there was no connection. The circuit diagram above shows switches in "0" position!

Switch A connected pin #10, switch B pin #11 and switch C pin #12 to any of the pins 1-9. The upper wafer (1) wound connect the pin in question through its contact (2) down to the lower contact (3) and to one of the 9 pins. The L-shaped contacts simply feed-though all nine pin connections from switch A, through switch B and C to all tube socket pins in parallel (all pins #1 together, all pins #2 together, and so on).

DIY PROJECT: Fixed adapter for selected power tubes

Say you need an adapter to test 4-pin directly heated triodes such as 300B and 2A3, EL12 and LS-50 power tubes, Russian 6C33C-B output triode, and PL519 pentode. However, your tester has none of the five sockets needed. The solution is simple. You'll need to wire six-core cable to an octal plug, using wiring and pinout for a 6L6 power tube. Why 6L6? Because every tester under the sun can test those.

Find a suitable box (plastic, metal, even timber). It doesn't matter, providing it's sturdy enough and can be opened (to access the socket pins and the wiring) and closed firmly. Mount the sockets onto the box and wire all cathodes together, all anodes together, all control grids together, etc.

Since PL519's anode is on the top cap, you'll need to have one wired. Also, 6C33C-B has two independent heaters that can be wired in parallel for 6.3V or in series for 12.6V operation. Wire them in series; otherwise, the 6.3V winding on your tester's heater transformer will burn down! The heater current draw at 6.3V would be 6.6A, and no tester that I know of could supply that much current. Even at 12.6V, most testers' transformers cannot supply 3.3A, so check first. You may have to power the heaters of that tube from a separate AC or DC power supply!

Since the octal plug is wired for 6L6, you must use 6L6 settings on your tester to test the tubes in question. Of course, the heater voltage will need to be changed to suit (5V for 300B, 40V for PL519, 6.3V for EL519 and EL-12, 12.6V for LS-50 and 6C33C-B). Also, the "Shunt," "Plate," "Load," "Sensitivity," setting (usually a rheostat shunting the meter to obtain the right reading) will need to be set experimentally.

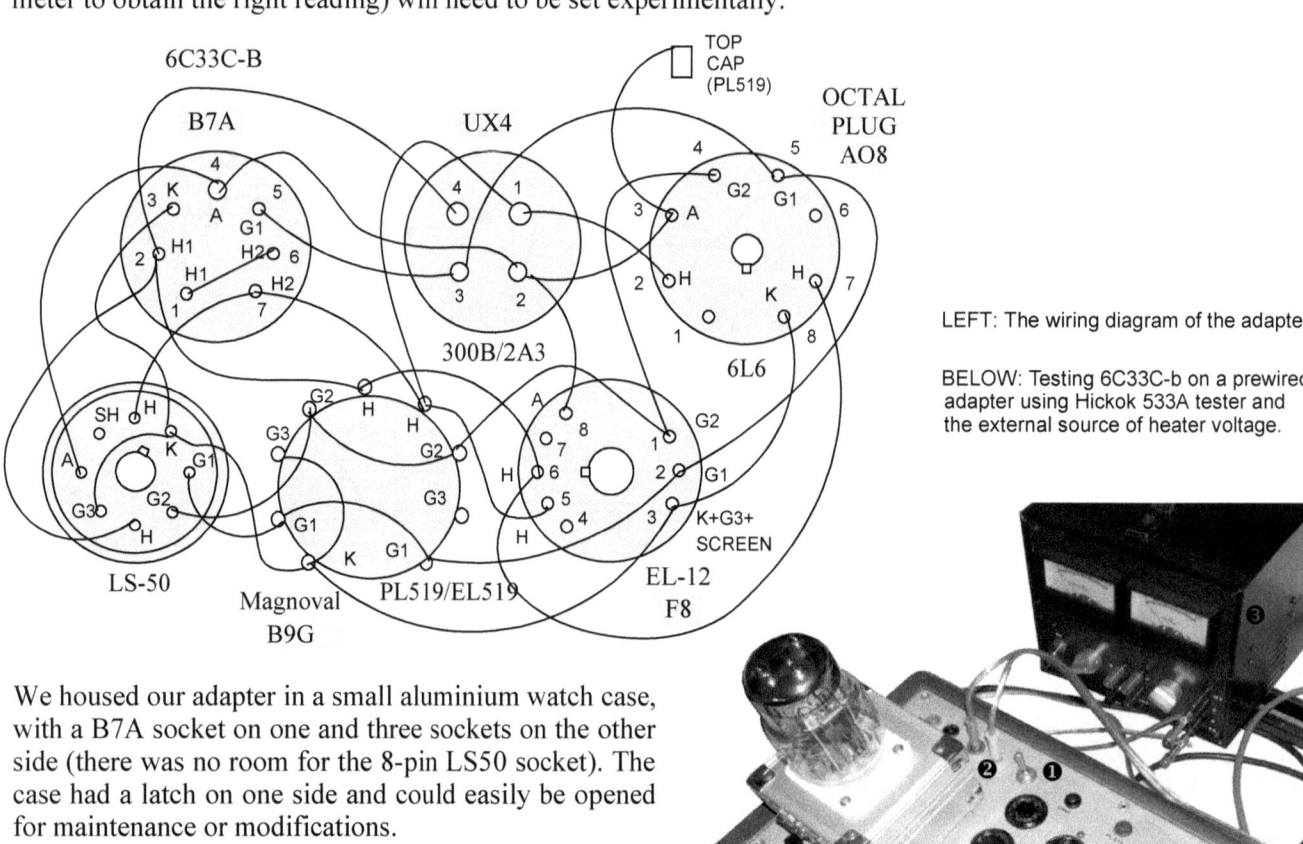

LEFT: The wiring diagram of the adapter

BELOW: Testing 6C33C-b on a prewired adapter using Hickok 533A tester and the external source of heater voltage.

We housed our adapter in a small aluminium watch case, with a B7A socket on one and three sockets on the other side (there was no room for the 8-pin LS50 socket). The case had a latch on one side and could easily be opened for maintenance or modifications.

The photo on the right shows it being used to test a 6C33C-B triode on the Hickok 533A tester. The tester's lousy transformer could not supply enough heater power, so we installed a heater changeover switch (1) and two banana sockets (2) for the external heater supply connection.

In this case, our DIY high voltage power supply (3) provided the external heater voltage.

Build your own USU - Universal Switching Unit

DIY PROJECT

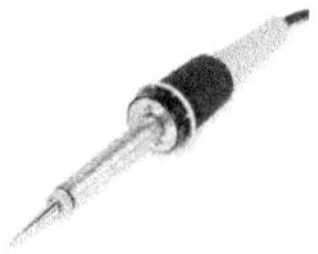

If you need to install more than a couple of additional sockets into your tester, a more elegant and long-term solution is to build your own switching adapter. Since we didn't need Compactron and similar 12-pin sockets, 10-pin controls were installed in this prototype switcher, one rotary switch for each pin.

Standard 12 - position rotary switches were used, but only 6 positions were needed, so every second position (marked "OFF") leaves that pin disconnected - sweet! Electrodes are (clockwise): two heater pins H1 and H2, cathode, control grid (G1), anode, and screen grid (G2). The output goes to an octal plug (wired as 6L6), so the switcher can be used with any tester (1).

With a dedicated tester, you can wire the output permanently to it, with no need for plugging and unplugging.

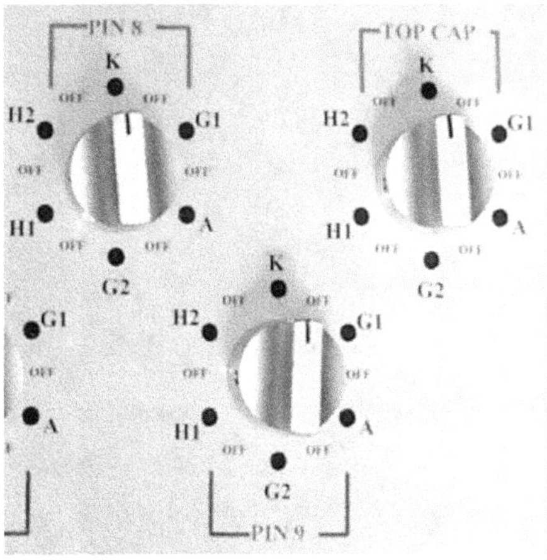

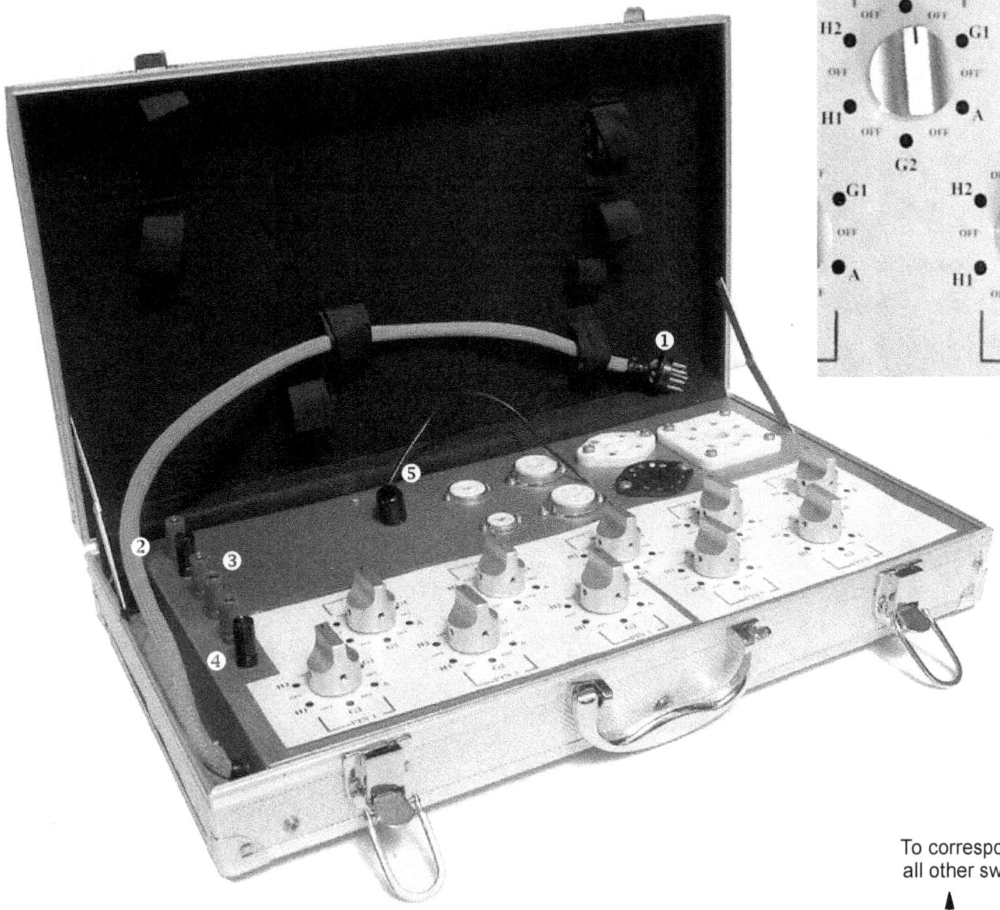

The case was a nice aluminium storage box of a BBQ toolset, and the top panel was a steel top cover salvaged from an old DVD player and trimmed to size with an angle grinder. The markings were laser printed in color, laminated, and glued to the steel fascia.

The top red & black sockets (2) are for measuring plate current, the middle red & green sockets (3) give you access to anode and cathode to measure plate voltage, while the bottom green & black sockets (4) are connected to the cathode and G1 so that you can measure grid bias voltage. There is a top cap (5), connected as pin 10.

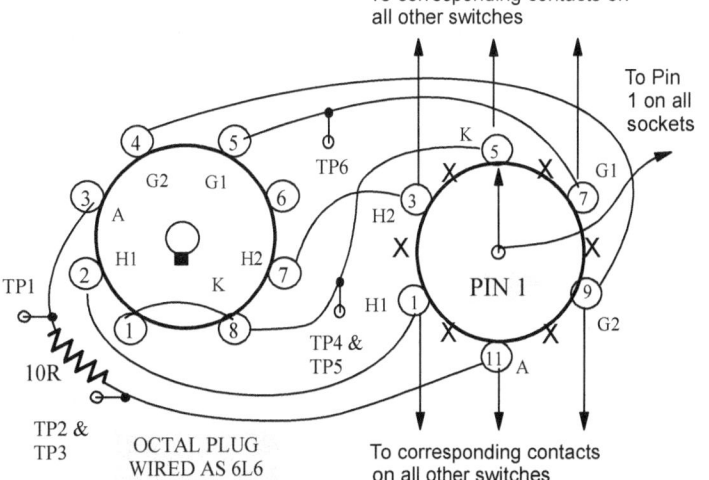

ABOVE: Wiring diagram of the Universal Switching Unit. Only one switch (for pin 1) is shown. All other switches are wired in the same manner.

OTHER AUDIO-RELATED BOOKS BY IGOR S. POPOVICH
Available from Amazon, Barnes & Noble, Book Depository and all other major online bookstores

Audio Tests & Measurements: How to Test Electronic Components, Audiophile & Guitar Amplifiers and Loudspeakers Using Modern and Vintage Test Instruments
ISBN: 978-0-9806223-9-3

- TEST INSTRUMENTS, ERRORS, LIMITATIONS & SAFETY ISSUES
- SIGNAL SOURCES, TRACERS, POWER SUPPLIES AND FILTERS
- MULTIMETERS - TYPES, OPERATING PRINCIPLES AND FUNCTIONS
- OSCILLOSCOPES - HOW THEY WORK & HOW TO USE THEM
- TESTING PASSIVE ELECTRONIC COMPONENTS (RESISTORS, CAPACITORS & INDUCTORS)
- TESTING AUDIO AMPLIFIERS AND PREAMPLIFIERS
- DISTORTION MEASUREMENTS
- TRANSFORMER TESTS & MEASUREMENTS
- LOUDSPEAKER TESTS & MEASUREMENTS
- TRANSISTOR TESTERS AND CURVE TRACERS
- TESTING VACUUM TUBES (VALVES)

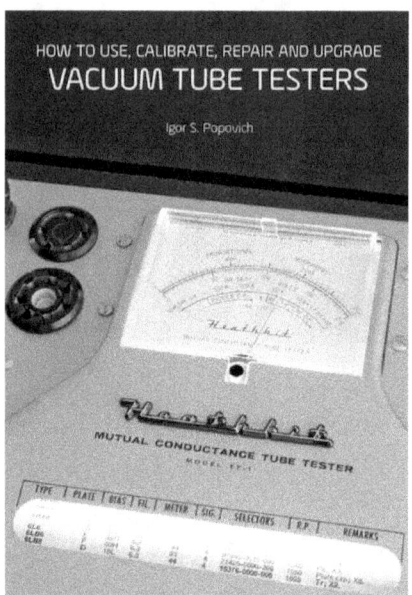

How to Use, Calibrate, Repair and Upgrade Vacuum Tube Testers
ISBN: 978-0-9806223-7-9

- HOW VACUUM TUBES WORK
- TESTING & MATCHING VACUUM TUBES
- EMISSION TESTERS
- GRID CIRCUIT TESTERS
- DYNAMIC CONDUCTANCE TESTERS
- PROPORTIONAL MUTUAL CONDUCTANCE TESTERS
- HICKOK-TYPE TESTERS
- TRUE MUTUAL CONDUCTANCE TESTERS
- REPAIRING & UPGRADING VINTAGE TUBE TESTERS
- TESTING & MATCHING TUBES WITHOUT A TUBE TESTER

Transformers For Tube Amplifiers: How to Design, Construct & Use Power, Output & Interstage Transformers and Chokes in Audiophile and Guitar Tube Amplifiers
ISBN: 978-0-9806223-8-6

- PHYSICAL FUNDAMENTALS OF MAGNETIC CIRCUITS AND TRANSFORMERS
- FILTERING CHOKES (INDUCTORS WITH DC CURRENT)
- TRANSFORMER MATERIALS, CONSTRUCTION METHODS AND ISSUES
- MAINS (POWER) TRANSFORMERS
- PHYSICAL FUNDAMENTALS OF AUDIO TRANSFORMERS
- SINGLE-ENDED OUTPUT TRANSFORMERS
- PUSH-PULL OUTPUT TRANSFORMERS
- SPECIAL MAGNETIC COMPONENTS: LOW POWER INPUT, PREAMP OUTPUT & DAC OUTPUT TRANSFORMERS, TRANSFORMER VOLUME CONTROL
- INTERSTAGE TRANSFORMERS, GRID & ANODE CHOKES
- OUTPUT AND INTERSTAGE TRANSFORMERS FOR TUBE GUITAR AMPS
- TRANSFORMER TESTS & MEASUREMENTS

TESTING & MATCHING TUBES WITHOUT A TUBE TESTER

10

- SIMPLE OHMMETER AND MULTIMETER CHECKS
- EMISSION AND TRANSCONDUCTANCE TESTS
- DIY CURVE TRACER FOR MATCHING TUBES

SIMPLE OHMMETER AND MULTIMETER CHECKS

The ultimate test of any tube is inside its amplifier. Tube testers, even the more elaborate ones, suffer from both false positives (a tube tests "bad," but it works in the equipment) and false negatives when a tester passes a tube as good, but the tube then fails to work properly in an amp. So, unless you buy & sell tubes commercially, you don't need a tube tester for amplifier repair and construction work.

These three tests range from the simplest, using ohmmeter only, to the more complex tests with the tube-under-test in a hot state (heaters energized).

Ohmmeter checks in the cold state

These cold checks are necessary but not sufficient to proclaim a tube to be OK. An analog ohmmeter should show a very low reading (a few ohms) when measuring the resistance of the tube's heater. If the ohmmeter shows infinite resistance, the heater has burned out, and you can throw that tube away. Between all other electrodes, the ohmmeter should show infinite resistance.

Pairs of electrodes physically close to each other are more likely to show a fault, for instance, heather to the cathode, control grid to the cathode, and screen grid to the anode. However, even anode to cathode test can show low resistance due to foreign material stuck in the mechanical structure of the tube.

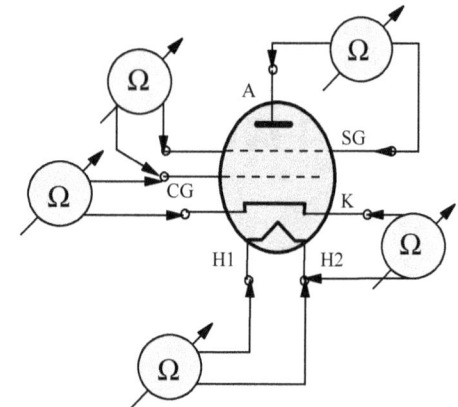

ABOVE: Apart from the very low ohm reading across the heater pins, all other ohmmeter checks should show infinite resistance.

Hot ohmmeter checks

Most cheap tube emission checkers use a similar "hot" test. Connect a small mains transformer to supply the rated heater voltage. If the heater needs 6.3 V@3A, a 20 or 30 VA transformer will be sufficient. You can use a universal adjustable DC power supply instead.

Connect an analog ohmmeter between the cathode and the control grid. As shown, you don't even have to connect the anode to the grid. Connect the + lead to the anode or control grid and the negative lead to the cathode.

The ohmmeter should show some deflection; how much will depend on the type of tube tested, its internal resistance, and other parameters. If you get no indication, reverse the ohmmeter's leads.

Use the highest range on your meter, for instance, 10k or 100k. Most multimeters use an internal 9V battery for one or two of the highest resistance ranges, but only a 1.5 V battery for lower ranges and 1.5V may not be enough to test a tube.

Digital ohmmeters and multimeters, which work on a different measurement principle, are not suitable for this test.

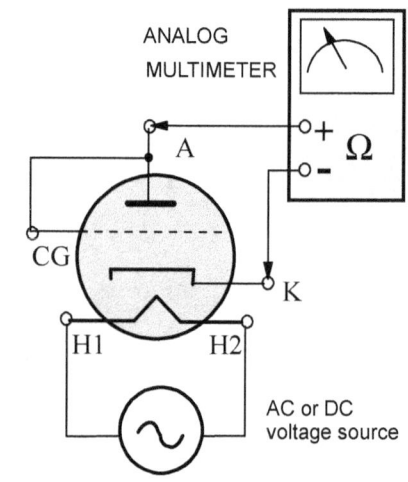

ABOVE: 9V from the ohmmeter's internal battery is enough to cause a current to flow through a hot tube, which indicates emission levels from its cathode.

Hot checks with a function generator and a multimeter

The low voltage power supply provides a DC heating voltage (adjustable and readable on the power supply voltmeter). A function generator provides a sine or square voltage fed between the tube's cathode and the grid/anode. A transformer of suitable voltage and current rating can be used for AC heating instead of the DC power supply.

A DC milliammeter or a multimeter set on DC current range displays grid/anode current. The tube's grid should be safe at such low test voltages (most function generators provide 7-10V_{RMS} signals). Some higher-spec models may even go up to 30V.

If you are still worried about damaging the tube-under-test due to too much current flowing through the grid, strap the grid to the cathode instead.

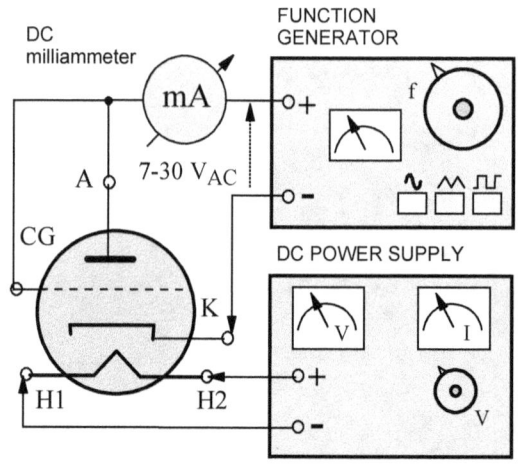

ABOVE: A function generator as a signal source plus an AC or DC heater supply, and you have a simple yet effective emission tester!

TESTING & MATCHING TUBES WITHOUT A TUBE TESTER

EMISSION AND TRANSCONDUCTANCE TESTS

Even the best tube testers are limited in their features and range of tubes that can be tested. The most noticeable is the small range of tube socket types, but that is easily expanded through adapters or add-on sockets.

The next limitation is the low heater power that their meager power transformers can supply. Tubes requiring high heater power (such as 6C33C-B, GM70, 211, 813, and 845) cannot be tested. An external heater supply can be hooked up temporarily (crocodile clips to tube pins), or a tester can be modified (a simple switch and a couple of terminals added) to enable external heater AC or DC voltage to be brought in. During those tests, the changeover switch will automatically disconnect the tester's internal heater circuit.

The most serious drawback of all but a few laboratory testers, one that easy modifications cannot overcome, is the limited range of grid bias, anode and screen voltages available, and low anode current capabilities. Even a decent tester such as Triplett 3444 only goes up to $V_A=V_S=250V_{DC}$ and 50mA of anode current!

The solution is to hook up your own test setup. At least two DC- and one AC-power supply (for the heater) is required for plotting curves with only negative grid bias voltages, plus two DC voltmeters and one DC milliammeter for anode current. An additional DC power supply is needed to be able to bias the grid positively unless you are happy to manually reverse the V_G polarity.

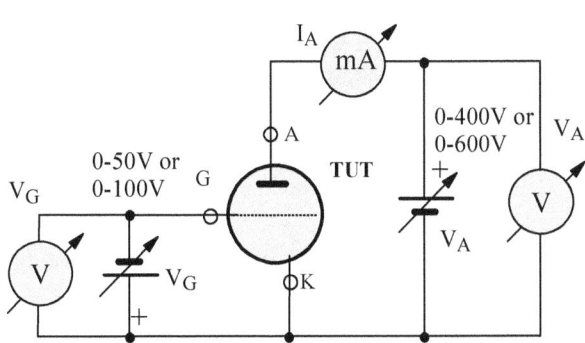

The circuit for recording transfer, anode and constant current curves of triodes. Only negative grid voltages are obtainable with two DC power supplies. A third power supply is needed for the heater circuit.

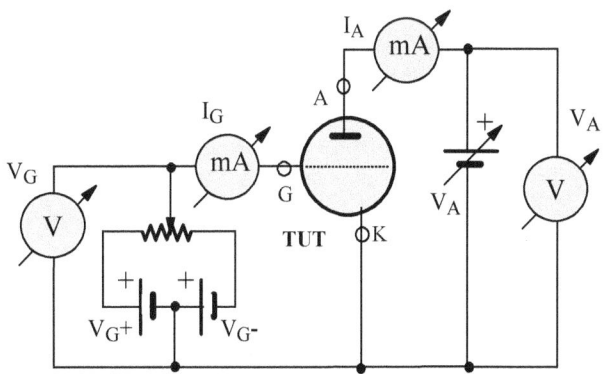

A more versatile circuit for recording characteristic curves of triodes. Both positive and negative grid bias voltages are obtainable with three DC power supplies.

The grid-shift method

We have already seen a few vintage commercial testers that used a variant of the grid-shift method as their operating principle, namely AVO and Taylor. It seems UK tester designers disliked the American way of feeding an AC test signal into the tube's control grid and using the AC component of anode current to drive the Gm meter through some sort of filter and amplifier.

Why pay $2,000 or more for a vintage AVO or Taylor tester when a better (higher test voltages, higher heater and anode current capability, higher precision) grid shift Gm tester can be hooked up in a couple of minutes?

You will need a high voltage (HV) power supply, at least $0-400V_{DC}$ 125mA, a source of bias voltage 0-100V, rated at few mA only (since you will not test tubes in class A_2 with grid current flowing), a source of heater voltage and a good quality DC ammeter or multimeter.

The source of heater voltage could be a low voltage (LV) DC power supply, say 0-40V, 0-5A or 0-7A (better), or a transformer with multiple secondary taps and a selector switch.

This can also be an old emission tester with a decent-sized power transformer, used here only to supply heater voltages.

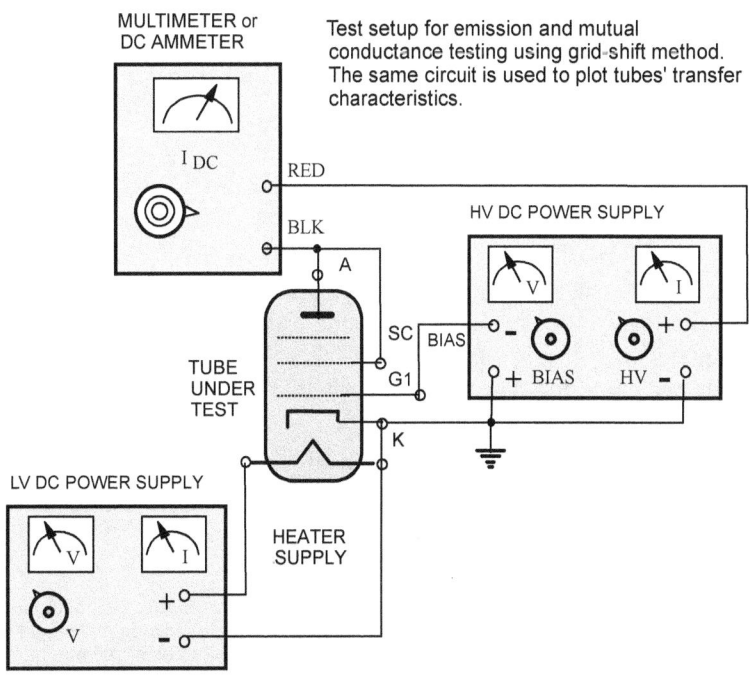

Test setup for emission and mutual conductance testing using grid-shift method. The same circuit is used to plot tubes' transfer characteristics.

$V_A=V_S=300V_{DC}$ $V_{G1}/V_{G2}=-14.0V/-18.4V$	I_{A1}	I_{A2}	ΔI_A	Gm [mA/V]
Valve Art 1	54.9	37.2	17.6	4.02
Valve Art 2	53.8	36.0	17.8	4.04
Valve Art 3	53.5	36.6	16.9	3.84
Valve Art 4	54.5	36.7	17.8	4.04
JJ 1	61.6	43.5	18.1	4.11
JJ 2	63.3	45.5	17.8	4.05
EH 1	60.8	43.6	17.2	3.91
EH 2	55.9	39.3	16.6	3.77
EH 3	61.9	45.3	16.6	3.77
EH 4	64.3	46.7	17.6	4.00

ABOVE: Results of testing two "matched" quads and a "matched" pair of EL34 pentodes using grid shift method

In this illustrative experiment, we bought ten brand new EL34 pentodes on eBay as matched pairs and quads, strapped them as triodes ($300V_{DC}$ anode/screen voltage), adjusted bias at -14.0V, and measured DC anode currents. Then, we increased the negative bias to -18.4V and recorded anode currents again.

If you want a precise Gm in one point, the bias shift should be very small, 0.1 or 0.2V. However, that requires a precise adjustment of bias, and many DC power supplies aren't capable of such fine-tuning.

Also, the change in anode current would be correspondingly small, and sensitive mA meters would be needed.

Finally, unless all DC supplies were closely regulated, the mains voltage would fluctuate during the test and affect results, so you would never be sure if a change in anode current I_A is due to a change in bias or due to mains fluctuations. Surprisingly, that makes test results of large grid shift changes more reliable and accurate!

In terms of matching, the Valve Art quad of Chinese manufacture was reasonably matched in terms of anode current, 53.8mA to 54.9mA, but in terms of Gm three of the four tubes were very closely matched (4.02-4.04-4.04mA/V) while the fourth one was weaker, at only 3.84mA/V!

The JJ pair was also reasonably matched, both in terms of I_A ([63.3-61.6]/61.6 = 0.0276 or 2.76%) and Gm ([4.11-4.05/]/4.05= 0.0148 or 1.48%).

The ElectroHarmonix quad's current was matched to (64.3-55.9)/55.9 = 0.15 or 15%, meaning it wasn't matched at all, while Gm results were closer, (4.0-3.77)/3.77 = 0.061 or 6.1%.

The grid signal method

The setup, in this case, is the same as for the grid shift method, with an addition of a grid audio signal of a certain frequency f_0, provided by a function generator or any other signal source.

An accurate true-RMS AC voltmeter or multimeter measures the voltage drop the AC component of anode current produces on a load resistor.

The AC voltmeter/multimeter must be capable of measuring signals at your test frequency f_0. If you use 1-5 kHz, most digital multimeters will have their upper -3db way higher and will be fine (quality multimeters will go up to at least 30kHz). The el-cheapo ones go up to 3-4 kHz, so stick to a 1kHz test signal in that case.

You can use a decoupling capacitor in series to eliminate DC current from the AC voltmeter's input, but these already have an input capacitor for that purpose; just make sure its voltage rating is at least 600V.

To protect the output transistors of the function generator from high bias DC voltages, add a series capacitor C_S. If your test frequency is between 1 and 5 kHz, its reactance at that frequency must be negligible, less than 10Ω, meaning its capacitance should be 160-220nF.

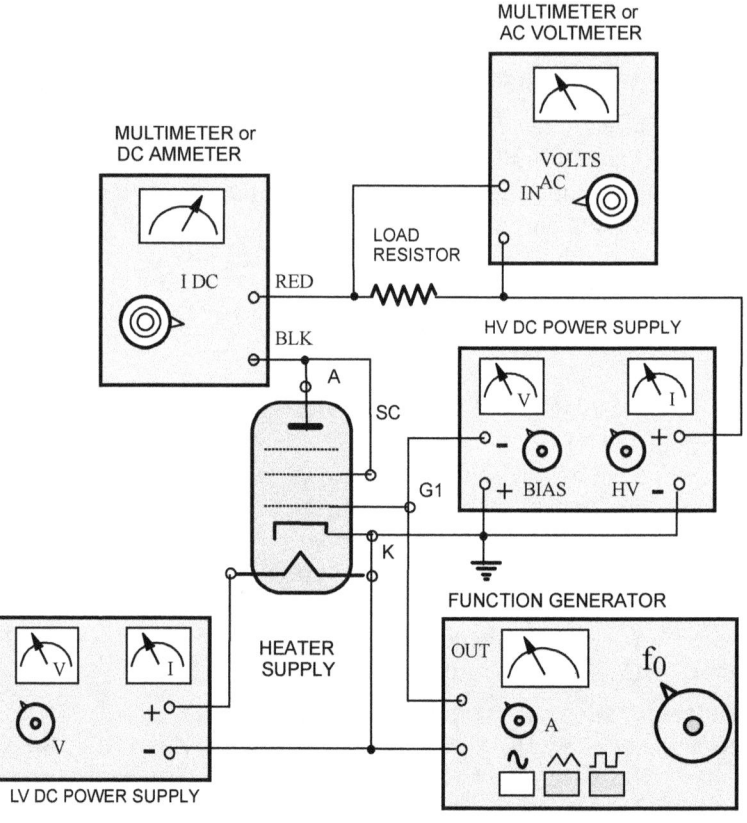

ABOVE: Test setup for mutual conductance and emission measurement using a dedicated grid signal source (such as a function generator), a load resistor in the anode circuit and an AC millivoltmeter as an indicator of transconductance.

TESTING & MATCHING TUBES WITHOUT A TUBE TESTER

Vintage high voltage regulated DC power supplies

Vintage high voltage regulated power supplies are available for sale online but are limited to 125 mA and 400V. They all use the same series regulated design with two triode-strapped 6L6 tubes in parallel, so it doesn't matter which one you get (Eico, Heathkit, PACO, Knight).

The bias voltage is adjustable 0 to -150V, but since an ordinary potentiometer was used, it is impossible to adjust it precisely; as soon as the knob is moved a tiny bit, the voltage increases or reduces two or three volts. Thus, these crude instruments aren't suitable for precise measurements. For bias supply, you are better off using a precisely regulated low voltage DC supply 0-25V or 0-50V if you are testing directly heated triodes requiring very high negative bias, such as 300B and 2A3.

If you plan to do lots of testing, you should build your own better HV power supply. You can go as high in voltage and current capabilities and other functionality as you need or want. Ours can go up to 600V and 300 mA.

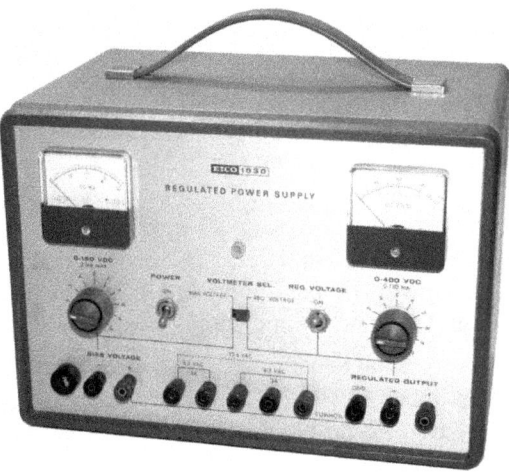

RIGHT: Vintage high voltage regulated power supplies such as Heathkit IP32 (left), Heath Zenith IP2717 or Eico 1030 (right) are useful vintage workshop instruments, although limited both in voltage/current range and the precision of their adjustments.

Measuring & plotting the transfer characteristics of a tube without a tube tester

We wanted to design a tube guitar amplifier that would operate at relatively low anode and screen supply voltages of 50-60V_{DC}. After eliminating most other power pentodes, PL508 worked happily at such low voltages.

Tube catalogs and data sheets don't contain any data or characteristics for low voltage tube operation. So, no matter what power tube you decide to use, you need to do a little experiment to determine the required bias of the output stage. A transfer curve of the tube at the DC voltages in your circuit needs to be plotted, meaning with +60V on the anode and +55V on the screen grid in our case.

You'll need the heater, anode, and screen grid power supplies, a 9V battery, 1k to 5k LIN potentiometer, and two multimeters (below). One will measure the grid voltage (bias), on "DC Volts," 1-10V range, the other will be connected as mA-meter, in series with tube's anode. It could also be placed in the cathode circuit if you prefer.

Start at 0V on the CG and record the reading of the mA-mater, in this case, 58mA. Then keep lowering the negative grid voltage in -1V steps by adjusting the potentiometer and record the anode currents. Once you have your table with data, you can draw the TX curve if you wish, as illustrated.

RIGHT: Transfer characteristics of PL508 pentode with +60V anode and +55V screen grid voltages
BELOW: The test circuit for plotting transfer curve (I_A versus V_G) of an output tube

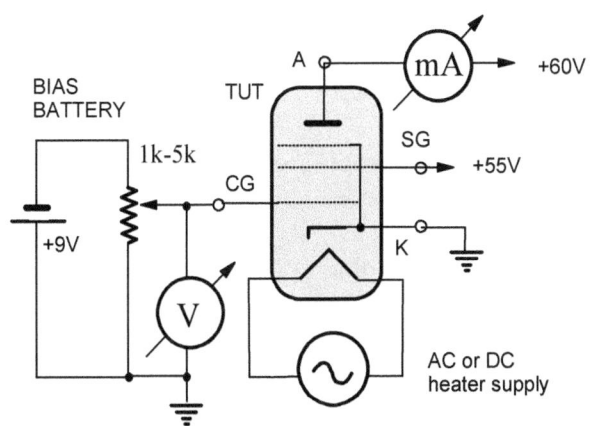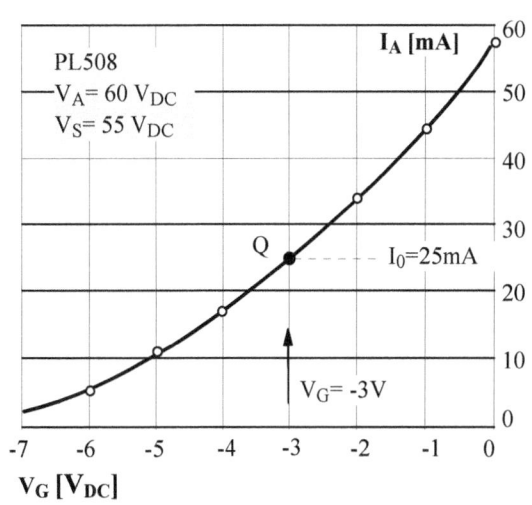

With a zero bias, the maximum anode current is only 58mA, so at such low anode/screen voltages, the tube is safe from overheating and destruction in any case. The cutoff is just below -7V, so we can choose the quiescent point at any point around the middle of the curve to ensure the maximum possible input signal swing, which is 7V peak-to-peak or around $7/2.8 = 2.5$ V_{RMS}!

DIY CURVE TRACER FOR MATCHING TUBES

As with most things, you have more than one option. You can use existing pieces of test gear and put them together in the right way or construct your own curve tracer from scratch.

Option #1: Using a commercial semiconductor curve tracer

Heathkit models IT-1121 and IT-3121 are budget curve tracers for semiconductors. However, they can be used as a step-generator for a tube curve tracer. They can test FETs, and these, just like tubes, require voltage steps on the gate (analogous to control grid in a tube). Similar models include B&K 501A and Leader LTC-905. Eico 443 cannot test FETs (has no voltage steps) and is unsuitable for this application.

B&K 501A uses pulsating rectified sinewave of up to a maximum 100V peak for the collector supply (will become anode supply), as does Leader LTC-905. The voltage steps for testing FETs needed for testing tubes are 0.05V, 0.1V, 0.2, 0.5, and 1V, with a maximum of 6 steps displayed and an accuracy of +/-4%.

Leader LTC-905 has voltage steps only up to 0.5V; it can display 7 of them with +/-5%, so a mixed bag.

Heathkit IT-1121 and IT-3121 can display up to 9 voltage steps (1) with 1V maximum, but its sweep voltage can be as high as 200V (2) and supply up to 200mA.

The horizontal sensitivity is switchable between 100mV and 50V per division; vertical sensitivity (4) ranges from 0.5 V/div to 200 V/div, so it is the most capable of all the budget curve tracers mentioned.

The heater supply can be a separate power supply or an existing tube tester.

You also need a source of sweeping plate voltage. Commercial curve tracers already have such an output (collector power supply), but if you use your own DIY staircase generator, any tube tester that uses full-wave rectified sine wave for plate voltage, like Hickok or B&K models, can provide such a pulsating sweep voltage.

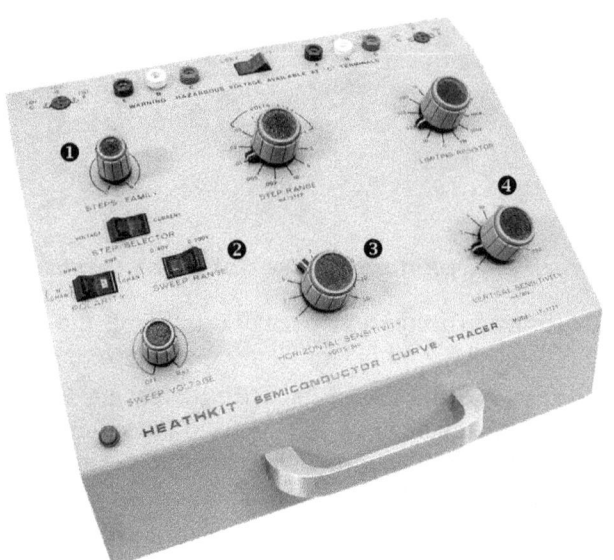

You can tap into plate voltage on the octal socket between pins 3 (+ or anode) and pin 8 (- or cathode). The sweep voltage goes to Channel 2 (X- or horizontal deflection) on the oscilloscope, while the input to the vertical channel (Y- or vertical deflection) is taken from the cathode resistor, between cathode and GND or COM.

The oscilloscope's time-base is switched off; it must be in "X-Y" mode.

The "Base" or "Gate" (for FETs) output of the curve tracer supplies the stepped voltage waveform to the control grid, while its "E" or "Emitter" terminal is grounded.

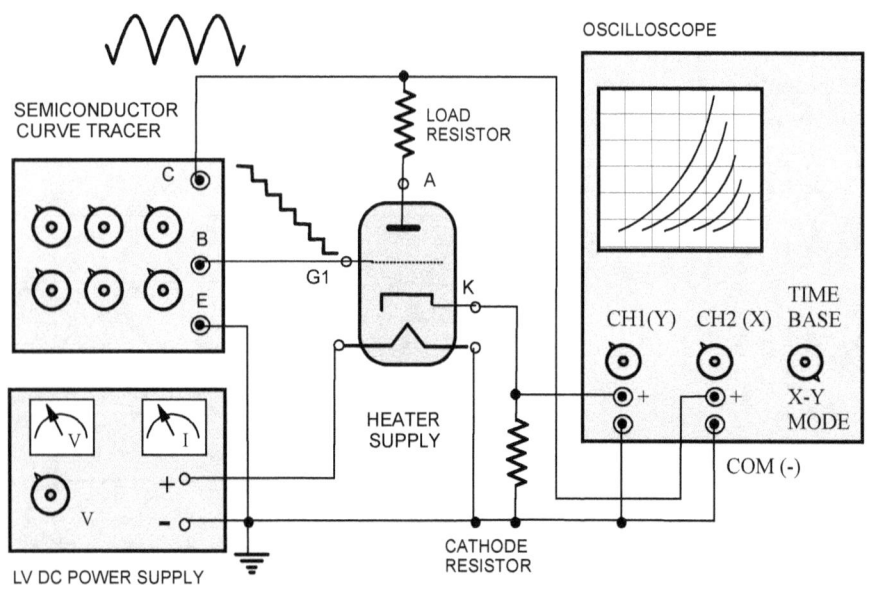

ABOVE: Test setup for displaying triode's anode (plate) characteristics on an oscilloscope. AC heating voltage can be used instead of the DC power supply.

Option #2: Build your own curve tracer for preamp tubes

The simplest schematics for a staircase generator we could find was designed by Daniel Metzger and published in the August 1971 issue of Electronics World. It works very well.

Its steps are 1 Volt in amplitude, so it is not suitable for curve tracing of power tubes where a larger step size would be needed.

The tube biasing circuit is fairly simple; all you need is half-a-dozen resistors, a 3-pole double-throw switch (for selecting one of the two triodes you want to test), and a tube socket, Octal, Noval, 7-pin mini, RimLock, etc., depending on the tubes you want to test.

You could add a switching circuit so tubes with different pinouts could be tested and a variable cathode and anode resistors to change the value of the cathode bias resistors.

We wanted to illustrate the principle here, not to overcomplicate the schematics with multiple plate voltages, various switchable heater voltages, or similar options you can easily add yourself.

The 6.3 V winding provides heating for the tube under test. If you need more than one heater voltage, you can use a heater supply from a commercial tube tester.

The half-wave rectified, filtered, and stabilized 15V supply powers up the transistorized staircase generator.

The full-wave rectified mains voltage provides the sweep voltage for the anodes. In our case, the amplitude is 100V.

With multiple secondaries, you can add a voltage selector switch and change the amplitude of the sweep, say 100 - 150 - 200 - 250 Volts.

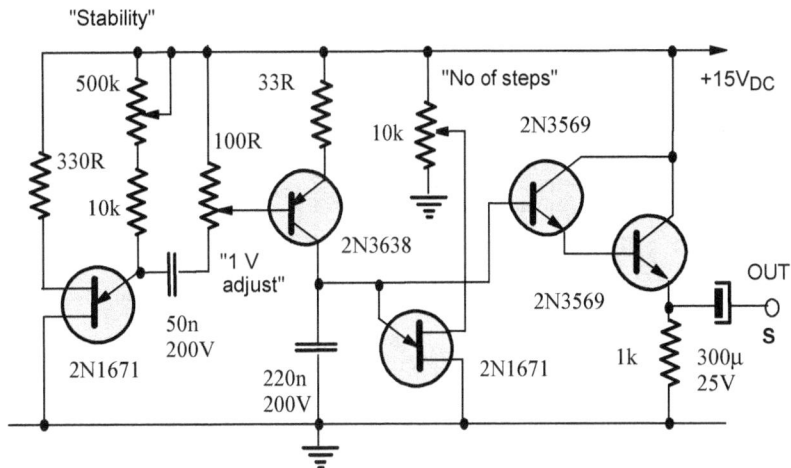

ABOVE: A simple solid state staircase generator, source & copyright: Daniel Metzger, Electronics World, August 1971

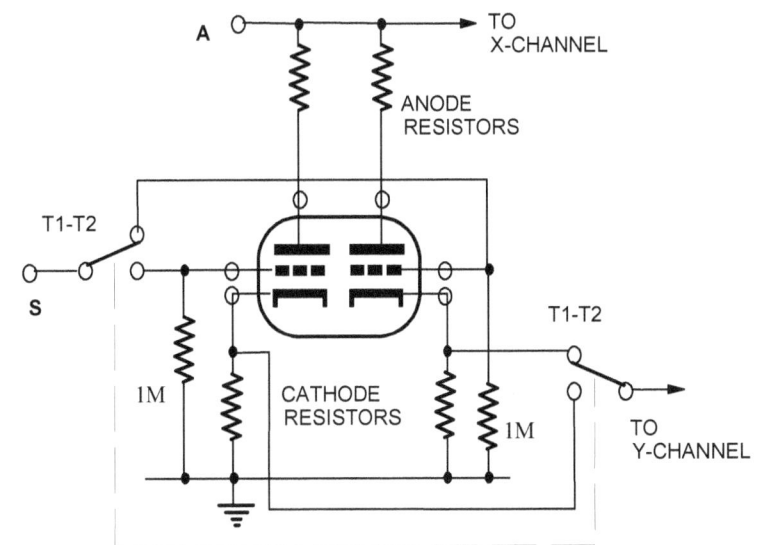

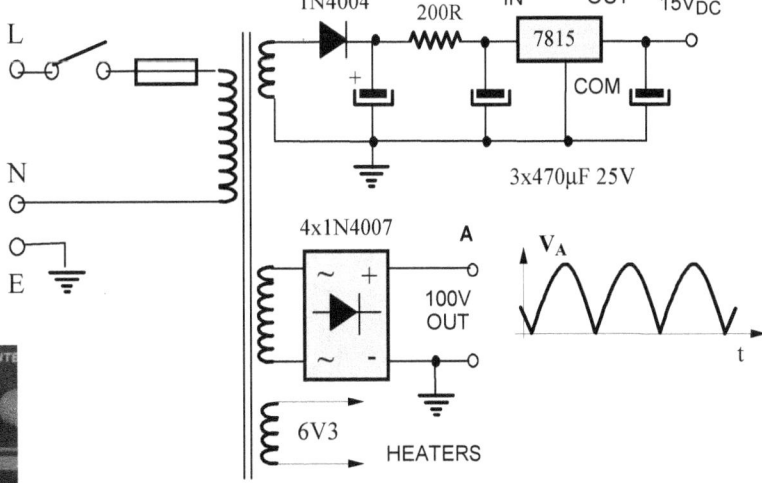

ABOVE: One possible power supply configuration for the curve tracer

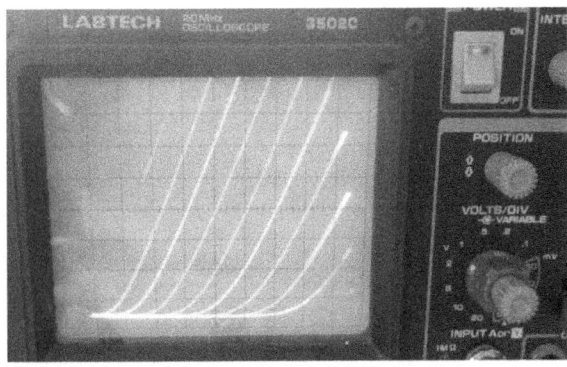

LEFT: A family of ten plate curves for 6DJ8 duo triode displayed on an analog CRO. Each V_G step (one curve) is -1V. Half of the 0V bias curve on the left and one curve (-9V) on the right are missing on the photo due to the synchronization effect between CRO sweep frequency and digital camera's operation.

INDEX

A

Accuracy, 23
Adding plate current measurement capability, 103, 119, 123, 126, 131, 138
Adding rectifier balance control to Hickok bridge testers, 130
Adding signal calibration control to Hickok bridge testers, 130
American Scientific Development Company TV-20, 60
Analog meter,
    protection, 87
    shunts, 163-164
    symbols, 165
Anode dissipation, 15
Anode resistance, 12, 17-18
AVO MIII Valve Characteristics Meter, 85-87

B

B&K 500, 103-104
B&K 550, 105-107
B&K 650, 107-108
B&K 675, 108-113
B&K 600 & 606, 54
B&K 607 & 667, 56
B&K 610 test panel, 116-117
B&K 625, 55
B&K 700 & 707, 114
B&K 747 115-116
Barkhousen's equation, 18

C

Calibrating tube testers, 24-25, 120, 122
Calibration module
    for Sencore testers, 52
    for Hickok bridge testers, 107
Cathode, 10
Checking tubes with an ohmmeter, 172
Conar 221, 223 & 224, 41-43
Contact troubles, 159
Curve tracers, 176-177
Cutoff point, 17

D

Dynamatic DM456, 128
Dynamic conductance testers, 62
Dynamic resistance (of a tube), 13

E

Eico 625, 37
Eico 635, 37, 59
Eico 666 & 667, 73-77
Electric shock, 157
Electron flow versus current flow, 10
Elettra Provavalvole tester, 45-46
Emission testers, 36

F

False positives and negatives, 5, 153
Fault location methods, 159-160
Ferrite beads, 177
Fixing Gm range mismatch, 129

G

Gassy tubes, 26
Gas test, see: Grid current test
Grid circuit testers, 48
Grid current test, 25, 56, 59, 71, 102
Grid emission, 26, 102
Grid shift gas test, 26
Grid shift Gm test, 89-90, 154, 173
Groove Tubes' performance rating system, 29-30

H

Heathkit tube checkers, 38-41
Heathkit TT-1, 134-138
Hickok bridge, 96
Hickok testers, 98-102
High voltage DC power supplies for tube testing, 175

I

Increasing anode current metering range, 144

J

Jackson tube testers, 72-73
JAN (Joint Army-Navy), 14

K

Knight KG-600, 38-41

L

Lafayette TE-21, 38
Life test, 53
Line adjustment rheostats, 37
Line control adjustment, 69, 76
Low bias tube testing, 34

M

"Magic eye" tube, 149
Mains circuit safety upgrade, 158-159
Matching tubes, 23, 27-30
MaxiBurn, 154
Maxi Matcher II, 154
Maxi Preamp Digital tube tester, 154
Megaohmmeter leakage test circuit, 41, 57, 67

M, cont.

Mercury emission testers, 22, 59
Mercury Gm testers, 118-122
Metering circuits, 52, 57, 82, 134-135, 140, 145
Metrix 310CTR, 88-90
Micromho, 17
Mutual conductance, 16, 17, 21
    under reduced voltages, 22
    triode versus pentode, 22
    changes with anode current, 28

N

Noise & microphony test, 125

O

Orange DIVO VT1000, 152-154

P

PACO 650 & T-62, 57-58
Perveance, 15
Plate resistance, see: anode resistance
Plate dissipation, see: anode dissipation
Precise 111, 123-127
Precise 116, 122-123
Precision (PACO) 640 & 660, 43
Proportional mutual conductance testers, 80

R

Replacing analog meters, 163
Replacing selenium rectifiers, 140, 150, 162
Replacing tube rectifiers with silicon diodes, 12
RCA WT-110A, 147-149

S

Safety rules & precautions, 157-159
Scale resolution, 23
SECO 78, 88 & 98, 48-50
SECO 107, 149-152
Sencore automatic biasing circuit, 146
Sencore TC28 ("The Hybrider"), 52-53
Sencore "Mighty Mite" testers, 50-52
Sencore MU-140 & MU-150, 145-146
Shorts & leakage neon test circuit, 40, 44, 60
Sico 85, 63-64
Sico TV-12, 65-67
Simpson 330, 91-93
Simpson 1000, 73

S, cont.

Signal oscillator, 81, 82, 91, 141, 145, 147
Socket adapters, commercial, 166-167
Socket adapters, DIY, 168-169
Soldering, 165-166
Special quality (SQ) tubes, 10, 14, 21
Sylvania 139 & 140, 68-69
Sylvania 219, 220 & 620, 69-71

T

Taylor 45D, 83-85
Testing
    bipolar transistors, 161
    electrolytic capacitors, 58, 161
    potentiometers, 162
    silicon diodes, 160
    triodes with low internal resistance, 135
    unlisted tubes, 142-143
    Zener diodes, 161
Thermionic emission, 10
Transconductance, see: Mutual conductance
Transfer characteristics, 13, 16, 28, 30, 32-34, 175
Triplett 2413 & 3414, 44
Triplett 3423, 81-82
Triplett 3444 & 3444-A, 139-144
Tube naming conventions, 13-14
Tube parameters, 20
Tube test certificates, 24
Tube testing without a tube tester, 172-177

U

Universal switching unit for any tube tester, 169

V

Variation of tube parameters, 20-21, 27
Vacuum diode, 11
VTVM (Vacuum Tube VoltMeter), 49, 57-58, 135, 145
Voltage amplification factor, 15, 17-18

W

Westmore 501, 139-144
Weston 798, 80-81
Weston 981, 134

www.ingramcontent.com/pod-product-compliance
Lightning Source LLC
LaVergne TN
LVHW080116250326
834688LV00040B/1157